Echoes **of Life**

Susan M. Gaines
Geoffrey Eglinton
Jürgen Rullkötter

scientific illustrations by Florian Rommerskirchen

Echoes of Life

What Fossil Molecules Reveal
about Earth History

OXFORD
UNIVERSITY PRESS

2009

Oxford University Press, Inc., publishes works that further
Oxford University's objective of excellence
in research, scholarship, and education

Oxford   New York
Auckland   Cape Town   Dar es Salaam   Hong Kong   Karachi
Kuala Lumpur   Madrid   Melbourne   Mexico City   Nairobi
New Delhi   Shanghai   Taipei   Toronto

With offices in
Argentina   Austria   Brazil   Chile   Czech Republic   France   Greece
Guatemala   Hungary   Italy   Japan   Poland   Portugal   Singapore
South Korea   Switzerland   Thailand   Turkey   Ukraine   Vietnam

Published by Oxford University Press, Inc.
198 Madison Avenue, New York, New York 10016

www.oup.com

Oxford is a registered trademark of Oxford University Press

Library of Congress Cataloging-in-Publication Data
Gaines, Susan M.
Echoes of life : what fossil molecules reveal about earth
history / Susan M. Gaines, Geoffrey Eglinton and Jurgen Rullkotter ;
scientific illustrations by Florian Rommerskirchen.
p. cm.
Includes bibliographical references and index.
ISBN 978-0-19-517619-3
1. Biomolecules, Fossil.   I. Eglinton, G. (Geoffrey)   II. Rullkötter, J.
III. Title.
QP517.F66.G35 2008
572'.33—dc22   2008019905

9 8 7 6 5 4 3 2 1

Printed in the United States of America
on acid-free paper

in memoriam

Guy Ourisson

March 26, 1926–November 3, 2006

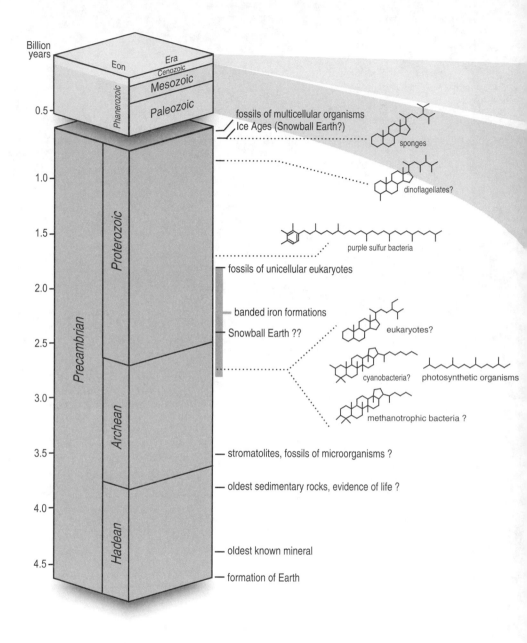

Fossil Molecules in Geologic Time

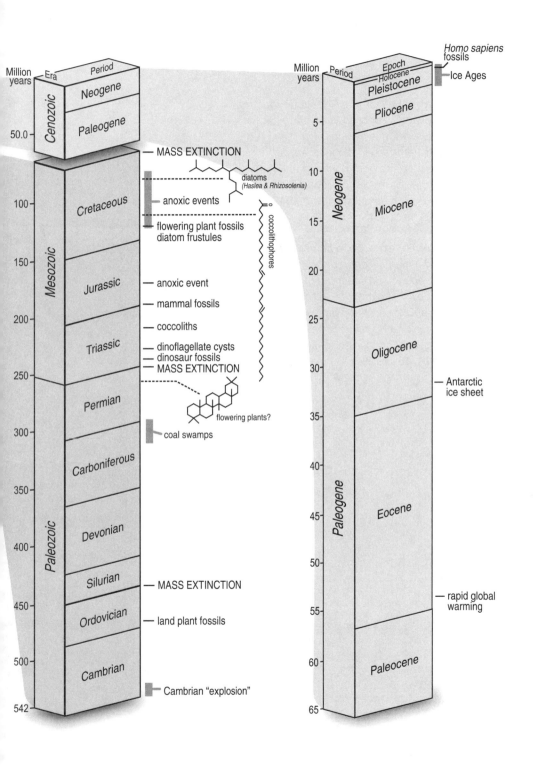

*It is in the atmosphere of [the First World War] that
I have approached a conception of nature, at that
time forgotten and thus new for myself and for
others, a geochemical and biogeochemical
conception embracing both nonliving and living
nature from the same point of view.*
—Vladimir Ivanovich Vernadsky, 1863–1945, Russian mineralogist
    From *American Scientist* 33 (1945)

*An organic solvent extract of a bituminous oil
shale showed a magnificent red color and the
characteristic absorption spectrum of the metal
complexes of pyrrole pigments.... The apparent
preservation of complicated pigment materials
through geological eras raises hopes that the same
may also be true of other types of molecules.*
—Alfred Treibs, 1899–1983, German chemist
    From *Angewandte Chemie* 49 (1936)

Preface

Seventy years ago, a German chemist identified certain organic molecules, extracted from ancient rocks and oils, as the fossil remains of chlorophyll—presumably from plants that had lived and died millions of years in the past. Outspoken against the Nazi regime, Alfred Treibs lost his university post and standing not long after this discovery. Though he proposed that other types of molecules might have left similarly recognizable traces, it would be another 25 years before his insight would be developed and the term "biomarker" coined to describe fossil molecules whose atomic makeup and molecular structures could reveal the activities, both past and present, of otherwise elusive organisms and processes. Over the past 50 years, hundreds of biomarkers have been identified in oceans and sediments, ancient rocks and oils, soils and coals, and in individual fossils. They are helping paleoceanographers to elucidate the temperatures and carbon dioxide concentrations of ancient seas, and climatologists to predict those of future ones; petroleum geologists to predict the whereabouts of oil; evolutionary biologists to understand cell evolution; paleontologists to determine when flowering plants evolved; microbiologists to understand the ecology of microbes in natural waters and sediments.... Biomarkers have provided evidence for Vladimir Vernadsky's prescient stipulation that the biosphere was inseparably linked to the geosphere and had been so since its inception. They are providing clues to some of the earliest forms of life on the earth, and as space exploration makes extraterrestrial rock samples available, biomarker analyses may help in detecting past or present life elsewhere in the solar system. *Echoes of Life* is the story of these molecules and how they elucidate the history of the earth. It is also the story of the scientists—once a small scattering of mavericks defying the dictates of their disciplines, now a large, thriving scientific community—who have learned to detect, measure, and interpret the molecular clues.

This is a book to be read for pleasure and contemplation as well as information and education, lying on a couch or sitting at a desk, depending on one's inclination. It is an attempt to unite 50 years of research, scattered across half a dozen disciplines and at least as many countries, into one scientifically and historically coherent treatise. Students and practitioners of organic geochemistry will, we hope, enjoy and learn from the panoramic view of their discipline and the history of its development, just as we have. Geologists, biologists, microbiologists, and oceanographers will gain a deeper understanding of and appreciation for the insights provided by organic chemistry. The extent to which molecular structure and function can reveal the history of the earth and evolution of biochemical systems will be a revelation for many organic chemists and biochemists and, we hope, an inspiration for their students. Indeed, undergraduate students and non-academic readers may find that *Echoes of Life* offers them a rigorous but exciting entrance into the natural sciences—chemical, geological, and biological—as well as an appreciation for science as a living, breathing beast that grows and develops, errs and triumphs. The book is based on a simple premise, all too often forgotten in the scientific literature but a banal fact of life for any working scientist: science is a narrative, a perpetually unfolding story, and despite its definitive rules of objectivity, human characters are integral to every scientific plot. With that in mind, the story of this book's genesis may be instructional.

In 2001, Meixun Zhao, then a professor at Dartmouth College in New Hampshire, told his British colleague and longtime collaborator Geoff Eglinton about an unusual novel. It included biomarker research among its main themes and told a story of discovery that was eerily reminiscent of one that had taken place in Geoff's own laboratory in Bristol—but it was set in California, and Geoff had never heard of the author. Susan Gaines, Meixun told him, was a graduate school friend he hadn't seen in ten years. Recently, she'd turned up as a guest speaker at Dartmouth, invited by a geology professor who'd been intrigued by her literary rendition of geochemistry. Perhaps, Meixun suggested, she could write the biomarker text that he and Geoff had been mulling over, something that could be used in courses for undergraduates like the one they'd been teaching together at Dartmouth. Geoff, Professor Emeritus at Bristol, was then on the adjunct faculties at both the Woods Hole Oceanographic Institution and Dartmouth. He'd drafted an outline for a sort of "biomarker casebook" based on his and Meixun's lectures. But when he read *Carbon Dreams* he started dreaming of something more ambitious: a popular science book about biomarkers that would bring the process of doing scientific research to life the way the novel had, replete with all its warts, tedium, beauty, and excitement.

When Geoff contacted me about writing a biomarker book, I was in California, deeply immersed in work on another novel. He invited me out to Woods Hole to give a presentation from *Carbon Dreams* and discuss his book idea. I'd read many of his papers while working on the novel and was intrigued and flattered by the invitation. I was genuinely fascinated by the biomarker work, I told him, but

I'd given up science a decade ago and committed myself to writing fiction. The sort of literary fiction I was attracted to had little commercial appeal, so the rest of my time was consumed by freelance editing and teaching. And to be honest, I didn't generally like popular science books. I found them somehow frustrating, the science too facile and watered down to be interesting. Thinking that was the end of it, I went back to California, and then headed south to Uruguay, where my novel-in-progress was set.

Meanwhile, *Carbon Dreams* was making the rounds in the scientific community. Jürgen Rullkötter at the University of Oldenburg in northern Germany was charmed by the unusual book, which his friend Keith Kvenvolden from the U.S. Geological Survey in California had sent him as a gift. Here was a mix of science and literature that he had never seen before! He would nominate the author as the "art-in-science" fellow at the nearby Hanse Institute for Advanced Study, which hosted scholars and scientists from around the world and was experimenting with an occasional creative artist. She could write another of these science novels, and it would be great fun!

Geoff, who had been one of the Hanse Institute's first fellows, had a similar idea—except that he was still thinking about his biomarker book. I'd told him, somewhat tongue-in-cheek, that if the book paid all my bills and subsidized writing a novel I might be interested, and he'd taken this as a green light to find funding and support for the project. All that was needed to nominate me for a Hanse fellowship was a collaborator at the University of Oldenburg or Bremen who was similarly inspired, and it so happened that he and Jürgen were working on a paper together....

So this is how the three authors of *Echoes of Life* came together: an organic geochemist in Oldenburg, a founding father of the biomarker concept from Bristol, and a marine chemist *cum* novelist who suddenly found herself commuting between California, Montevideo, and someplace called Delmenhorst in northwestern Germany. Out of the blue, when I had long forgotten about Geoff's curious proposition, I received an e-mail from a stranger named Jürgen Rullkötter, who said he had enjoyed reading *Carbon Dreams* and wondered if I would be interested in a fellowship at an Institute for Advanced Study in Germany. He said he had been talking to Geoff Eglinton about the possibility of me writing a book about biomarkers. Only now, after four years of intense immersion in the science I once abandoned, as we prepare this manuscript for publication, do I learn that Jürgen's first impulse had been to offer a Hanse fellowship for writing a novel—which would have been like manna from heaven for this writer. As it was, the manna was something less than heavenly, but still pretty sweet. All I had to do was spend 10 months over the next two years in Delmenhorst, wherever that was, give one talk, and start work on a book about biomarkers. I said yes....But what sort of book?

Not a popular science book, with an entertaining story and watered-down science. Not a dry textbook, with its efficient presentation of accumulated, finely honed knowledge. Rather, a narrative science book. We would present the science in the same way it was produced, as the product of rampant human curiosity that had been systematically indulged and constrained by the rules of the scientific method. We would make the book accessible, entertaining, *and* rigorous. Like its subject, *Echoes of Life* is geared to a wide spectrum of readers, but some minimal education in the sciences—an introductory course in organic chemistry, biology, or geology at some point in one's past—is probably prerequisite. We employ a loosely historical structure, divided by the ever-expanding applications of the biomarker concept in petroleum exploration, space sciences, oceanographic and climate studies, microbiology, and so on. Chemical descriptions often elude those not trained in the language of chemistry, but our historical narrative moves naturally from the simple to the more complex, and the chemistry is presented in the context of its applications: readers not versed in organic chemistry can absorb its language and basic premises while reading about the earth's changing climates or the first signs of life in its rocks, and no one has to wade through a tedious exposé on chemical nomenclature. All but the most general scientific terms are defined at first mention, and there is an extensive glossary at the end of the book. Figures and illustrations are fully integrated into the text, but readers can relocate them easily by referring to the list of figures at the end of the book. Those with a limited background in geology may find the opening figure ("Fossil Molecules in Geologic Time") useful for reference, and the appendix ("Biomarkers at a Glance") and closing figure ("A Biomarker-centric Tree of Life") provide a quick overview and reminder of molecular structures, names, sources, and biomarker uses.

I have pieced our story together from thousands of scientific papers and from researchers' memories of the evolution of their thinking. A few disclaimers are in order. We have strived for the utmost rigor in our presentation of organic geochemistry, with some minor compromises for accessibility and the narrative style. The many associated fields of science that we discuss are, of necessity, treated more superficially, and our cumulative expertise is more limited: we are likely to incur the wrath of specialists, though we have made every attempt at accuracy. The book's historical narrative is, however, openly and unabashedly anecdotal, used mostly as a vehicle for exploring the science. Certainly the interests and personal experience of the three of us have flavored the account. With one of the founding fathers of biomarker research—onetime Ph.D. or postdoctoral adviser to many of its current practitioners—as coauthor, it was only natural to begin with his story and perspective. I also engaged many of the book's main characters in long interviews and conversations, and their ideas and memories inform its narrative. I sought out many of them, but there is also a circumstantial component to these conversations, as whom I talked to depended to some extent on who happened through northern Germany while we were working on the book. For a more definitive outline of the history of organic geochemistry, we recommend two articles in the 2002 volume of *Organic Geochemistry*, Keith Kvenvolden's "History of the Recognition of Organic Geochemistry in Geoscience" (pp. 517–521) and J. Hunt, R. P. Philp, and Kvenvolden's "Early Developments in Petroleum Geochemistry" (pp. 1025–1052).

In the interest of readability, clarity, and literary aesthetics, we have resisted the temptation to cite every author of every work discussed and generally named only central "characters," often the leaders of dedicated biomarker laboratories or researchers whose work comes up repeatedly throughout the book. With the exception of a few key historical figures, we generally discuss work in peripheral fields without assigning credit. This leaves many coworkers, younger researchers, and students unacknowledged, even though they may have done the bulk of the work on a given project. We beg their pardon, but hope they will take some comfort in knowing their science has been integrated into a larger, cohesive panorama and their personal citations subordinated to the cause. We are confident that the chapter bibliographies at the end of the book will lead readers to their work and names. The bibliographies are selective, comprising only a fraction of the papers that inform the text, but these include representative papers from the recent literature and review papers with full reference lists from the older literature.

Echoes of Life is a somewhat unorthodox science book. For a more straightforward textbook on organic geochemistry, we recommend S. Killops and V. Killops's *Introduction to Organic Geochemistry* (2005). For an encyclopedic reference book on biomarkers and their many applications, we recommend the two-volume *The Biomarker Guide* (2005), by K. E. Peters, C. C. Walters, and J. M. Moldowan.

Acknowledgments

Dozens of members of the scientific community have contributed their thoughts, memories, and ideas to this enterprise, corrected and recorrected large chunks of text, and, in a very real sense, turned this book into a community project. They also made writing it a lot more fun than its first author had anticipated.

Roger Summons, John Hayes, Keith Kvenvolden, Kai Hinrichs, Marcus Elvert, and David Ward contributed above and beyond the call of duty, answering endless questions, pointing out research papers we had missed, and reading and commenting on multiple drafts of various chapters and figures. James Maxwell, Pierre Albrecht, Joan Grimalt, Jaap Sinninghe Damsté, Stefan Schouten, Jan de Leeuw, and Simon Brassell all spent hours explaining their work and its genesis, often commenting on long chapters of the manuscript and responding cheerfully to our persistent queries and challenges.

One of the most enjoyable interludes of this project for Susan was the weekend spent in Strasbourg at the invitation of Guy Ourisson. His enthusiasm, insight, wisdom, and passion for science inform this entire volume. Mandy Joye's infective excitement about the bizarre "bugs" she studies, perhaps aided by the free-flowing after-dinner wine at the Hanse Institute where she and Susan met, provided the impetus for the extensive microbiology chapter. Without Mandy's inspiration and encouragement, Susan probably would not have undertaken it. Antje Boetius also commented on portions of this text, offering crucial corrections and updates.

John Sargent, Pat Parker, Richard Evershed, Mike Moldowan, John Volkman, Walter Michaelis, Jeff Bada, Stu Wakeham, Steve Rowland, John Jasper, Ellen Hopmans, Paul Farrimond, Richard Pancost, Archie Douglas, Dick

Hamilton, Paul Philp, Chris Reddy, Tim Eglinton, Colin Pillinger, and Artur Stankiewicz all supplied us with reprints and literature lists when requested, and provided thoughtful answers to our questions about their work or corrected excerpts from the book, often at length.

We hope *Echoes of Life* offers some reward for their time and attention, without which it would surely be riddled with errors, large and small. That some errors remain seems almost unavoidable in a work of this breadth, but we hope such errors are small, and whether large or small, we know they are our own.

Thanks are due to the Organizing Committee of the Twenty-first International Meeting on Organic Geochemistry in Krakow, with special thanks to Artur Stankiewicz for facilitating Susan Gaines's attendance. Thanks also to Mark McCaffrey for arranging her attendance at the 2002 Gordon Conference on Organic Geochemistry, where the three authors gathered for the first time to discuss ideas for this book.

We would also like to thank Tim Knowles and Pete Smith for turning Geoff's sketches into the finished cartoons that adorn the chapter bibliographies at the end of the book, and Stephan Leibfried for serving as a nonscientist guinea pig for some of the figures and random bits of text. Geoff wishes to express his great indebtedness to his wife, Pam, for her patience and forbearance during the many hours he spent hidden away in his study, and we *all* wish to thank her for answering his pleas for "a little typing" as it occasionally spared us from having to decipher his handwritten notes. We are grateful to Ellen Immergut, a political scientist who became intrigued by the book's concept while she and Susan were fellows at the Hanse, and her father, Edmund Immergut, who directed it quickly and painlessly to its home at Oxford University Press while it was still, in fact, little more than a concept.

Without the support of the Hanse Institute for Advanced Study in Delmenhorst, this book would not exist. The Hanse provided generous fellowships and infrastructure for Susan and Geoff, and furnished the three of us with a comfortable meeting place for the first year of work. Its library retrieval service was essential, as it provided easy access to the old papers that much of the first half of the book relies on. The Hanse brings together a dynamic, international mix of scientists and scholars from different disciplines, and our spontaneous evening gatherings and conversations helped to shape this book in many subtle and not-so-subtle ways.

Contents

11. Thinking Molecularly, Anything Goes: From Mummies
to Oil Spills, Doubts to New Directions   259

———

Echoes **of Life**

1

Molecular Informants

A Changing Perspective of Organic Chemistry

*[T]he laws of mechanical affinity may be used for the most complete
physical separation of the substances soluble in certain fluids.
The green pigment of the leaves, the chlorophyll, is known to be
a mixture of pigments, the complexity of which was differently
estimated by different investigators. Chromatographic analysis is
called upon to settle finally this degree of complexity.... Amongst
the adsorption means I can provisionally recommend precipitated
CaCO₃ which gives the most beautiful chromatograms.*
—Mikhail Tswett, 1872–1919, Italian-Russian botanist and inventor of chromatography
   From *Berichte der Deutschen Botanischen Gesellschaft* 24 (1906)

Lodged in the earth's outermost layer, ephemeral scratch on a mineral skin,
life plays cards with a handful of elements—builds molecular extravaganzas
of carbon and hydrogen, oxygen, nitrogen, sulfur, or precious phosphorus, and
forms the pieces to the parts that, assembled, define it. When the game is over, the
cards reshuffled, the parts dismantled—membranes ruptured, shells dissolved,
bones ground to dust—a few of those organic molecules remain in the sediments
and rocks, bearing witness to the distant moments of their creation.

Imagine the most humble bit of life, a microscopic alga basking in the
sun-graced surface of the sea. Think of the tiny animal that grazes on the alga,
dismantling its parts, using the pieces and discarding the difficult-to-digest fats
and sturdy membrane lipids in tiny pellet-like feces that sink slowly into the dark
waters of the deep sea—a thousand meters, two, three, maybe more. Imagine the
bacteria that cling to the pellets as they settle onto the seafloor, zealous recyclers
of organic molecules, using some and transforming others, leaving them stripped
down or broken but still recognizable among the generic mineral bits of shell and
clay that accumulate, particle by particle, year by year, layer by layer. Dig down,
dig back, through meters and kilometers of sediments, through millennia and
epochs, and you'll find them yet, those molecular relics, testaments to that tiny,
light-loving bit of bygone life.

What do those molecules know, what do they have to say? Might they
remember their maker's name and environment, how that tiny alga lived and

died? Was it rich or poor, food plentiful or scarce, the water warm or cold? Perhaps there was a current from the south, or cold nutrient-rich waters upwelling from the deep. Maybe there was a drought in Africa and dry winds blew nutrient-laden dust over the Atlantic, the continent's misfortune a literal windfall for marine algae. Perhaps a meteor fell that year and the light went out of the sky, the temperature dropped suddenly, and the world died in a blink. Or perhaps it was an epoch of tectonic activity and volcanic turmoil, and the atmosphere was laden with CO_2 from the planet's molten depths—the earth's greenhouse fully insulated, temperatures soaring. Might those molecules tell a cautionary tale, perhaps even have some advice for us?

Nature's molecules can be fascinating in their own right, for their particular blend of form and function, the way the carbon atoms twist and turn, the rings they form, the tempting reactivity of a double bond or the weak link to an oxygen atom. For some chemists, an organic molecule is a challenging puzzle to be solved, or a beautiful sculpture to be duplicated, or improved upon, or put to use. But the organic chemist who looks at a rock—a dull gray-white outcropping of limestone, for example—and wonders about the molecules made by algae or bacteria 200 million years ago is likely to have another take altogether. These are chemists who hang out with geologists and biologists, oceanographers and climatologists. Microbiologists, and even molecular biologists. Paleontologists, archaeologists, and environmental scientists. They are thinking about time—distant past, recent past, present, and future. The organic chemistry they practice is a far cry from that laid out in the French chemist Marcellin Berthelot's definitive nineteenth-century treatise, based on man's ability to create and manipulate molecules: "Chemistry creates its object. This creative quality, resembling that of art itself, distinguishes it essentially from natural and historical sciences. The latter have an object given in advance and independent of the will and action of the scientist."* On the contrary, the organic chemistry of rocks, mud, and sea is both a natural *and* a historical science in which molecules are created, and manipulated over time, by Nature. It's a brand of natural history designed for the twenty-first century, when humankind is the most significant player in global cycles and molecules are our unwitting pawns. Ironically, it was developed in the latter half of the twentieth century, when it went against the prevailing trend toward specialization and divergence of disciplines that Berthelot's definition heralded.

If molecules can be witnesses, then the possibilities for what they have to tell us are endless. Not only can they tell us what the climate was like during the last ice age, but they can provide clues as to what the early humans were eating at the time; they can tell us about the advent of life on Earth, or the presence of life on Mars, about the source of pollution in a bay, or of the gasoline an arsonist used to ignite a fire. This idea of molecules as informants, as biological markers or "biomarkers" that carry information through time and space, is relatively recent, the product of largely extant generations of scientists, several of

---

* See *Chimie organique fondée sur la synthèse*, Mallet-Bachelier (Paris), published in 1860.

which are spanned by the three coauthors of this book. When I—the writer and lapsed scientist among us—was a graduate student in California in the mid-1980s amino acids extracted from ancient sediments were providing clues to the dinosaurs' demise and a British research group found that fossil lipid molecules could provide a record of sea surface temperatures during the last ice age. When my coauthor Jürgen Rullkötter interviewed for his first job with Germany's new Institute of Petroleum and Organic Geochemistry in the 1970s, chemists were just beginning to understand that the structures of sterane molecules extracted from mudstone might hold clues to the accumulations of oil hidden deep below the surface of the earth. And when our collaborator, main character, and inspiration, Geoff Eglinton, was a student in Manchester, England, back in the 1940s, chemists were chemists, geologists were geologists, and he was hard put to find anyone interested in small traces of organic molecules in rocks.

Like any good organic chemistry student, Geoff spent his time flirting with explosions in the laboratory and learning how to construct new and ever more exotic molecules from scratch. But as a youth he had read Darwin's *Voyage of the Beagle* and spent many an hour poring over natural history books, and he yearned to connect the disembodied chemistry he was learning in the laboratory to the "real" world outside. One day, when he was hiking in Derbyshire with the university mountaineering club, he noticed a stinky, rubbery, brown tar oozing out of the side of a cliff. At the time, Geoff says, it seemed really weird to him that this gooey brown stuff was leaking out of the dull gray limestone. The geologists in the club told him the stone was compressed marine sediments, made up of the calcium carbonate shells of tiny animals and microscopic algae—zooplankton and phytoplankton that had lived in the surface water of the ocean several hundred million years ago. Wondering if the brown goo contained traces of chlorophyll or other organic molecules made by these ancient organisms, Geoff scooped a bit into a tin and took it home.

"Didn't you know about Treibs?" I asked Geoff when he told me this story. A decade earlier, the German chemist Alfred Treibs had asked similar questions about deposits of petroleum and the oils that came out of certain types of sedimentary rocks. But no, Geoff says, he didn't know about Treibs's work. Petroleum was outside the realm of inquiry for most chemists at the time, something formed deep within the earth—the domain of geologists. When he showed his brown goo to his chemistry professor and expressed an interest in finding out what was in it, the professor just laughed and told him he'd better concentrate on getting his degree first, because it would be a complicated, if not impossible, task to separate such a complex mixture of compounds, let alone figure out what they were. It wasn't until Geoff was a lecturer with his own research group at the University of Glasgow, dutifully engaged in the enterprise of constructing and transforming organic molecules as he'd been trained to do, that he would find himself in a position to contemplate the composition of anything even faintly resembling the Derbyshire goo.

The Glasgow chemistry department of the 1950s and early 1960s was known for its synthetic organic chemists, but there were also a few distinguished natural products chemists, people who were searching for interesting molecules made by nature. They isolated compounds from exotic plants and fungi, following leads from biologists and anthropologists in an attempt to find compounds that were useful as pharmaceuticals, food flavorings, perfumes, or cosmetics. It was by no means a systematic attempt to compile a natural history of life's molecules, but as the catalog of chemical structures and their sources grew, that was what it amounted to—an aspect of the enterprise that Geoff found particularly attractive. Though he counted himself among the synthetic chemists, he was in charge of a new instrument that was particularly helpful in figuring out chemical structures, the infrared spectrometer, and the natural products chemists often brought him their new compounds for analysis. What, he wondered, was the ultimate fate of all these interesting compounds the organisms made? Were they entirely ephemeral, recycled to CO_2 as soon as the organisms died? Or did some end up buried and preserved in the sediments?

But the natural products chemists were no more interested in such queries than the synthetic chemists. Guy Ourisson, a French natural products chemist who occasionally collaborated with his counterparts in Glasgow, says that he had a geologist friend who pestered him for almost a decade to analyze rock and sediment samples, before he finally acquiesced. His friend reasoned that even if most sediments contained less than 3% organic matter, the earth had such a large mass of sedimentary rock that it would contain 10,000 times more organic matter than all the living organisms together. I recently came across a 1973 paper where Ourisson—known to me in the 1980s as the founder of one of the world's leading biomarker laboratories—quotes these same facts and exclaims, with the zeal of the convert, "Organic chemists should therefore mostly study rocks!" But in the 1950s and 1960s, he tells me, he found the concept entirely unappealing. He wanted to know about the molecules of life, and who could imagine that a rock would have anything important to say on the matter? Who would want to study the intractable stuff his geologist friend was offering, when there was a world full of plants and animals that only needed to be tossed into a blender and extracted with solvent to yield exciting molecular discoveries? It was challenge enough to isolate, purify, and determine the chemical structure of a compound in such an extract, let alone in the horrendous, impossible-to-separate mixture of compounds that must have accumulated in the sediments.

Geoff, nonetheless, was still curious. He wrote to his brother and asked him to send another sample of the Derbyshire tar, thinking he might give it a go. But when it arrived and he took another look at it, his reaction was much like Ourisson's. "It looked so unpromising," he says. "I didn't really know what to do with it." He stored it away, along with all the leftovers from the samples the Glasgow chemists gave him for infrared analysis, just in case. Meanwhile, the instruments that would eventually allow chemists to separate such an impossible mixture of organic compounds had just come on the market, and Geoff had acquired one for his laboratory in Glasgow. It would be another few years before it occurred to him that he might use the new gas chromatograph to analyze something like his stinky old tar, but he was, in fact, already headed in that direction.

The gas chromatographs that came on the market in the mid-1950s were based on chromatographic principles that organic chemists, in their ongoing quest to obtain pure substances for analysis, had been using and refining for almost half a century. They could grind up a plant or an insect and attempt to dissolve it in various liquids, using the fact that like dissolved like to extract different types of molecules into different solvents. Then they could evaporate off the liquids, reextract the residual solids, and eventually, if they were lucky, obtain a few pure crystals of a single component from the hundreds in the original mix. But chromatography allowed the systematic separation of Nature's messy mixtures into clans of molecules, and the clans into families, and even, eventually, into individuals. The Russian botanist Mikhail Tswett recognized the basic concept at the turn of the last century when he used column chromatography to separate the pigments in a leaf: he poured an extract of a leaf into the top of a glass tube packed with finely ground chalk, washed it down with solvent, and noted that the pigments moved at different speeds through the column, such that the green extract separated into bands of green, blue-green, yellow, and orange. The premise was prosaic enough: push a bunch of

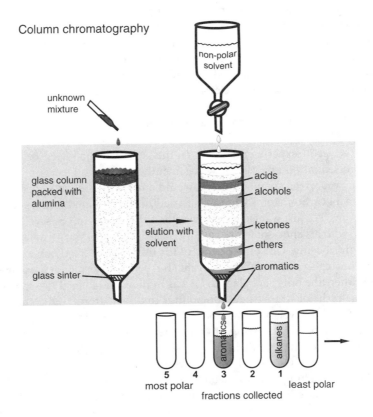

molecules through a column of fine-grained solid with a flood of solvent, and different types of molecules will move at different rates, like runners at a race who start out together but reach the finish line at different times—small and light move faster than large and heavy, graceful faster than awkward, solvent-lovers faster than solvent-phobes—depending on what solvent you use and what sort of stationary phase you pack in the tube. Alumina or silica packing clings to more polar or water-soluble molecules such as alcohols or acids; large, cumbersome, water-insoluble hydrocarbons move fast in solvents like hexane or toluene. Choose well, refine the process, and if you collect what comes out the end of the tube in increments, you have relatively simple mixtures of molecules with similar traits, or even pure compounds.

There are myriad variations on the theme. If, instead of a column, one uses a glass plate covered with a thin layer of silica, places a drop of the sample mixture near the bottom edge of the plate, and stands it in a pool of solvent, one has thin layer chromatography, where the solvent crawls upward and the compounds separate and stick to the silica along the way. Their positions can then be compared with those in a standard mix of known compounds, or the silica can be sliced up and the compounds analyzed separately. If one replaces the glass column or plate with a long, thin tube, and the silica or alumina stationary phase with oil or grease, and then vaporizes the sample mixture and pushes it through with a flow of gas instead of solvent, then one has a gas-liquid chromatograph, now colloquially known as a "GC." Here the mixture separates according to the molecules' size, volatility, and preference for the gas or liquid phase. Various sorts of detectors can be connected to the GC, so that as the organic compounds exit the tube they generate an electronic signal, which, in turn, makes a pen slide up and down on the moving graph paper of a chart recorder, thus indicating both the amount of compound and length of time it took to make its way through the tube.

The GC Geoff experimented with in the late 1950s had been designed for food and essential oil analyses. It was generally used for mixtures of relatively light, easily vaporized compounds that would move with the gas, molecules with fewer than 14 carbon atoms—but Geoff tried his out on anything he could get his hands on. "We just wanted to see what the thing could *do*," he says, sounding even now, with his shock of white hair and beard, like a child with a new toy he's been caught abusing. One of the perks of his infrared spectroscopy expertise—acquired during an unbearably hot summer he spent as a postdoc at the University of Ohio, where the only cool place to be was in the infrared facility—was that he had a huge cache of strange new compounds to choose from, things the designers of the GC had never even imagined. "We rooted around in there, picking out anything that looked interesting. It's a wonder we didn't poison ourselves, since half the time we didn't know what it was we were analyzing."

In these early GCs, the oil stationary phase was coated onto a fine powder and packed into a long, narrow metal tube that was housed in an oven or heated jacket. Geoff tried reducing the oil coating to a bare minimum, increased the flow of gas through the column, decreased his sample size, cranked the temperature up as high as he dared, and started pushing through bigger and bigger molecules.

He found he could separate mixtures of long-chain hydrocarbons up to 30 carbons long. He even tried running the 27-carbon cholesterol, not really expecting it to get through the column, with its bulky ring structure—and was astounded to see a peak appear on the chromatogram. "It was a rather fat one," Geoff says, "took a while to come through...but there it was. After that we started running all the big molecules we could find." The bulky cholesterol molecule is ubiquitous in animals, where it serves as a rigid insert in the otherwise flexible cell membrane. But it was its special role as a culprit in heart disease that made it a hot topic among biomedical researchers in 1959, when Geoff and his small research group published a description of their GC analysis. They received thousands of requests for reprints of the brief paper—and, more importantly, they received samples of all sorts of things to analyze with the new method.

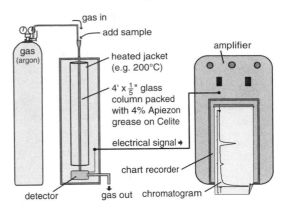

Gas chromatograph, 1950s

One very famous Cambridge plant biochemist who had spent a good part of his career studying the wax coatings of leaves offered Geoff his legacy: dozens of cigarette tins full of leaf waxes that had been extracted and recrystallized from hundreds of species of plants. Each wax contained a complex mixture of compounds and before the GC could separate them, they had to be passed through a simple alumina chromatography column and divided into families based on polarity. The individual compounds in each family could then be separated on the GC and tentatively identified from a comparison of their behavior—how long they took to get through the column at a given temperature—with that of known compounds in a standard mixture. The easiest family of compounds to analyze, and by far the most prevalent in all of the leaf waxes, was that of the normal alkanes. From most organic chemists' and biochemists' point of view, these were the most boring compounds imaginable: straight chains of carbon and hydrogen atoms, distinguished only by their different lengths, with no branches or interesting rings or flat double bonds, no oxygen or nitrogen or sulfur atoms to break the monotony.

Geoff remembers that one chemist who shared his laboratory in Glasgow for a time found it so incomprehensible that he finally paused by Geoff's lab bench and exclaimed: "*Why* are you people working on *n*-alkanes?! They don't *do* anything!!" Geoff says he mumbled something in reply, but he can't remember what. The man was right: the *n*-alkanes *are* boring chemically—they don't do anything. They have no reactive functional groups dangling like charms from the ends of their chains—no water-loving oxygen atoms as in the hydroxyl groups of the alcohols or the carboxyl groups of the acids, no ether links or sweet reactive esters

Normal alkanes: different representations
used to understand molecular behavior

methane, CH_4, the simplest alkane

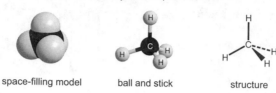

space-filling model   ball and stick   structure

nonacosane, $C_{29}H_{60}$, a typical long-chain *n*-alkane

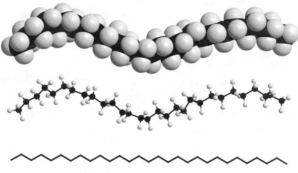

line drawing (carbon skeleton)

$$CH_3(CH_2)_{27}CH_3$$
condensed formula

to attract a water molecule or enzyme, not even a double bond or a ketone's carbonyl group. Even biochemically, once these leaf-wax alkanes are made, they just sit there on the outside of a leaf with their long sturdy chains lined up side by side and layer upon layer in wax crystals. With no reactive double bonds or other weak links for enzymes to attack, they are unattractive to bacteria and fungi, and with no oxygen atoms or functional groups, they are decidedly nonpolar and hydrophobic, completely immiscible with water.

About all one could say about the leaf-wax *n*-alkanes at the time was that their hydrophobia and general inertness made them particularly well suited to their job of protecting plant leaves from dehydration and disease: they kept water inside and pathogens outside. From a purely chemical point of view, they were decidedly uninteresting. But they were so prevalent, so easy to observe, that Geoff couldn't just ignore them. He was exploring, like the nineteenth-century

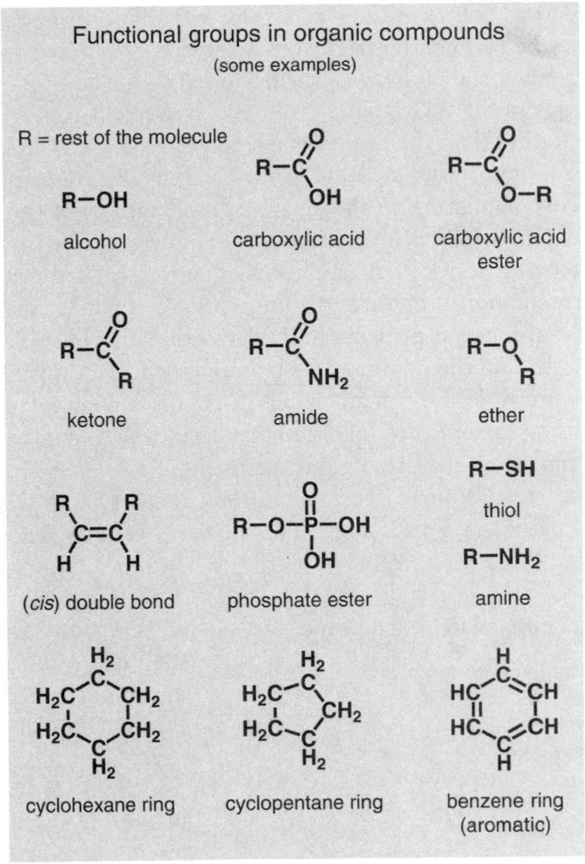

Functional groups in organic compounds (some examples)

naturalists who traveled the world describing and naming all the plants and animals they could find. Geoff had embarked with his chromatographic ship to explore the molecular world of those same creatures, and he couldn't very well ignore all the dull brown finches and sparrows just because the tanagers were more beautiful or the eagles more spectacular. Perhaps, like the Galapagos finches that inspired Darwin's monumental observation that "one species had been taken and modified for different ends," these boring *n*-alkanes would prove more enlightening than the fantastic ring structures Geoff's colleagues were building and isolating.

Usually, it's the thrill of conquest or challenge of complicated new structures that attracts organic chemists: like a bird or a beautiful woman, a molecule is more interesting and desirable for being exotic or elusive, or, as one prominent Swedish organic chemist was fond of saying, "a tough nut to crack." The *n*-alkanes offered no structural challenges and were easy to identify: their peaks in the chromatograms appeared in order, shortest chains first, longer later, and one only needed to compare their positions with those of a standard mix of *n*-alkanes that one had synthesized from scratch, a relatively easy task. But as Geoff looked at his chromatograms, which showed the makeup of the entire mixture of *n*-alkanes in the wax from a given leaf, he began thinking of molecules in a new way. Rather than the beauty of a single character in the series, for they were all rather homely, what stood out in such an analysis were the patterns of peaks in the chromatogram: in the arrays of *n*-alkanes from the leaf waxes, it was the different *distributions* of the same set of molecules, the variations within this one homologous series, that caught his attention. Each chromatogram was a graphic representation, a picture or a fingerprint of a particular species, characterized by the absence or presence and relative amounts of the homologues. It was these patterns and the relationships between them that was interesting, and that would eventually lead Geoff and a handful of his contemporaries into a new way of considering the mixtures of organic molecules they obtained from plants and animals, and, ultimately, rocks and sediments.

A chemist friend who worked in the food industry told Geoff that such patterns were being used to classify the mixtures of fatty acids—which are simply long-chain *n*-alkanes with a carboxylic acid group at the end—in different animal fats and vegetable oils, and it occurred to Geoff that the patterns of peaks in his chromatograms of leaf wax *n*-alkanes might follow the taxonomy of the plants. They might provide new information about differences and similarities between species, even about the evolutionary pathways that led to their differentiation. To test this idea, he needed waxes from species that would show a clear evolutionary pattern, something like Darwin's Galapagos finches. He needed a large group of closely related genera of plants, species that had evolved in an isolated environment with a minimal amount of differentiation.... What he needed, of course, was an island.

Gas chromatograms of leaf wax alkanes

1960 vintage chromatogram

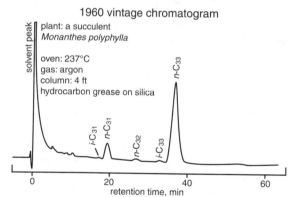

Modern chromatogram with well-resolved peaks

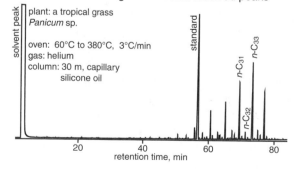

"We had a postdoc in the lab," Geoff says, "who'd come from this way out place, the island of Gomera, in the Canaries. You can't think of a place farther from the center of science at the time. He told us they used a whistling language to communicate across the island's hills and valleys, because there were so few phones.... I think that's what got me interested in the Canaries, to tell the truth, this fellow's stories." Once he started thinking about it, however, the Canary Islands seemed just the thing. They were relatively undeveloped, so there probably wouldn't be a lot of introduced species; the botany had been well studied; and they were volcanic and geologically young, so the species would be recently differentiated and tightly grouped.

Of course, it wasn't an easy thing to mount a scientific expedition to such a place, but a chance encounter at a university banquet in Cambridge around this time presented an opportunity: Geoff found himself sitting next to the laboratory manager for the chemistry department, who, it turned out, had been involved in providing instruments to the only prominent organic chemist on the island of Tenerife, at the university where Geoff's Gomera postdoc had been trained. Over dinner, the Cambridge laboratory manager told Geoff that the chemist had funding, but couldn't get new instruments past the local customs office. He suggested that Geoff, as a foreigner, could take a new GC into the country with him and then resell it to the chemist, bypassing customs. In return, the Tenerife chemist would pay part of Geoff's project expenses and provide laboratory facilities. So that's what Geoff did: he got a grant from the Carnegie Foundation, took a leave from teaching at Glasgow, and set off with his young family and his Ph.D. student, Dick Hamilton, on a scientific-*cum*-smuggling expedition to the wilds of Tenerife.

Tenerife in 1960 was not extensively developed. Nor was it, to Geoff's dismay, particularly wild. He had chosen the perfect group of native plants for study, a subfamily of drought-resistant succulents that have particularly thick waxy leaf coatings. In a recent article, botanists had even compared its evolution to that of the Galapagos finches. Unfortunately, most of Tenerife was under cultivation, and the pristine native flora had long been overrun by introduced banana and palm trees. This was a far cry from the isolated Galapagos! But once word got round that the British scientists were looking for native vegetation, the locals guided them into the island's isolated wild canyons, and the botanist at its magnificent Botanical Gardens offered to identify their finds.

Analysis of the *n*-alkanes was relatively easy: dip the leaves in chloroform to extract the waxes; pour the extracts into chromatography columns packed with alumina; collect the least polar fraction, composed mostly of *n*-alkanes and their one-branched brothers, the *iso*-alkanes; and then run this hydrocarbon fraction on the GC. The amount of each alkane could be determined from the size of its peak in the chromatogram. Geoff and his student prepared histograms that showed the relative amounts of different homologues for each species, and then they compared them, looking for patterns that might link and differentiate the taxonomic groups.

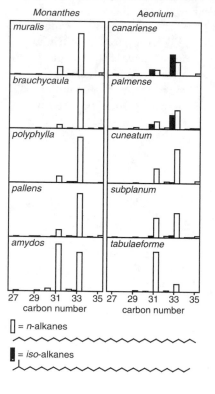

Comparing alkane histograms, 1962
Two genera of Canary Island succulents

There were patterns, but they were not as distinct as one might have hoped. Each species did indeed have a characteristic distribution of alkanes, an individual fingerprint that might allow one to identify an otherwise hard-to-distinguish species. But the patterns didn't reveal the evolutionary links or common ancestors of the related species within a genus.

"Were you disappointed?" I asked Geoff the other day, a little disappointed myself, as I looked at histograms from the Canaries expedition. Geoff shrugs off the question, but I wonder if he really remembers or if it's only now, in hindsight, that he recognizes the importance of the other, seemingly mundane observations in his 1962 paper. The attempt to use leaf wax hydrocarbons as a taxonomic tool was only marginally successful, but the study revealed some of their essential properties as molecular chroniclers—as biomarkers—before the concept even existed: that the *n*-alkanes range from 25 to 35 carbons long, with the 27-, 29-, and 31-carbon compounds particularly prominent; that those with odd numbers of carbon atoms dominate the mixture by a factor of 10 over even-carbon-number chains; and that no matter how old the plant or where it grew or what time of year one chose to analyze it, the species fingerprints remain true, the *n*-alkane distributions nearly constant.

This family of long-chain *n*-alkanes with odd numbers of carbon atoms would turn out to be specific to land plants, found not only in their leaf waxes, but in their stems, flowers, and pollen. The odd-over-even preference is easily explained by our knowledge of how organisms build unbranched carbon chains, a core biosynthetic process that appears to be common to all forms of life. Unbranched chain structures are assembled two carbon atoms at a time: acetic acid is combined with malonic acid, losing a carbon to form a four-carbon acid, which then latches onto another malonic acid, losing a carbon to form the six-carbon acid, and so on. Accordingly, the fatty acids and the alcohols—biosynthesized via a simple reduction of the acids—have carbon chains with primarily *even* numbers of carbon atoms. The *n*-alkanes like those in the leaf waxes, however, are produced by snipping the carboxyl carbon off the end, which results in chains with *odd* numbers of carbon atoms.

The same physical qualities that make these long, water-phobic hydrocarbon chains such good leaf protectors also make them resistant to breakdown in the environment once the plant dies. Once they hit the dirt, so to speak, they go where it goes, eroded by rain, washed into the sea by rivers, blown out to sea with the dust— indeed, small particles of wax are even blown directly off the surface of leaves and carried long distances on the wind. The presence of long-chain *n*-alkanes in marine sediments, far from their origin in both time and place, thus records

Biosynthesis of unbranched carbon chains: odd vs. even numbers of carbon atoms

the movements of wind and water from land to sea, and even, if one looks more closely at the carbon atoms in their chains, the ecosystems and climates where the plants that made them formed. Some four decades after Geoff's Tenerife smuggling exploit, he and Jürgen are once again engaged in a study that has them poring over histograms of leaf wax alkanes. This time, their extracts come not only from the leaves of living plants, but from sediments buried beneath the floor of the South Atlantic Ocean; this time, the patterns tell a story that spans nearly 200,000 years of changing climate and vegetation across an entire continent. But more of that later. First, we need to take a look at the intervening decades of molecular exploration and discovery by an ever-expanding legion of rock-, mud-, and sea-loving organic chemists, among others.

2

Looking to the Rocks
Molecular Clues to the Origin of Life

[I]t happens also in chemistry as in architecture that "beautiful"
edifices, that is symmetrical and simple, are also the most sturdy.
—Primo Levi, 1919–1987, Italian chemist and writer
   From *The Periodic Table* (1984)

Every chemical substance, whether natural or artificial, falls into
one of two major categories, according to the spatial characteristic
of its form. The distinction is between those substances that have
a plane of symmetry and those that do not. The former belong to
the mineral, the latter to the living world. . . . I could not indicate the
existence of a more profound separation between the products born
under the influence of life and all others.
—Louis Pasteur, 1822–1895, French chemist and microbiologist
   From *Oeuvres de Pasteur (1922–1939),* Vol. 1 (1922)

When Geoff started analyzing leaf waxes, he hadn't consciously been looking for compounds in living organisms that are likely to survive in sediments, attacking his Derbyshire question from its other end, as is now a standard ploy in biomarker studies. But when he returned from the Canaries in 1961, it began to dawn on him that this was, in fact, what he had done. It also became apparent that he was not alone in his curiosity about the fate of organic compounds in things like rocks and tars. In fact, he had some very good company.

Two prominent Nobel chemists gave talks in Glasgow that year, both posing fundamental questions about the stuff of life. And both, Geoff found, shared his maverick interest in the organic chemistry of rocks and oils. Sir Robert Robinson sought to resolve the apparent paradox of petroleum, which appeared to have formed from the buried detritus of plants and algae that had been subjected to high pressures and temperatures, and yet also contained organic compounds that were quite different from any known in plants and algae. And Melvin Calvin, who had discovered how CO_2 is converted to organic molecules in photosynthesis, talked about the origin of life. Experiments done in the past decade had given new meaning to such questions, by showing that simple organic chemicals could form spontaneously under conditions similar to those that were likely to have prevailed on the early, prebiotic earth. The mystery of life's origin suddenly seemed approachable, and discussions had shifted from the purely theoretical and philosophical toward the empirical, and from the paleontological to the chemical. On the contemporary earth, life has something of a monopoly on making organic compounds. But if amino acids and sugars, the basic building blocks of living things, were produced by other than living things on the early earth, before the advent of life, then one could imagine the world before photosynthesis, before bacterial chemosynthesis. One could imagine some form of chemical evolution that predated Darwin's biological evolution and launched the simplest, most primitive of bacteria. And one could look to the oldest rocks for evidence.

Calvin immediately embraced Geoff's idea of using the new gas chromatographic separation techniques to facilitate the hunt, and he invited Geoff to spend a sabbatical year working in his laboratories in Berkeley. At a time when the natural sciences were growing and dividing into ever more disparate new disciplines, Calvin had gathered scientists from all camps—biologists, biochemists, organic chemists, and physicists—into a specially designed laboratory building that was a model of interdisciplinary collaboration. Geoff says it was a wonderfully creative environment when he was there in 1963. The building was round and set up like a doughnut, with the labs in a ring around a meeting area in the middle, and no doors, so that everyone always knew what everyone else was doing. "The only problem," he says, "was that the well-equipped chemistry lab Calvin promised me was completely empty. There wasn't even a bloody test tube!" Once he got over the shock, however, starting out with a blank slate turned out to be not such a bad thing. It so happened that a local company had started producing commercial GCs, so he acquired the most up-to-date instrument. And a student from someone else's project heard about what he was doing—or supposed to be doing—and showed up one day to help. Ted Belsky turned out to be indispensable to the whole undertaking. He'd served a stint as a mortuary technician in the U.S. Navy and was very practical, the sort who got things done: he begged and borrowed equipment from around the Berkeley campus, and Geoff soon found himself with a fully furnished lab.

Calvin had acquired samples from some of the oldest known sedimentary rock formations, and he was impatient for Geoff to start analyzing them. So far, a clear history of life as told by fossils went back only about 600 million years, to the advent of multicellular organisms a little before the beginning of the Cambrian period. The earlier strata bore so few legible clues to history that paleontologists had delegated them to geologic prehistory and referred to the entire expanse as the *Precambrian*, undivided by the eras, periods, and epochs that they meticulously assigned based on the fossil assemblages in more recent strata. Only in the past decade had radiometric dating techniques revealed that this prehistory corresponded to seven-eighths of the earth's existence, and it was just in the past few years that geologists and paleontologists had begun finding indirect, circumstantial evidence that life may have been present for most of it.

The earth's oldest minerals had been dated at about 4.5 billion years, and a sedimentary formation in Greenland was dated as 3.8 billion years old. Other sedimentary rocks in Africa, the northeastern United States, and Canada ranged between 1 and 3.5 billion years old. All of them contained organic matter, and many had laminated, mushroom-shaped "stromatolite" structures that looked very much like some that grow in shallow waters on the contemporary earth. In modern stromatolites, the distinctive laminated layers of mud and calcium carbonate are formed by photosynthetic uptake of CO_2 and excretion of a sticky substance by cyanobacteria. Were these ancient stromatolite structures formed by similar, oxygen-producing photosynthetic bacteria?

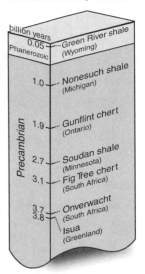

Ancient rocks, 1960s

The rocks Calvin wanted to analyze spanned the period from 1 to 3.7 billion years ago. Paleontologists had just developed new microscope techniques that allowed them to detect the fossils of microorganisms, and there were recent and very tentative reports of fossil bacteria in some of these rocks: tiny marks that appeared to be coated with residual organic matter and bore a close resemblance to the shapes of contemporary bacteria. In younger rocks, the cell wall and detailed structure of the bacteria might actually be apparent with an electron microscope, but in these ancient Precambrian rocks it was unclear whether the marks were the remains of organisms, or simply serendipitous marks left in the matrix of the rock by water, or gases, or mineral crystallization.

If there had been microorganisms, then perhaps they had left molecular remains. Maybe the rocks contained recognizable remnants of biological molecules, or clues of early life-forms with different biochemical systems. Or maybe they contained traces of prebiotic organic chemicals like those formed during experiments that simulated conditions on the early earth. But how could one recognize primordial biological molecules if they were radically different from contemporary ones? For that matter, how could one distinguish a biological from an abiotic or prebiotic organic molecule?

There was a conscious naiveté to the assumptions that had to be made as Calvin, Geoff, and the small team they assembled tried to track life backward through geologic time, an unavoidably circular logic. They were trying to find out when and how biochemical pathways had changed over time and, ultimately, how they had come into existence. And yet, of necessity, they were looking for molecules they could recognize, molecules that resembled those they knew from contemporary life-forms. The assumption here was that even as life-forms evolved and changed dramatically, the materials used in building them had remained constant over hundreds of millions of years. Such an assumption mirrored nineteenth-century geology's premise that the geological processes now at work within the earth have operated with general uniformity over immensely long periods of time: it was Charles Lyell's Uniformitarianism applied to biochemistry. Geology has now outgrown Lyell's brand of all-encompassing Uniformitarianism, but it was doctrinaire in the 1960s when Geoff and Calvin started dabbling in geology.

Microfossils in the Gunflint chert
1960s microphotographs

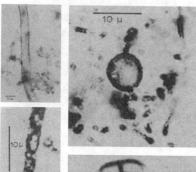

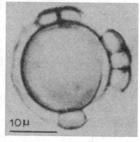

Its basic tenet—that the present is the key to the past—was crucial in understanding the geological record, and has proven valid in many, if not most, instances. To what degree it would hold true in the case of biochemical pathways remained to be seen, and the chemists adopted it not as doctrine or even hypothesis, but as a provisional supposition, which they questioned at every turn.

Though Calvin wanted Geoff to leap right in and start analyzing the Precambrian rocks, Geoff was skeptical. For one thing, the rocks had been sitting on the shelf for years, and Calvin kept going in and picking them up to look at them. They were likely loaded with contaminants from his hands and the environment. And even if they could cut off the outside layers and get a pristine sample, Geoff balked at beginning their hunt with rocks about which so little was known, where the organic matter was so sparse and difficult to detect, and the possibility of contamination with organic compounds from later deposits so high. Had the sediments formed in a lake or marsh, near the coast, or in the deep sea? How deeply had they been buried, how much had they been heated within the earth, and how long? It seemed to him that all of these factors would have had an effect on the structures of any organic compounds they found, and on the degree to which they might or might not resemble their biological precursors. How could they recognize clues to past life-forms in such ancient rocks, when they didn't even know what they looked like in younger rocks? Which of life's molecules survived the first onslaught of bacterial breakdown in the sediments, and how were they transformed? For that matter, obtaining and analyzing an extract from a rock was not quite the same as dipping leaves in chloroform—shouldn't he practice a little before starting in on Calvin's precious Precambrian collection?

Geoff was still getting set up and considering other options when a geologist he'd been talking to in the earth science department came dashing into his half-furnished lab, excitedly waving a single sheet of paper. It was just a one-paragraph abstract from a recent conference, but it was, as his colleague surmised, exactly what Geoff was looking for: a couple of scientists at the U.S. Bureau of Mines had tentatively identified two hydrocarbons in a sample of 55-million-year-old shale from the Green River Basin in the western United States. Geologists had been looking for oil in the area since the 1920s, so quite a lot was known about the basin's geologic history, which was apparently relatively simple. Much of the Great Plains region was covered with shallow freshwater lakes during the Eocene epoch, and the rock that the Bureau of Mines scientists studied was a shale made up of compressed, solidified sediments that had formed in these lakes. The shale contained large numbers of fossils and a relatively high content of organic matter, and it was old enough that sediment bacteria would long since have eaten or transformed the most reactive organic compounds, leaving only those that were stable on a geological timescale of tens of millions of years. But it had never been deeply buried and subjected to the high pressures or temperatures that were likely to completely destroy most organic molecules. Most promising of all, the two compounds identified, though transformed by bacteria and time, retained certain structural idiosyncrasies that linked them firmly to source molecules in living organisms, and were just the sort of molecules Geoff would need if he were going to track life back into Calvin's ancient Precambrian rocks.

In his initial experiments, Geoff used the Green River shale to develop dependable extraction and separation procedures, confirming and extending the results of the Bureau of Mines scientists. The organic matter in the rock was the chemical debris from a wide range of organisms and contained a much more complex mixture of compounds, in more minute quantities, than anything he'd ever analyzed—and most of that was bound up in an insoluble, unextractable, and impossible-to-analyze organic matrix. You could grind up the rock, extract it with every imaginable solvent, even dissolve the mineral in acid, and end up with everything in solution *except* its organic matter, some 90% of which remained like a bad joke in the bottom of the flask, a fine brown or black powder that to this day defies chemists' attempts to fully characterize it. Geologists had given it the Greek name "kerogen" because it produced a waxlike oil when heated. No one knew quite what it contained or how it had formed. It appeared to be made of all sorts of organic molecules glued together in an apparently random manner by bonds so strong that it took excessive heat or a strong oxidizing agent to break them. Of the 35% organic matter in Geoff's Green River shale sample, some 95% was insoluble kerogen. He did what any good organic chemist would do, and focused his attention on the 5% that he could get into solution: after slicing off the rock's outer surface to remove contaminants, he ground it to a powder, heated it in solvent, and then threw away the insoluble residue. The result was a surprisingly yellow-green solution that he figured must be full of pigments, probably from algae that had lived in the lakes.

Gratifying as it was to actually be able to *see* pigments left in a rock 55 million years ago, for the moment Geoff concentrated on the alkanes, which were easier to analyze and, with their complete lack of double bonds and functional groups, most likely to have persisted in older rocks. After all, Calvin's rocks were almost 50 times as old as the Green River shale! He poured the extract onto an alumina column, which retained all the pigments and other relatively polar compounds, and collected the first fraction to wash through. This contained only the colorless alkanes, but it was still a much more complicated mixture than anything he'd analyzed in the leaf waxes, so he split it into two fractions by pouring it into a molecular sieve, a mineral that catches the slim *n*-alkanes in its pores and separates them from all the alkanes that have branches or bulky rings. If he then ran each of these simplified mixtures on the GC, he could separate many of their individual compounds.

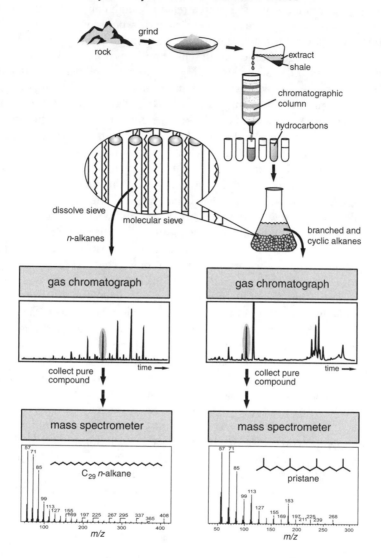

Analysis of hydrocarbons in ancient shales

Once he had the compounds separated, he needed to figure out what they were. This was no problem for the *n*-alkanes, which he could identify from their positions in the gas chromatogram. The branched and cyclic alkanes were more difficult, however, as there were myriad structural possibilities, and Geoff wasn't sure what to expect. Even when he could synthesize a structure that matched up with an unknown peak in the chromatogram from the shale extract, he couldn't be positive they were the same, as different alkanes could easily exhibit similar behavior in the GC column, and their peaks might well overlap. Luckily,

Al Burlingame, a young analytical chemist who worked with Calvin, had set up an instrument that showed great promise as a tool for determining the molecular structures of organic compounds, a state-of-the-art mass spectrometer that required only a trace of sample. When Geoff's separated compounds came out the end of the GC column with the gas, they cooled and condensed into a few drops of liquid. If he stood there patiently with a little vial, he could collect the purified compounds and then give them to Burlingame to analyze on the mass spectrometer.

The mass spectrometer is a complex instrument based on a rather juvenile impulse: the best way to see how something is put together is to break it apart. A high-energy electron beam breaks the molecules into charged fragments, or ions, that are then sorted and registered by a detector according to their masses. If the mass of a fragment is known, then the atomic weights of the elements—12 mass units for carbon, 1 for hydrogen, 16 for oxygen, and so forth—can be used to figure out how many carbon, hydrogen, oxygen, or other atoms it has. Some of the molecules in a sample will only be nicked by the electron beam, which just knocks off an electron and produces a molecular ion, so you can calculate the molecular weight of the intact molecule. Others will be hit more directly and break at the structure's weak spots, yielding large quantities of the fragment that corresponds to the most stable part of the molecule—the base ion—and lesser quantities of various other fragments. It's like dropping a tray full of wineglasses: a few will escape unscathed, some will shatter completely, but most will break into three pieces—the round cup, the stem, and the thick-glassed base. In a few cases the stem may break in half, or the round part may shatter, but the thick-glassed base will usually stay intact. By examining all these fragments one can, in principle, figure out what an intact wineglass looked like, even if one has never seen an intact wineglass. A branched alkane, for example, usually breaks next to the branches. The mass spectra of two such alkanes in the Green River shale exhibited a very distinctive pattern of fragmentation and showed that they were both open chains of carbon atoms with branches composed of a single-carbon methyl group at every fifth carbon atom, and a total of 19 or 20 carbon atoms—and confirmed the structures of the compounds pristane, with 19 carbon atoms, and phytane, with 20, that the Bureau of Mines scientists had proposed.

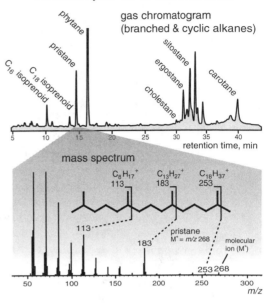

Identifying pristane in the 50-million-year-old Green River shale

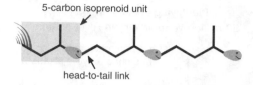

5-carbon isoprenoid unit

head-to-tail link

It just so happened that the distinctive five-carbon pattern that made the mass spectra of these two compounds so readily recognizable also linked them firmly to biological parent compounds and made them particularly promising as indicators of life in the ancient rocks that Geoff was setting out to analyze. The carbon skeletons of the family of compounds with this pattern, known collectively as isoprenoids, are constructed by stringing these five-carbon branched segments together like beads, joining them head to tail into longer chains and rings, which form the basis for a wide variety of lipid molecules found in all forms of plant, animal, and bacterial life.

It's a universal method of construction, and it results in a recognizable and idiosyncratic architectural style that is built into the most enduring part of the molecule—its carbon skeleton—and has a very good chance of surviving the ravages of decay, time, pressure, and even heat.

The two isoprenoids that were of such interest in the Green River shale didn't look like anything Geoff had identified in the leaf waxes, but they were, nonetheless, quite familiar: they looked for all the world like the lopped-off tail of the chlorophyll molecule. In fact, it was the pared down version of chlorophyll's other piece, its recalcitrant heart, that Alfred Treibs had identified in coal and in oils from even older shales back in the 1930s. Treibs had noticed a certain familial resemblance between the blood-red pigment molecules he extracted from the oil and the chlorin structure at the heart of the green chlorophyll molecule, and had proposed a chemically probable pathway to get from one to the other.

Though Geoff was unaware of Treibs's work when he first turned his attention to rocks, it wasn't long before he realized that he had been unwittingly following its lead and that a small tribe of chemists was scattered around the world doing likewise, defining a field they called organic *geo*chemistry. Many of them were working for oil companies that forbid them to publish their results, so the organic geochemistry literature was scarce and often not known to organic chemists working in universities and research institutes. But a few key observations did make their way into the wider scientific community during those early years, published either by petroleum geochemists who managed, by sheer force of will, to sidestep company policies, or by geochemists at the enlightened Mobil Oil, which actually encouraged its scientists to publish. Keith Kvenvolden, who worked for Mobil, says that the management had figured out that when its scientists' work went unpublished, there was no outside scientific scrutiny to verify it.

While Geoff had been in the Canary Islands noting the odd-over-even carbon number preference of long-chain *n*-alkanes in leaf waxes, a couple of oil company geochemists in Texas had published a paper describing similar patterns in sediments and rocks. The *n*-alkanes in Geoff's Green River shale analysis also showed this distribution: a predominance of compounds with an odd number of carbon atoms, with the 27-, 29-, and 31-carbon compounds particularly prominent. But whereas none of the plants Geoff had analyzed contained any *n*-alkanes

Geological transformation of chlorophyll *a*

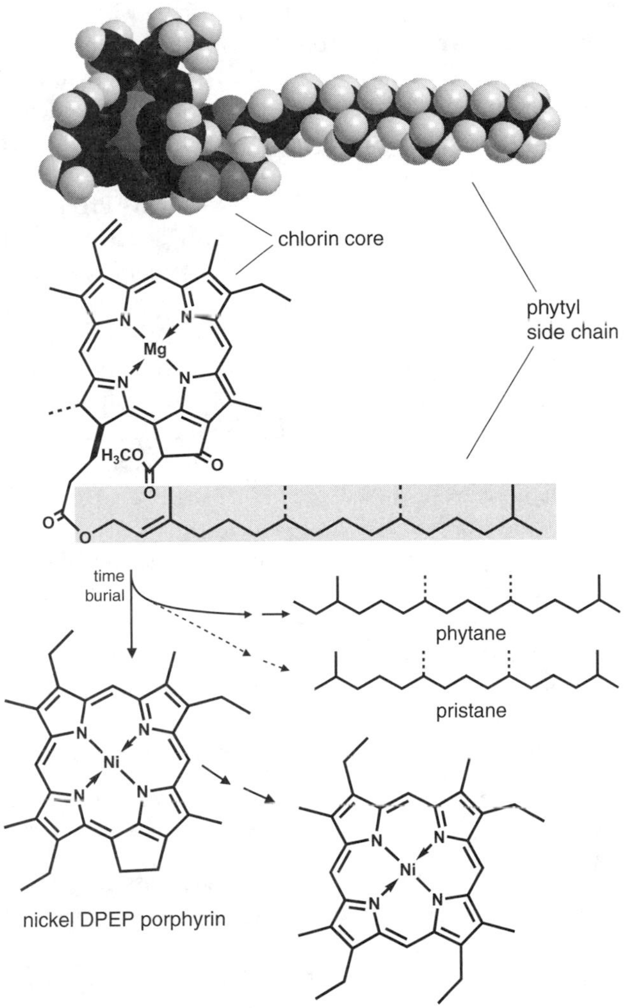

chlorin core

phytyl
side chain

H₃CO

time
burial

phytane

pristane

nickel DPEP porphyrin

nickel etioporphyrin

fewer than 25 carbon atoms long, the shale contained the 17-carbon *n*-alkane in almost equal abundance with the long-chain compounds. Had the *n*-alkanes in the Green River shale come from land plants, washed into the lake from the surrounding watershed? Or from the algae growing in the lake? Or both? What about the 17-carbon *n*-alkane that was so prevalent? Had that come from the algae? Or, perhaps, from bacteria living in the sediments? Very little was known about the lipid constituents of either algae or bacteria. For that matter, this short-chain *n*-alkane might not have been synthesized as such by any organism. It might have been a chemical degradation product of some other compound altogether. These questions about the precise sources of the compounds extracted from the shale would remain unanswered for years to come. But there was little doubt that the pattern of alkane abundance and the prevalence of the isoprenoids, pristane and phytane, in the Green River shale constituted a record of life in North America 55 million years ago, even if scientists couldn't yet decode the details. With the hope that similar patterns might provide a record of life in rocks formed during the long prehistory of geologic time, when fossil-forming organisms were scarce or absent, Geoff turned his attention to analyzing Calvin's Precambrian rocks.

Copper mining in northern Michigan had provided access to the gray and black shales of the one-billion-year-old Nonesuch Formation, which were so rich in organic matter that drops of oil seeped from their pores. Unlike the Green River shale, which was loaded with fossils of plants and animals, the only physical fossils that had been found in these rocks were submicroscopic bits of organic material in the shape of microbial filaments or spores. Geoff began by extracting the alkanes from the oil, and found that the *n*-alkane distribution was quite different from what he'd found in plants and what the Texas geochemists had found in contemporary sediments. Like the Green River shale, the Nonesuch contained an abundance of the 17-carbon *n*-alkane, but in the Nonesuch only *n*-alkanes with fewer than 20 carbon atoms were present, and these included both odd and even carbon-number homologues, with little preference for the odd. Where did these short-chain *n*-alkanes come from? Nothing like them was known in organisms, and indeed, they seemed to defy what was known about the biosynthesis of straight-chain compounds. The Texas petroleum geochemists had found similar distributions in extracts of crude oil from much younger rocks. Could such compounds have been generated by some purely chemical, nonbiological means in the earth's crust?

The one-billion-year-old Nonesuch shale
Gas chromatograms of alkanes, packed column, 1964

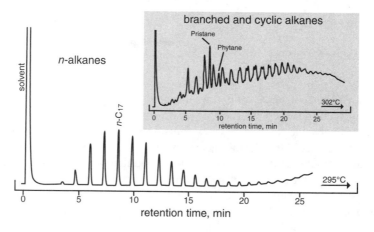

Chemists trying to simulate the chemical reactions that might have occurred on an early, lifeless earth had generated hydrocarbons from inorganic chemicals by dissolving iron carbide in hydrochloric acid, or by applying an electrical discharge to frozen methane. But these chemical processes produce thousands of different combinations of carbon and hydrogen, rather than the discrete selection of *n*-alkanes Geoff found in the Nonesuch shale oil. This alone implied that these hydrocarbons derived from some more discriminating biological reactions. There was, however, another, more direct and informative indication of past life in the one-billion-year-old rocks: the branched alkane fraction of the extract was loaded with pristane and phytane.

As Geoff and his colleagues were nearing completion of the Nonesuch study, Calvin had a visit from Phil Abelson, the editor of *Science* and director of the Carnegie Institute's Geophysical Laboratory in Washington, D.C. Abelson was interested in the chemistry of organic compounds on the early earth, and had himself been trying to determine how long the distinctive components of life such as proteins and amino acids could survive when they were trapped in the mineral matrix of a fossil shell. Geoff still remembers showing him a chromatogram of the branched alkanes in the Nonesuch shale, pointing out the pristane and phytane peaks and mumbling that they might be of interest—at which point the rather large, imposing senior scientist suddenly looked up from the chromatogram, fixed him in his gaze, and exclaimed heartily, "You bet!"

Abelson told Geoff to hurry up and submit the Nonesuch study for publication, which turned out to be timely advice: unbeknownst to either of them, Warren Meinschein, a geochemist at Esso Oil who somehow managed to keep a foot in academic science—and publish his results—had been thinking along similar lines and working on the Nonesuch shale with a couple of Harvard paleontologists. Abelson published Meinschein's paper back to back with Geoff's in the July 1964 issue of *Science*. Both reported the identification of fossil organic molecules in the Nonesuch shale—but different ones. Geoff and crew had identified the acyclic isoprenoids phytane and pristane, and Meinschein and his colleagues had found porphyrins, the complex ring structures that Treibs first identified in oil. It was a sensational finding, but they needed more information, more knowledge, before it could be verified, let alone teach them anything about Precambrian life. Were the porphyrins and isoprenoids remnants of chlorophyll—its chlorin heart and phytyl tail—that had served to harvest light and energy for organisms that lived more than a billion years ago? Or had they come from some other source? Longer isoprenoid chains had just been identified in the membranes of certain salt-loving microbes, and phytane and pristane could, conceivably, be fragments of such chains, perhaps formed in the membranes of some yet-undiscovered but ubiquitous bacteria. Here they were, trying to recognize the molecular remnants of ancient life, but they didn't have a complete inventory of life's molecules, let alone an understanding of how those molecules are transformed over time in the rocks—the central conundrum of the whole enterprise.

In 1964, as Calvin and Burlingame continued their analyses of the Precambrian rocks in Berkeley, Geoff returned to Glasgow and began to confront that conundrum. He enlisted the help of the only two Glasgow chemists who had experience with complex mixtures of hydrocarbons, largely from doing medical research aimed at identifying carcinogenic compounds in lubricating oils used by industrial workers and machine operators. One of those chemists, Archie Douglas, got interested in how the carcinogenic hydrocarbons formed to begin with and had already started working with petroleum geochemists. An analytical wizard by all accounts, Douglas was pivotal to assembling a laboratory that was up to the challenges posed by geologic materials of the sort Geoff wanted to study. Within the year, the fledgling organic geochemistry group had acquired two other essential resources—one in the form of a disgruntled graduate student, and the other, a newly minted analytical instrument. James Maxwell tells me that he'd started his doctoral work in an organic synthesis laboratory, but his hands were covered with horrible skin diseases from the chemicals and he was shopping for another project. "I looked like Frankenstein!" he says, 40 years later. He was not particularly interested in rocks when he queried Geoff about working in the organic geochemistry group…but he was easily seduced by the research papers Geoff directed him to read, which included the Nonesuch papers. As for the laboratory's star instrument, it had been designed by biomedical chemists in Stockholm, but it was precisely what one needed to decipher the chemical structures of unknown compounds in complex mixtures extracted from rocks: a gas chromatograph linked directly to a mass spectrometer, immediately christened with the acronym GC-MS. Now, as the compounds came through the GC column, they were routed one after the other directly into the mass spectrometer for structural analysis. Not only could Geoff's group now separate the mess of compounds in the mixtures extracted from rocks, it now had a much better chance of identifying what those compounds were. The group could, in principle, expand its horizons beyond phytane, pristane, and the *n*-alkanes.

When he'd first started his career at Glasgow, Geoff had shied away from natural products chemistry because the field was growing so fast and producing such a huge body of literature. "You had to spend all your time reading and trying to catch up," he says. But now he found himself not only reading the burgeoning natural products literature, but scanning the works of biochemists, paleontologists, and analytical chemists, even hunting down the elusive petroleum geochemists, in his search for relevant information, which, it seemed, could be tucked away in the most obscure journals. "It was hard to know what I should be reading and who I should be talking to," he says. "It was bewildering…but exciting."

During this period in the mid-1960s, a handful of petroleum geochemists were attempting to follow the geological fate of a class of lipids that, in addition to being the biosynthetic precursors to the *n*-alkanes, are much more prevalent in living organisms than the *n*-alkanes: the straight-chain fatty acids. These compounds had been extensively studied for years, partly because of their crucial roles in organisms, partly because of their commercial value for the food and detergent industries, and partly because the carboxyl group at the end of the carbon chain makes them easy for chemists to work with. In organisms, fatty acids are an important means of storing energy, and they are the main ingredient of cell membranes. They come in both saturated and unsaturated versions, the latter with double bonds between some of the carbon atoms; both the chain length and the number and placement of double bonds vary among different groups of organisms. The carboxyl group is bound by an ester linkage to form the phospholipids that make up most cell membranes, and Geoff had found that the leaf waxes contained long-chain fatty acids bound by ester linkages to alcohols.

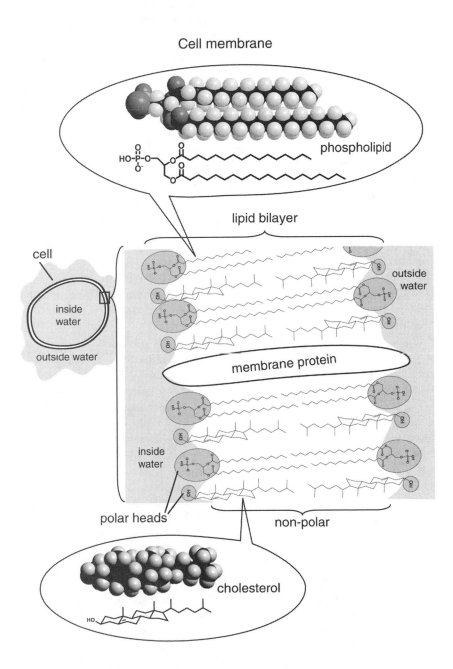

Such ester bonds are easily hydrolyzed and cleaved, however, so as soon as the organism dies and begins to decompose, free fatty acids are released. The carboxyl group at the end of the carbon chain in these free acids makes them more polar and soluble in water, more chemically reactive, and easier for bacteria to consume than the alkanes. And yet some fatty acids appeared to be surprisingly persistent. Petroleum geochemists found them in shales and crude oils, and for a time even postulated that fatty acids play a pivotal role in petroleum formation. Abelson and others found them in the Green River shale, and Kvenvolden found them in even older Cretaceous-age rocks.

Like the *n*-alkanes, the fatty acids in the oils and ancient rocks posed a number of enigmas that couldn't be explained by their known occurrence in organisms. Though the only known biosynthetic pathways for fatty acids produced mostly acids with an even number of carbon atoms, and these were by far dominant in organisms, the distributions of fatty acids in older oils and rocks didn't seem to reflect this. It appeared, rather, that the even-over-odd carbon number preference in fatty acids was progressively erased, over tens of millions of years, just as was the odd-carbon-number preference in *n*-alkanes. Kvenvolden and his colleagues at Mobil came up with a plausible, even elegant, chemical reaction scheme that very neatly explained this observation, as well as the general excess of short-chain *n*-alkanes found in petroleum: fatty acids reacted with each other in the sediments, producing acids with an odd number of carbon atoms and *n*-alkanes with an even number. Not until years later would it become apparent that the process is not so tidy, that the distributions of *n*-alkanes and fatty acids in sediments, rocks, and oils are influenced by the two most difficult to analyze aspects of the geologic record: the insoluble kerogen, and the multifarious and ever-elusive microbes.

Back in the 1960s, while Kvenvolden was at Mobil in California trying to figure out where the compounds in oil came from, and Geoff and James Maxwell were in Glasgow beginning analyses of the fatty acids in the Green River shale, Max Blumer at the Woods Hole Oceanographic Institution in Massachusetts was finding some of the mysterious short-chain *n*-alkanes in samples of phytoplankton collected from the surface waters of the ocean. Blumer was another wayward organic chemist whose early fascination with fossils had diverted him briefly into the petroleum industry, and from there to Woods Hole. In one of the first forays into the organic chemistry of the marine environment, Blumer was attempting to trace the pathways from organic compounds made by marine organisms to the organic compounds in ocean sediments. The *n*-alkanes made by the phytoplankton apparently remained unchanged as they passed through the food chain, because Blumer found the same distributions in zooplankton and in detritus filtered from the water—even bacteria didn't seem to be particularly fond of the *n*-alkanes. This all made chemical sense, as it's hard to cleave a single carbon–carbon bond when it's part of a straight chain, with no functional groups or other weak links. And yet, when Blumer and his students analyzed extracts of recently deposited ocean sediments, they found significantly different distributions of *n*-alkanes and concluded that they must have had some other source or sources—but what?

Meanwhile, Pat Parker, a chemist at the Texas Marine Institute, was doing similar sorts of studies in the rich, stinky mud flats and lagoons around the Gulf of Mexico, trying to track the fate of fatty acids in the first stages of their journey from organism to sediment. The laboratory next to Parker's was occupied by a distinguished microbiologist who was studying the thick mats of photosynthetic bacteria growing in the shallow waters around the institute, and Parker managed to interest him in the enterprise and obtain samples from his cultures of cyanobacteria and algae. Extracts of these organisms contained simple mixtures of both saturated and unsaturated fatty acids, dominated by the 16- and 18-carbon compounds—which was precisely what they found when they analyzed extracts from the surface layer of sediments in the lagoons. The unsaturated acids disappeared quickly, however, and were absent from the deeper layers of mud, presumably because the double bonds made them more accessible to bacterial scavengers in the sediments. No one had ever actually observed such scavengers in sediments, but one assumed they were living in the top layers just as they did in soil or anyplace else where organic detritus accumulated. If something disappeared quickly, it was reasonable enough to blame the invisible bacteria. But the Texas group also found a series of unusual acids that seemed to *increase* in the sediments and didn't resemble any of the compounds they'd isolated from the cultures of algae and cyanobacteria. Commonly known as *iso-* or *anteiso*-compounds, these were fatty acids with a single methyl group branch on the second or third carbon atom from the end.

A review of the biochemical literature revealed that the *iso-* and *anteiso*-acids had been identified among the lipids of heterotrophic bacteria—just the sort of scavengers that might be making a last-ditch effort to recycle organic matter before it was forever buried and turned to stone. Most microbiologists were too busy studying the kind of bacteria that make people sick to get excited about something found in a bucket of mud, but Parker's distinguished neighbor was intrigued enough to isolate cultures of bacteria from the mud itself. Sure enough, when Parker prepared a solvent extract of these bacteria and ran the fraction with the fatty acids through the GC, he found the same *iso* and *anteiso* acids he'd found in the mud itself. On the one hand, it was exciting to find the source of the acids, like finding the treasure in a treasure hunt after only a few clues.

On the other hand, it brought home just how complicated it was going to be to read the molecular cipher in the rocks. Not only did you need to think about what the enigmatic sediment bacteria *ate*—or more precisely, what they disdained to eat and left intact or slightly transformed—but you had to consider what they *produced*, as well.

Common fatty acids

unsaturated

saturated

iso- and *anteiso*-fatty acids

iso

anteiso

Members of Calvin's lab had also started analyzing extracts of algae and bacteria, looking for the sources of the compounds they were finding in ancient rocks. Not much was actually known about the microscopic inhabitants of natural waters and sediments, but even a limited sampling from the organisms that botanists and microbiologists had managed to isolate and grow in culture provided a potential explanation for some of the hydrocarbons that were being found in sediments and rocks. The long-chain compounds that were so prevalent in land plants, and missing from the Precambrian rocks, were entirely absent from the algae and bacteria analyzed—as might be expected, since organisms that live in water would have no need for long, hydrophobic molecules to protect them from drying out. The 17-carbon n-alkane that was absent from higher plants, but a prominent component of the organic matter in both the Green River and the Precambrian rocks, turned out to be the most abundant straight-chain alkane in the algae and photosynthetic cyanobacteria. All of the bacteria analyzed, including heterotrophs, contained odd-carbon-number short-chain n-alkanes ranging from 11 to 24 carbon atoms, with the 15-carbon compound in relative abundance. In a few cases, though, they also contained the corresponding even-carbon-number compounds, a finding that cast doubt on the reputedly dominant and universal biosynthetic pathway for making straight-chain compounds and caused confusion about the sources of these alkanes in oils and rocks... until someone figured out that these compounds had come not from the organisms under study, but from the plastic hoses used to bubble air into the water while they were growing. The plastic was, of course, made from petroleum, and while it would be many years before anyone figured out how these short-chain n-alkanes were generated in the petroleum, one thing was becoming clear: petroleum hydrocarbons were everywhere in the man-made environment, and they were confounding attempts to understand the presence of biological lipids and their decomposition products—which included many of the same petroleum hydrocarbons—in nature.

One of the main tasks confronting the scattered cadre of organic chemists who were trying to analyze geologic materials in the 1960s was that of developing experimental protocols that eliminated contamination—plastic containers, impure solvents, car exhaust drifting in through an open laboratory window, pollutants in the ocean and lakes. When it came to analyzing Precambrian rocks the problems were even more insidious. Unlike the Nonesuch shale, most of these rocks contained such small quantities of organic compounds that the merest trace of contamination could render results completely meaningless. Indeed, much of the excitement and funding for such work was coming from NASA, which was preparing for its Apollo 11 mission: not only would men be landing on the moon in 1969, but they would be bringing moon rocks back to Earth for chemical analysis. Those analyses would include a search for minute traces of organic compounds, and NASA was offering generous funding for anything that might provide a knowledge base, including much of the Precambrian work and Burlingame's analytical facilities in Berkeley. One of the most frustrating contamination problems that Geoff had faced with the Precambrian rocks was simply that they'd been sitting around on Earth for so long that they had become contaminated by organic matter that had seeped in from later deposits. Calvin had maintained that the ancient rocks were impervious, until Geoff did a simple experiment: he soaked a representative chunk of rock in water containing long-chain fatty acids that had been prepared for biochemical studies and "labeled" with radioactive carbon atoms. As feared, when he analyzed the rock, he found that the labeled compounds had made their way deep into its middle. The Nonesuch shale had been relatively easy to work with, and he was able to compare the hydrocarbon content of different components and layers of the rock, thus verifying that the compounds identified were indeed indigenous to the one-billion-year-old shale. But verifying that the minuscule amounts of hydrocarbons in older, more altered rocks actually derived from organic matter that was deposited with the sediments that formed the rocks was a difficult, if not impossible, task—one of the reasons Geoff focused the Glasgow efforts on learning about the fate of organic constituents in younger and more genial geologic material. In this younger material, which ranged from 5,000-year-old lake sediments to Geoff's all-time favorite rock, the 55-million-year-old Green River shale, Maxwell identified a series of acyclic isoprenoid acids—including a substantial amount of pristanic acid, which Blumer also found in his analyses of zooplankton from the marine environment.

Blumer's studies were showing that the zooplankton that grazed on microscopic algae in the surface waters of the ocean contained phytol, the alcohol produced from cleavage of the phytyl tail of chlorophyll, as well as pristane and pristanic acid, which could easily have derived from the phytol via oxidation and loss of a carboxyl group. All three compounds were present in the extracts of zooplankton and in surface sediments, but phytane, which has the same carbon skeleton as phytol, was absent from both, and Geoff's group was finding that it only appeared in older deposits, apparently produced by the slow reduction of the hydroxyl group in phytol as the sediments aged. The Green River shale, which had spent 55 million years aging, contained phytane, pristane, and their acid versions, but lacked phytol.

Even with this bread-crumb trail of evidence from chlorophyll to phytol to the series of isoprenoid acids Maxwell was finding, and with plausible chemical mechanisms to produce the corresponding alkanes, pristane and phytane, the origin of the acyclic isoprenoids in ancient rocks and oils was far from certain. Luckily, though they are only slightly more complex than the *n*-alkanes, these molecules are asymmetric and contain a subtle extra source of information in the way their atoms are oriented in space. The tetrahedron formed by a carbon's four bonds is symmetrical when any two of the attached atoms or groups are the same, as in the *n*-alkanes, where every carbon is bound to at least two hydrogens. But if all the atoms or groups attached to a carbon atom are different, as they are at the branch carbons where the methyl groups are attached in the middle of phytane and pristane, then an asymmetric, or chiral, center is created and the same molecule can exist as either of two mirror-image configurations, or stereoisomers. These stereoisomers are identical except that, like a left hand and a right

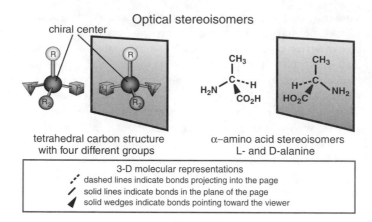

Optical stereoisomers

chiral center

tetrahedral carbon structure
with four different groups

α–amino acid stereoisomers
L- and D-alanine

3-D molecular representations
dashed lines indicate bonds projecting into the page
solid lines indicate bonds in the plane of the page
solid wedges indicate bonds pointing toward the viewer

hand, they cannot be superimposed one upon the other. They are, then, chemically indistinguishable until they interact with other asymmetric molecules, at which point they react differently, much as when two people try to shake hands right to left instead of right to right.

The amino acids that make up proteins and enzymes are asymmetric—with a chiral carbon atom bonded to an amino group, an acid group, a carbon chain, and a hydrogen atom—and one of the most crucial, and inexplicable, steps in the origin and evolution of life is the adoption of just one configuration from each of the possible pairs. Accordingly, biochemical reactions, which depend on enzymes, distinguish between otherwise identical pairs of stereoisomers and are very specific about which ones they make and use—unlike abiotic chemical reactions. Nineteenth-century chemists recognized this peculiar "one-handed" aspect of life when they shined polarized light through solutions of plant extracts and found that certain extracts bent the rays to the right or left. Indeed, it was this optical activity and a simple bit of geometric reasoning that led to the realization that the four atoms attached to a carbon must lie at the corners of a tetrahedron and ultimately, if you wanted to understand the molecules of life, you needed to know not just what is bonded to what, but how the molecule's atoms are oriented in space—its stereochemistry.

The phytyl tail of chlorophyll is made with a single, specific configuration at each of its two chiral carbons—in other words, as only one of four possible versions. Any phytol released by cleaving the tail would presumably have the same configuration at those two carbons, and it was likely that any pristane or phytane that came from phytol would retain it. Maxwell says that he still remembers a 1967 paper from Calvin's group that noted what a major breakthrough it would be if one could correlate the stereochemistry of the isoprenoids in a geological sample with that of their supposed biological precursor, namely, the phytyl tail of chlorophyll. "I took that paper of Calvin's literally," he tells me. At the time, it seemed a sure way of tracking a fragment of isoprenoid skeleton through hundreds of millions of years in the sediments....It also seemed a near-impossible task. How could they possibly separate and determine the configurations of four chemically identical or nearly identical stereoisomers of phytane or pristane, two configurations at each of the two chiral carbons of interest? Perhaps they'd have a chance with the isoprenoid acids found in younger rocks. Acids allowed you to do things with them in the lab. You could form an ester by tacking an alcohol onto the end, and if you used a single stereoisomer of an alcohol with its own chiral carbon, then the esters formed would be more distinctive than the acids alone and, hopefully, easier to separate. Help with the project came from a strange quarter: the Fisheries Research Board in Halifax, Canada, where a couple of biochemists were trying to measure the amounts of phytanic acid in fish oils. A rare genetic defect in humans can cause a dangerous buildup of this acid in the body, and the biochemists were developing a

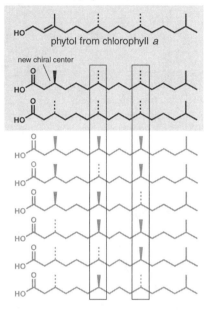

Phytol, and the stereoisomers of phytanic acid

phytol from chlorophyll *a*

new chiral center

method of checking for its presence in foodstuffs, trying out a new GC column, which, they discovered quite by accident, separated the acid ester into two peaks.

Working with the Canadians, Geoff and Maxwell used phytol from chlorophyll to prepare pristanic and phytanic acids, thus obtaining only the stereoisomers with the same configuration as in chlorophyll—two stereoisomers per acid, due to the creation of a new chiral center in the acids. Sure enough, these were precisely the two stereoisomers—the only two—that they detected among the isoprenoid acids in their extract of Green River shale, the clearest indication yet that the pristane and phytane in the shale came from chlorophyll.

Did this mean that all of the pristane and phytane they found in geological samples had come from chlorophyll, that there was no other significant source? What of the phytane that had just been reported in a chert from South Africa that was more than three billion years old? Was it possible that they were looking at evidence not only that life was in full swing, but that something as biochemically complex as photosynthesis was already the dominant form of production? Could Maxwell somehow extend his techniques to determine the stereochemistry of isoprenoid alkanes in Precambrian rocks, where the acid forms had long since disappeared?

In the mid-1960s, fueled by NASA funding and encouraged by the success of the Nonesuch shale analyses, there was a small flurry of studies on older and older rocks. Paleontologists found what appeared to be the fossils of photosynthetic bacteria and algae, and chemists, cynical about what looked to them like a few odd marks in the rock, nevertheless looked for their own clues, and continued to find them: in the 1.9-billion-year-old chert from the Gunflint Formation in Ontario, the 2.7-billion-year-old Soudan shale from Minnesota, and the 3.1-billion-year-old Fig Tree Formation in South Africa....Pristane, phytane, and a preponderance of 17-carbon *n*-alkane all implied that photosynthetic microbes had been present. In the Soudan shale, Meinschein tentatively identified ring compounds that seemed to be related to sterols like cholesterol—known constituents of the cell membranes of more complex eukaryotic organisms. But when did life *begin*? Were there any rocks old enough to have recorded a time *before* life imposed its order, a time when the only organic compounds on Earth comprised the random assortment made by abiotic chemical reactions?

Such was the excitement of the time that when Calvin's group extracted a 3.7-billion-year-old sample from the Onverwacht Formation in South Africa and showed him the first chromatogram, he wondered if he might actually be seeing the transition from abiotic to biotic production. There was a large hump of hydrocarbons, presumably so myriad and various in structure that they couldn't be separated, and, rising from the hump, a number of tiny, well-defined peaks corresponding to the alkanes they'd been seeing in contemporary sediments and living matter. "This is precisely the kind of spectrum one might expect from a mixture of abiogenic materials (the continuum) and biogenic materials (the superimposed peaks)," Calvin wrote in his 1969 treatise on the chemical evolution of life. After giving a brief inventory of the caveats, he went on to say, "we may indeed have found a period in time, 3700 million years ago, during which the transition between the abiogenically developed organic substrate was being converted by the newly developed autocatalytic replicating chemical systems which give rise to life."

But Calvin's excitement was premature, and his speculation remained unsubstantiated. Even as analytical techniques improved, facilitating the search for more complex and informative organic molecules, doubts were growing about the veracity of what they were finding in any of the Precambrian rocks older than the Nonesuch. Thomas Hoering, a colleague of Abelson's at the Carnegie Institute, had long been intrigued by the signs of past life in Precambrian rocks, but was keenly skeptical of his own and others' results. He was always looking for unequivocal evidence that the organic compounds they contained were biogenic and did in fact hail from the early years of earth history when the rocks formed—and not from the dust filter in the lab's heating system or the ink on the newspaper the rocks were wrapped in by the geologists who'd collected them, or most difficult of all, that the compounds hadn't leached in from later deposits such that the remains of organisms that lived a few million or hundred million years ago were being confused with those of the first organisms to leave their mark on Earth. Geochemists from the era consistently tell me that it was a 1967 paper of Hoering's that finally deflated their hopes of getting meaningful results from rocks that were more than two billion years old, and indicated that much of what they'd found was, as Keith Kvenvolden puts it, "bogus."

Kvenvolden himself had detected amino acids in several Precambrian samples and was, at first, excited to see that they matched those used by organisms in their proteins, providing evidence that "complex organisms were in existence more than 3.1 billion years ago," with an amino acid composition that had "not changed significantly since very early in earth history." But when he repeated the analysis with a system that could separate the amino acid stereoisomers—an easier task than separating hydrocarbon stereoisomers—he immediately realized that the amino acids he'd found in the Fig Tree Formation couldn't possibly have been made by organisms 3.1 billion years ago. Abelson had just discovered that amino acids in fossil shells gradually convert from the biological isomer—whose configuration is designated L by chemical convention—to the nonbiological D isomer as the fossils age. The conversion was slow, but not so slow that it couldn't be studied in the laboratory, and the basic mechanism had already been elucidated: the hydrogen at the chiral carbon was removed as a naked proton, without its electron, leaving a negatively charged planar intermediate. A proton could then reattach to either side of this flat intermediate with equal probability, producing either the L or the D isomer.

Isomerization of α-amino acids

Abelson suggested that one might be able to figure out the precise rate of this process and use the relative amounts of the two isomers to date fossils—but only relatively young ones. The reaction is reversible, with the protons constantly removed and reattached and the molecules inverted and reinverted, until at some point an equilibrium is established where the rate of conversion from D to L balances that from L to D and there is no observable change in the relative amounts of the two isomers. Such an equilibrium mixture, with the two isomers present in near-equal amounts, is obtained after a few tens of millions of years even at low temperatures. If there were amino acids left from organisms that had lived during the Precambrian, *billions* of years ago, they would surely be present as a 50:50 mixture of L and D isomers, what is known as a racemic mixture of the two mirror images. And yet the amino acids Kvenvolden had extracted from the Fig Tree rock were all in the L configuration that life uses. Clearly, they had leached in from later deposits, and their evidence of primordial life was, indeed, "bogus."

The chemical record of the origin and early evolution of life that Calvin had hoped to find seemed to have been pushed off the map of the rock record—at least for the time being. "People got frightened and overreacted," says Maxwell, who spent 1967 as a postdoc in Calvin's lab. "No one wanted to risk their careers by making bogus discoveries." Maxwell did in fact develop a method that allowed him to distinguish the stereoisomers of pristane and phytane in geological samples, but he laughs when I ask why he never tried it on the Precambrian rocks. "You want me to be cynical? At that time? Because of the problems of contamination in those rocks, I got frightened off." Instead he focused his energies on much younger organic-matter–rich oil shales, where the stereochemistry of isoprenoid alkanes would lead him to some unexpectedly practical discoveries—and, along the way, make it clear that knowing the stereochemistry of phytane in the ancient Precambrian rocks would not, in fact, have told him anything about its origin.

By the end of the 1960s, as interest in the Precambrian rocks waned, the small community of organic geochemists that had assembled around their study turned its attention and newly forged analytical methods to the more attainable goal of learning about the molecular remnants of life in younger rocks, sediments, and petroleum...and, at the same time, vied for a chance to analyze some samples that were even more dubious, and more seductive, than the Precambrian rocks. If it was hard to resist the mystique and romance of three-billion-year-old rocks that might contain clues to the origin of life on Earth, it was even harder to resist a hunt for signs of life or its precursors on other planets—if only for the thrill of seeing and handling their rocks.

3

From the Moon to Mars
The Search for Extraterrestrial Life

*Does this carbonaceous earthy material truly contain humus or a
trace of other organic compounds? Does this possibly give a hint
concerning the presence of organic structures in other planetary
bodies?*
—Jöns Jakob Berzelius, 1799–1842, Swedish chemist, referring to his analysis of the
    Alais meteorite, which fell in France in 1802
    From *Annalen der Physikalischen Chemie* 33 (1834)

*In biogeochemistry we have to consider that life (living organisms)
really exists not on our planet alone, not only in the Earth's
biosphere. It seems to me that this has been established beyond a
doubt, so far, for all the so-called terrestrial planets, i.e., for Venus,
Earth and Mars. At the Biogeochemical Laboratory of the Academy
of Sciences in Moscow…we identified cosmic life as a matter for
current scientific study already in 1940. This work was halted due to
the war, and will be resumed at the earliest opportunity.*
—Vladimir Ivanovich Vernadsky, 1863–1945, Russian mineralogist
    From *American Scientist* 33 (1945)

"But did anyone really expect to find anything?" I ask Geoff, as he shows me
the canister that had contained his sample of moon dust from the 1969 Apollo
11 mission. "Well, no," he replied, "we didn't think there'd ever been life on the
moon. But we didn't know. We thought there might be organic compounds."

And why not? People had been finding organic compounds in meteorites
for more than a century, and no one was quite sure where they'd come from or
how they'd formed. In 1834, the Swedish chemist Jöns Jakob Berzelius noted the
high carbon content of a meteorite that had fallen in southern France a couple of
decades earlier. Meteor showers in Europe were described as early as 1492, and
their extraterrestrial provenance had been documented in 1803, when the distinguished French physicist Jean-Baptiste Biot featured among the scores of citizens
who witnessed the stones falling from the sky above the village of l'Alsace. But
the source of the carbon compounds Berzelius and others found in meteorites
would remain controversial far into the next century.

Another carbonaceous meteorite fell in Hungary in 1857, and the eminent chemist Frederick Wöhler—Berzelius's student, and the first to show that one could create carbon compounds like those made by organisms from inorganic substances in the lab—found organic compounds that he was convinced were of extraterrestrial biological origin. A decade later, Marcellin Berthelot found what he called "petroleum-like hydrocarbons" in a meteorite that had fallen near Orgueil, France, in 1864. He postulated that the hydrocarbons had formed abiotically from reaction of metal carbides with water, but in the next few years there was a spate of meteorite treatises in which the fossils of an astounding assortment of exotic extraterrestrial creatures were described in minute detail. Louis Pasteur had just presented his famous experiment showing that a protected, sterile medium remained devoid of life ad infinitum and debunked the popular theory that life could burst spontaneously into being from nonliving matter, but now the debate shifted to the possibility that life on Earth had originated with live cells or spores delivered by meteorites from space. Finally, in an attempt to put the debate to rest, Pasteur applied his meticulous analytical techniques to the Orgueil meteorite and found it to be indisputably sterile and incapable of generating life.

Nearly a century later, three well-respected chemists, including Warren Meinschein, stirred things up again. Using the newly developed techniques of gas chromatography and mass spectrometry, they analyzed samples of the Orgueil meteorite from a natural history museum in France, where they had been kept in a sealed jar since 1864. At a meeting of the New York National Academy of Sciences in 1961, the group reported finding distributions of hydrocarbons like those in terrestrial sediments, which they took as evidence of biological activity on the meteorite's parent asteroid or planet. Microbiologists then weighed in with reports of "organized elements" that resembled algal microfossils and even more assertive claims that they had evidence of life's existence in regions of the universe beyond Earth. But the work was immediately challenged and ridiculed, and it soon emerged that the hydrocarbons had come from pieces of coal, the "organized elements" were ragweed pollen and rush seeds, and the meteorite had been broken apart and glued back together. Apparently, back in the 1860s, when the debate about spontaneous generation and extraterrestrial life was raging in the scientific and popular press, someone had added "evidence" to the meteorite, never imagining that the sample would be sealed in a jar and the hoax would play out a hundred years in the future.

Hoax notwithstanding, the 1961 endeavor inspired a rekindling of interest in carbonaceous meteorites, whose chemical composition turned out to be radically different from that of the majority of meteorites. But even analyses of virgin pieces of the Orgueil and other meteorites were inconclusive: though they confirmed nineteenth-century chemists' findings of hydrocarbons, interpretation was still plagued by an inability to distinguish earthly contamination, albeit accidental, from extraterrestrial organic material.

And what about the moon? In 1969 chemists had their first chance to examine extraterrestrial rocks that hadn't been lying around on Earth for decades or even years, but were, rather, brought here under meticulously controlled conditions of cleanliness. If organic compounds were generated abiotically in meteorites or on the early earth, then couldn't they have been generated on the moon as well? Or perhaps they'd formed in space, or on other planets, or in asteroids, and then been delivered to the moon by meteorites. One might even speculate that the molecular precursors of life on Earth had arrived in a similar fashion.... Maybe the moon contained traces of abiotic organic compounds like the ones Melvin Calvin had hoped to find preserved in the earth's oldest rocks. Calvin, of course, was among the first to speculate about such possibilities. Others were more skeptical.

Keith Kvenvolden says that the only scientists who *really* thought they'd find much in the way of organic compounds on the moon in 1969 were astronomers. They were, it seems, thinking rather simplistically from a chemical standpoint: you've got hydrogen, carbon, and energy, so you ought to have hydrocarbons. "They thought the moon was paved in asphalt," he says, and then adds, with a grin, "Rational people, like geologists, were skeptical." Spectral analyses of the lunar surface from previous unmanned Apollo missions indicated that it contained little or no carbon. Nonetheless, Kvenvolden, a rational geologist by training, was lured away from his comfortable job at Mobil Oil to work in NASA's Ames Research Center in California. "There was one chance in history to work on the first lunar samples," he says, "and I just couldn't miss it. We could do any experiment and it was world news."

NASA orchestrated the distribution and reporting of results from the first lunar samples with Hollywood theatricality and suspense. Despite the scientific consensus that the moon was unlikely to harbor life, NASA designed the Lunar Receiving Laboratory, or LRL, in Houston with the express purpose of quarantining the astronauts and samples. One young chemist from Al Burlingame's Berkeley lab had just read Michael Crichton's novel *The Andromeda Strain*, and when he got to Houston to help prepare the LRL and saw the elaborate measures NASA was taking to protect the earth from dangerous extraterrestrial microbes, he got so scared that he abandoned his job and left Houston before Apollo 11 landed. As Kvenvolden points out, the LRL was hooked into the city's main sewage system, so it's hard to gauge if it was all just part of the show, or if it was a serious, and genuinely inept, precaution. "With NASA, there was always this undercurrent about the possibility of extraterrestrial life," Kvenvolden says. "It attracted public interest, and support for the program."

Cindy Lee, who was a graduate student in Jeff Bada's then-nascent amino acid laboratory at the Scripps Institution of Oceanography, says it was an exciting time to be a student. Big, brash questions about the chemistry, origin, and extent of life in the solar system were on the table—and that was in large part due to NASA's funding and support. It was no easy task to sustain public interest and funding for such basic scientific questions, so perhaps NASA's theatrics can be forgiven in the name of science. From some of the stories the Apollo 11 researchers tell, it would appear that NASA organizers themselves had been reading Crichton's novel. The principal investigators from each proposal were required to travel to Houston and retrieve their samples in person, and NASA gave them such dire warnings about thieves that they resorted to all sorts of James Bondish antics to transport them home. Lee recalls a Scripps party where a slightly ine-briated Caltech physicist, who was en route from picking up his ration of moon, rose from his seat and, with great solemnity, dropped his pants to show how he'd taped the tin of lunar dust to his leg so it couldn't be stolen.

Flouting the usual rules of scientific discourse and openness, NASA swore all the scientists to secrecy about their results until January 5, 1970, when they were to gather in Houston for the first Lunar Science Conference and report them simultaneously. They were forbidden not only from talking to the press, but also from talking to each other. "We got our samples last," Geoff says, "so we had a late start and had to work fast. The hardest part was getting the aluminum sample canister open! They'd sealed it so well we had to use a lathe to get the bloody thing open." By this time, Geoff and James Maxwell had moved to the University of Bristol and set up their Organic Geochemistry Unit, where the meticulous, ultraclean, and sensitive techniques they'd refined for the analysis of ancient rocks were standard procedure. "We were looking for any identifiable organic compounds—amino acids, porphyrins, biomarker hydrocarbons....Any signs that there'd been life on the moon at some time."

"And?" I ask, eager to get to the punch line.

"And then there was the Lunar Science Conference in Houston. *Science* had exclusive rights to all the papers, so Phil Abelson was presiding. It was hair-rais-ing, because we didn't know what the others had found out. We were worried we'd report we couldn't find anything, and then someone else would get up and say they'd found something we'd missed. Everyone was petrified of this!"

"So?"

"Of course there was nothing there."

As suspected, there were no identifiable biomarkers above the low back-ground contamination levels, no organic compounds at all except the smallest hydrocarbon: methane gas was found trapped in the mineral matrix of the dust. John Hayes, a postdoc in the Bristol lab at the time, says he was amazed when they found methane. He'd worked with Kvenvolden at Ames for a year, but was so convinced that the moon would contain no carbon that he hadn't even wanted to analyze lunar samples. In fact, he says, laughing, he'd applied for the Bristol appointment thinking he could escape the moon hullabaloo, only to be greeted by Geoff with the "wonderful news" that they were about to receive lunar sam-ples. For his part, Geoff was excited, not because he thought they'd find signs of life, but because he was expecting they might find methane. This had occurred to him when he was working at the Lunar Research Laboratory, listening to the physicists talk about the solar wind that constantly bombarded the moon with a stream of high-energy ions—including carbon and hydrogen, which he imagined might become embedded in the lunar minerals and react to form methane. "We were absolutely chuffed when we found all that methane," Geoff says, still sound-ing pleased almost 40 years later. And if there was metallic iron on the moon, as one might expect since it was covered with tiny craters from meteorites, and most of the meteorites that had landed on Earth were composed of iron, then the carbon should also react with iron to produce a sort of lunar steel....

This was all very interesting and led to an exciting new line of research in Bristol, but what about biomarkers for life? Wasn't that what all the hullabaloo was about?

Well, yes, Geoff admits, that's what they were supposed to look for, but there just wasn't anything to be found. "We kept with the Apollo program through Apollo 15, but mostly because of this methane and iron carbide work."

Sifting through the old papers, I find one by Kvenvolden's Ames group that describes the identification of porphyrins in lunar dust, like those found in petroleum. "We had this scientist in the lab who could find porphyrins in anything," he quips, when I ask him about it. "But it turned out to be contamination." Others reported finding amino acids in some of the Apollo samples, but it turned out that they had been collected from an area near the spaceship, where the dust had been contaminated by traces of rocket exhaust.

While the Apollo program chemists were still busy analyzing their bits of moon and preparing their reports for the first Lunar Science Conference, a much more interesting example of extraterrestrial organic chemistry had made a spectacular but unheralded landing on Earth. On September 28, 1969, a meteorite exploded in the atmosphere over Australia, and fragments rained down around the town of Murchison. In the wake of the Apollo moon drama, the meteorite fall was a bit anticlimactic. But when Kvenvolden received fragments of the stone at the Ames laboratory, he was excited: it was clearly a carbonaceous meteorite like the one from Orgueil; the fragments had been retrieved by residents and sent to NASA almost immediately after their fall, minimizing chances of terrestrial contamination; and the Ames laboratory was perfectly poised for their analysis.

Kvenvolden and his colleagues found a complex suite of hydrocarbons, but it was completely lacking in the organized, enzyme-directed structures and distribution patterns produced by organisms. They also found significant amounts of amino acids, including some of life's building blocks for proteins and enzymes, but with none of life's insistence on left-handedness. The mixture was racemic, with D and L configurations present in equal measure, and it was dominated by a number of strange amino acids that had never been found in organisms or, for that matter, anywhere on Earth except a chemistry laboratory. Indeed, the distribution of amino acids in the Murchison fragments bore a remarkable resemblance to the distributions produced during simulations of the abiotic chemical processes hypothesized to have taken place on the early earth. Here, finally, in this wayward bit of space debris, they'd found solid evidence of what chemists since the 1950s had suspected must exist somewhere in the solar system, what Calvin had been hoping to find in the oldest Precambrian rocks: organic compounds that had been produced not by organisms in their natural habitats or by chemists in their laboratories, but by purely abiotic chemical processes somewhere in space.

Thirty years of scrutiny and rescrutiny of the Murchison meteorite has only reinforced the NASA group's conclusion. More than 70 amino acids have been identified, most of them nonexistent in the earth's biota. The soluble organic matter consists mostly of polycyclic aromatic hydrocarbons, or PAHs, which are produced in quantity by abiotic reactions in early earth experiments, and appear to be ubiquitous and abundant in the dust and gas clouds of interstellar space. PAHs can also be generated by burning or heating organic matter on earth, and they are found in petroleum and as man-made pollutants in the atmosphere, but they are not known to be biosynthesized directly by any organisms. Next in abundance are the carboxylic acids, some of which are made by organisms. But like all of the groups of organic compounds found in the meteorite, the distribution of acids exhibits none of life's carefully constructed patterns: for any given number of carbon atoms, all possible isomers are present in at least trace amounts—a structural randomness that enzymatically mediated biosynthesis would never produce. Contamination remains a problem when analyzing meteorite samples, but the presence of "extraterrestrial" amino acids, as well as evidence of abiotic syntheses, helps to distinguish uncontaminated from contaminated samples, and careful analysis of several other carbonaceous meteorites has revealed remarkably similar arrays of compounds.

The carbonaceous meteorites are among the oldest objects in the solar system, dated by radiometric methods as more than 4.5 billion years old. They contain an elemental composition that is little changed since the first accretions of matter from the solar nebula and are thought to be pieces of asteroids produced during collisions with planets or other asteroids. The provenance of the organic compounds in them remains the subject of study and speculation, but now the focus is on the nature of the chemical reactions and conditions that generated them in interstellar space. Such primordial compounds would have been destroyed on the nascent earth, as its surface was covered with molten rock. But as the planet cooled, some 3.8–4.1 billion years ago, a constant hail of tiny meteors and cosmic dust, much more intense than it is today, would have reintroduced them. Regardless of whether life arose from such interstellar organic compounds, or whether it made use of organic compounds that were synthesized on the early earth as it cooled and water condensed to form oceans, one thing is clear: the carbon-based chemistry that laid the foundation for life was available throughout the solar system.

Has any other planet besides the earth ever provided enough shelter to preserve complex organic compounds, as well as the energy sources and water conducive to chemical evolution and the emergence of life? Clearly the moon is lacking. And according to results from the 1975 Viking mission to Mars, the red planet is equally lacking. Along with a number of automated biological experiments designed to detect life, the Viking included a GC-MS—a miniature, automated rendition of the combined gas chromatograph–mass spectrometer that Geoff had acquired for his Glasgow lab—which detected no organic compounds at all in samples taken from the top 10 centimeters of soil on Mars. That the surface of Mars should lack life is not surprising, as current temperatures are too low for liquid water to be present, and it is subject to intense ultraviolet solar radiation. But what about abiotic organic compounds? The planet is continually bombarded by meteors, and has been since its birth. Unlike the moon, where organic compounds in a meteor are most likely destroyed on impact, Mars has a generous enough atmosphere to cushion the meteor's fall, and many organic compounds should survive, just as they do on Earth. The hypothesis that best explained the Viking results was that high concentrations of strongly oxidizing chemicals such as hydrogen peroxide in the surface soil rapidly destroyed any organic compounds. NASA's official stance was that the Viking missions "neither proved nor disproved the existence of life on Mars," but the consensus in the wider scientific community was somewhat less ambivalent: a living cell clearly could not survive in such an environment. Of course, the Viking sampled a few grams of soil from the surface at a single site, and the theoretical oxidizing layer should be restricted to the surface. And the 1970s-era mass spectrometer was not sensitive enough to have detected truly minute traces of organics. All of this left open the possibility that microbial life exists or existed beneath the surface, in some more protected environment, or at some time in the planet's distant past. In 1996, when a NASA geologist made the sensational claim that he and a group of colleagues had found both chemical and physical fossils in a meteorite from Mars, the news was received with a mixture of excitement and skepticism, not to mention a feeling of déjà vu among many older scientists.

The meteorite had been found lying on the ice in Antarctica in the 1980s, but only recently had geologists established its origin, placing it among a dozen meteorites whose composition linked them with some certainty to Mars. Other evidence indicated that this particular emissary had landed on Earth some 13,000 years ago, that it had apparently been ejected from the surface of Mars by a large asteroid impact about 16 million years ago, and that it was composed of material that was almost as old as the planet itself. The evidence of past life consisted of submicroscopic, rod-shaped mineral deposits and crystals similar to those fabricated by some bacteria, and the presence of possibly biogenic organic matter. What followed was yet another replay of the extraterrestrial life debates of the mid-nineteenth century and the Orgueil meteorite fiasco of the 1960s.

NASA, which had been suffering from budget cuts and was desperately in need of public support for its programs, called a press conference to publicize the results even before the paper was published. Scientists of all stripes were quick to challenge the conclusions, but the public was immediately seduced by the prospect of life on Mars, and politicians weren't far behind. "I am determined that the American space program will put its full intellectual power and technological prowess behind the search for further evidence of life on Mars," said President Clinton in a statement shortly after NASA's press conference. Meanwhile, geologists, biologists, and chemists went to work to test the evidence.

At Scripps in California, Jeff Bada was skeptical of the way results from the organic analyses had been interpreted. The compounds that were construed as evidence of life were PAHs, whose provenance is ambiguous even when they are found in earth rocks. Unlike the molecular fossils that organic geochemists have discovered over the past 40 years, PAHs contain little specific structural information that links them to compounds made by organisms. Some PAHs are found in petroleum and in shales rich in organic matter, and may indeed be the distant transformation products of polycyclic isoprenoids from plants and bacteria.

But PAHs can also be formed by a number of abiotic reactions and are a common constituent of carbonaceous meteorites and cosmic dust. Of course, just to complicate matters, they are also ubiquitous atmospheric pollutants, produced by burning anything from forests to gas. Bada, who had spent much of his career studying the isomerization reactions of amino acids in fossils, decided to look for some slightly less ambiguous and more informative sign of past life in the Martian meteorites.

Bada and his colleagues analyzed the amino acids in two meteorites found in Antarctica, including the one the NASA researchers had studied. They found a distinctly earth-like suite of amino acids, predominantly in the L configuration and lacking the "extraterrestrial" amino acids that had been identified in the Murchison or in abiotic syntheses. In fact, the pattern of amino acids in the meteorites looked suspiciously like the one they found when they analyzed ice from the Antarctic landscape where the meteorites had been discovered. Like the pattern of amino acids Kvenvolden had found in his Precambrian rock samples 30 years earlier, it conveyed one not particularly surprising bit of information: during its 13,000 year residence in Antarctica, the meteorite from Mars had become contaminated with earthly substances.

In the years since the 1996 study, the other evidence of life in the Martian meteorite has been similarly revealed as either equivocal or ambiguous: the purported microfossils of bacteria are simply minute fragments of clay, and though the NASA group continues to defend its conclusions about the mineral formations, most scientists who examined them agree that they are more readily explained by simple inorganic chemical reactions. Nevertheless, the controversial paper and the debate it triggered may well have rescued the search for extraterrestrial life from scientific oblivion. The question of whether life *could* ever have existed on Mars is being revisited, now in the light of irrefutable evidence that water did indeed exist on the surface of the planet at some time during its history. Recent discoveries of bacterial communities thriving in extremely hostile environments on Earth—buried beneath thousands of meters of solid granite bedrock, or deep within the Antarctic ice sheet, or in the harsh chemical milieu of hot springs—have added a new dimension to the query and led many scientists back to the questions they thought the Viking had answered: *Did* life ever exist on Mars? Might it still? In the late 1990s, the combination of rekindled scientific interest and public excitement translated into renewed funding for Mars science programs, with both NASA and the European Space Agency making plans for new missions on the surface of Mars and the return of samples to Earth.

Researchers working to define sampling strategies for Mars are faced with the same fundamental question that plagued Geoff and Calvin as they planned their analyses of Precambrian rocks 40 years ago: what should we look for? The supposition of biochemical Uniformitarianism that facilitated the initial search for molecular fossils in ancient rocks has turned out to be surprisingly sound: the fundamental building blocks of life have not changed much in more than three billion years of earth history. We know of no reason why these should necessarily be the same in extraterrestrial life, if it exists—biology as yet boasts no universal laws like those of physics and chemistry. But physics and chemistry can nevertheless constrain the search: we can suppose that life anywhere in the solar system would depend on the chemistry of carbon for the same reasons it depends on carbon on Earth; that water is required as the milieu in which that chemistry functions; and that life of any form harvests energy and creates chemical disequilibrium. We can look, then, for simple patterns in the distributions of organic compounds, like Geoff first saw in his leaf wax *n*-alkanes. And, of necessity, we can make the provisional assumption that Martian life's primary biochemical pathways of carbon assimilation and metabolism would be similar to the ones on Earth.

Bada, of course, has his hopes set on amino acids. Together with a large team of NASA-funded researchers from several institutions in California, he has been designing miniaturized analytical systems that can both detect and determine the chirality of tiny traces of amino acids. The system, which is many times more sensitive than the Viking's GC-MS, is scheduled for inclusion on a European Space Agency mission that plans to analyze samples from up to two meters below the surface of Mars, in the hope of detecting what the Viking may have missed— whether it be signs of present or past life, or of the ancient interstellar organic chemistry evident in meteorites. As molecular fossils on Earth, amino acids are not particularly informative, partly because they are so mobile in the sediments and readily consumed by bacteria, and partly because the same set of 20 amino acids is common to virtually *all* life on Earth. But in the search for extraterrestrial life this universality is precisely what we need: we don't want to limit our search to a specific form of life. Amino acids are one of the most essential and abundant ingredients of the simplest bacterium. Though they are readily made by abiotic synthesis in interstellar space, their distribution in living things on Earth is distinctive and idiosyncratic. One might expect that it would be equally distinctive, though perhaps different, in Martian life-forms; likewise, if amino acids are to work as building blocks, linked into chains that form the precise sorts of structures that life requires—coils and sheets of a consistent three-dimensional form, enzymes that fit with a specific substrate—then they would have to be all of the same configuration, either predominantly L, as on Earth, or all D. Under the current frigid, dry conditions on Mars, such a chiral preference could persist in fossil Martian organic matter for millions of years and might indeed provide evidence that the planet was once host to organisms of one sort or another. However, if the Martian climate had been warmer and wetter for more than a few million years, amino acids from any organisms that existed before or during that period would be fully racemized, and it would be much more difficult to distinguish them from amino acids that had been created by abiotic chemical processes.

Other researchers are focusing on biomarkers with the distinctive architectural style of the isoprenoids, whose five-carbon isoprene units, linked head to tail, signal a probable biogenic origin *and* are sturdy enough to resist chemical breakdown for billions of years. Such a focus presumes an even higher level of uniformity between extraterrestrial and terrestrial biochemistry, as most of these compounds require a more sophisticated enzymatic system to fabricate than do amino acids. But they are also essential components of one of life's fundamental structures, the formation of which some theorists propose was a first step in the origin of life: its container. The cell membrane confined life's ingredients in proximity, segregated them from their environment, and generally provided the conditions necessary for a system of self-propagating reactions to develop. Long-chain acyclic isoprenoids make up the cell membranes of some of Earth's most widely adapted microorganisms, and hopanoids, which are more complex cyclic compounds based on the same architecture, comprise integral components of most bacterial membranes. In fact, as a group, isoprenoids are the most abundant organic molecules preserved in earth rocks. So why not on Mars? If there's anything at all to the presumption of solar-system-wide biochemical Uniformitarianism, then wouldn't such ubiquitous, persistent molecular fossils also have been preserved on Mars? Indeed. But some microbiologists are coming at the question from the other side, looking for outliers, for aberrations in the Uniformitarian norm: rather than searching on Mars for the most universal and persistent molecular fossils derived from terrestrial biochemistry, they are looking on Earth for environmental analogues to Mars—frigid temperatures, high exposure to damaging ultraviolet light, extreme aridity, caustic soil—and studying the biochemistry of terrestrial bacteria adapted to live under such conditions.

Most, though not all, researchers are still convinced that there is presently no life on Mars...but the more we learn about the adaptability of life here on Earth and about the chemistry of the solar system and beyond, the more plausible it seems that carbon-based life has evolved, or is evolving, or will evolve, somewhere besides Earth. Whether the search for evidence is worth the resources invested is a hotly debated issue in the scientific community, as well as in the political domain—but the temptation to search is almost impossible to resist. The exciting questions about the origin and distribution of life in the universe that chemists laid on the table and began to test experimentally with their first meteorite analyses in the 1860s, and revived in the 1960s, and then again in the 1990s, are destined to return ad infinitum, it seems, until they find some definitive answers.

4

Black Gold

An Alchemist's Guide to Petroleum

The fascinating problem of the origin of liquid petroleum, with which we must associate natural gas, mineral waxes, and asphaltic materials, is primarily of interest to geologists and chemists, but its solution would have a wider impact, since it would throw some light on certain aspects of the early history of the earth and the first steps in the evolution of living forms.
—Sir Robert Robinson, 1886–1975, British chemist
  From *Nature* 199 (1963)

For many of us who studied and came of age in the last two decades of the twentieth century, there was nothing more prosaic, lacking in romance, and less worthy of our scientific curiosity than petroleum. The basic questions about its composition and origin had been answered, and it was no longer one of Nature's secrets luring us to discovery, but rather the dull stuff of industry and business, money and technology. Some of us even imagined, naively, that we would witness the end of the age of fossil fuels: they were the bane of modern man, the source of pollution, environmental disaster, and climate change that threatened to disrupt ecosystems and civilizations around the entire globe. Finding new reserves, we reasoned, would only forestall the inevitable, or exacerbate the havoc. But when Jürgen joined Germany's government-funded Institute of Petroleum and Organic Geochemistry in 1975, there was still a sense of mission in finding new reserves. The energy crisis of the early 1970s had created a heightened awareness of the value of fossil fuels and the need for conservation, but the accepted wisdom remained that oil was the key to the future and well-being of civilization. And the chemistry, it seems, was anything but banal—it was, in fact, leading not just to a better success rate in finding new reserves of oil, but also to a new understanding of life that no one had foreseen.

Certainly for Geoff and the generations of organic chemists that came before him, the oils that occasionally seeped out of a crack in a rock, or came spouting out of the earth if one drilled a hole in the right place, were as intriguing as the life some said they came from. Liquid from a solid, organic from mineral, black or brown or dark red, it was as if blood were oozing from stone, an enigma that inspired inquiry from scientists long before it found its place among man's most coveted commodities. What was it made of, where did it come from, how had it formed? Geologists in the early nineteenth century thought it was synthesized directly from the elements at high temperature and pressure within the earth. Yet people had known since the Middle Ages that certain dark, flaky sedimentary rocks produced a similar substance when heated. Such oil shales were used to produce lamp oil in the eighteenth and nineteenth centuries, and their flammable nature was known to Native Americans in the Green River region, who burned the rocks for heating. In the mid-1800s, the first chemists to describe the peculiar molecular nature of the matter made by living things wondered about these oils that oozed out of rocks and collected beneath the surface of the earth. Their investigations revealed substances similar to those in organisms—molecules made of carbon and hydrogen bound with strong covalent bonds in countless variations—and yet the oils contained far less oxygen, nitrogen, and phosphorus than the material of life.

Since the earliest recorded history, people had been making use of the oil at natural surface seeps, or stumbling on subterranean accumulations of oil and gas while mining for salt, using the oil to boil off the brine. By the end of the 1700s, the Burmese had dug hundreds of wells around one such surface seep, producing tens of thousands of tons of oil a year and supplying the local economy with preservative for wood buildings, caulking for boats, and a rather smoky type of lamp oil. But it wasn't until the mid-1800s, when the first big underground oil reservoirs in North America and Russia were discovered by prospectors—who, like the Burmese, had drilled wells where they found a telltale seep of oil at the surface—that geologists began to note some commonalities in the geological setting of the deposits.

Observing that petroleum was always found near ancient sedimentary basins, they began to suspect that it came from the organic detritus of marine organisms that had been deposited with sediments in shallow seas. But accumulations of oil weren't found in these sediments, or even in the dense, organic-matter–rich shale that formed as the basin filled and the sediments were compressed into rock—they were found in porous sandstone or carbonate formations. The oil seemed to be produced in one place and then gradually migrate to and accumulate in another, something the geologists could explain with a bit of physics and observation: oil moved upward through the fractures and pores of rocks until it was blocked by an impermeable layer or collected in the inverted bowl created by an upward fold of the strata.

But how was the oil expelled from the shales, and, for that matter, precisely how did it form to begin with, from what? Was it deposited as part of the plant and animal debris and then somehow concentrated by this migration process? Was it made from that debris by bacteria in the sediments? And what, precisely, did it *consist* of?

"We didn't know what molecules were in there," Geoff says, and until coupled GC-MS came on the scene in the 1960s, it was almost impossible to find out. Nevertheless, a small cadre of analytical chemists made a go of it in the early part of the twentieth century. Many of them worked for the government-funded American Petroleum Institute, characterizing a single reference sample of crude

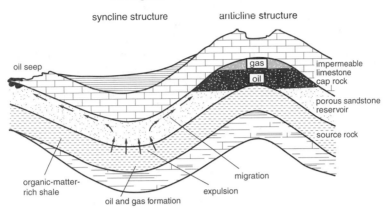

Oil migration and accumulation

syncline structure   anticline structure

oil seep   impermeable
                                        limestone
                           gas   cap rock
                           oil
                                        porous sandstone
                                        reservoir

                                        source rock

organic-matter-   migration
rich shale
                           expulsion
       oil and gas formation

oil from a well in Ponca City, Oklahoma. Over a period of almost 40 years, using a painstaking combination of distillation and column chromatography to separate fractions containing compounds of different size and boiling point, they identified more than 275 individual compounds—but it was clear that this was only a tiny fraction of what their single sample contained. Even their most exhaustive separation procedures usually resulted in complex mixtures of unknown structures. A good portion of the oil comprised a high-boiling, nitrogen-, sulfur-, and oxygen-containing complex macromolecular mixture that was as impossible to characterize as the insoluble kerogen that made up the bulk of the organic matter in sedimentary rocks. Most of these so-called "asphaltenes" were removed from fuel oils early in the refining process and used to make lubricating oils and asphalt. The lower-boiling fractions of the oil seemed to contain hydrocarbons of every size and shape, far beyond what was found in living things: branched and unbranched chains, five- and six-membered rings glued together like puzzle pieces into larger, more rigid molecules, and a plethora of aromatic compounds, with their flat, hexagonal benzene rings.

Indeed, so inscrutable was Nature's alchemy, and so extensive, that debate about the origin of petroleum raged well into the twentieth century, and even the basic premise that its hydrocarbons derived from compounds made by living things was called into question again and again.

The nineteenth-century chemists who first puzzled over the black oils and tars found oozing out of the earth had noted that their extracts possessed what Pasteur recognized as one of the most idiosyncratic qualities of living matter, the telltale mark of its one-handed enzymes: they deflected the rays of polarized light in one direction. "This optical activity was a connection to biology," Jürgen says, as Geoff and I sit in his Oldenburg office, flipping through one of his meticulously ordered collections of old papers. "And then, of course, you are interested. What is in there that can be tied to biology?" It was this seemingly alchemical connection

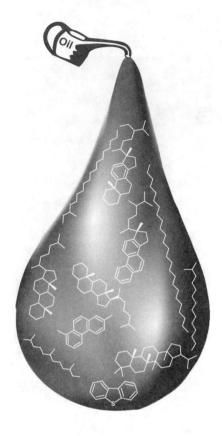

to life in a substance so far removed from life that had caught chemists' fancy from the first, and even now I detect a hint of understated excitement in Jürgen's voice as he poses his deceptively simple question. It was the same question Alfred Treibs was asking in the 1930s when he noted a structural resemblance between the red porphyrin pigments he'd extracted from petroleum and the green chlorophyll from plants. And that Geoff, Melvin Calvin, Keith Kvenvolden, and their assorted colleagues were asking about the ancient Precambrian rocks in the 1960s, and about meteorites and lunar dust, the one Jeff Bada and others are even now asking about Martian rocks. "Obviously one was curious about the origin of the stuff," Geoff says, leaning across the desk to see what Jürgen is hiding in his binder of papers.

But when I query the Strasbourg natural products chemist, Guy Ourisson, about the intellectual thrills of petroleum geochemistry, he rolls his eyes and tells me that he was never really interested. "You weren't?" I ask, surprised. Some of the most compelling and useful information about petroleum formation has come from his laboratory and students. He shakes his head. "No," he says, with a complicit little smile. "Not really." But then he considers for a moment and adds, "Well, maybe I'm cheating a little." My probing has reminded him of a conversation he had some 50 years ago, when he first started at the University of Strasbourg. "In 1954, when I was a candidate for a job here I had to meet the then-dean of science, who was a botanist.... And he asked me a number of questions, in particular, two that baffled me. He asked, 'Are you interested in the origin of life?' and 'Are you interested in the origin of petroleum?' I think in both cases I told him, 'Oh that's too complicated!' Interesting questions, but I didn't have any way of knowing where to start."

Apparently these two lines of inquiry, which seem worlds and philosophies apart to me—the one lofty, impenetrable, and plush with philosophy, the latter mundane and utilitarian—existed side by side in the scientific imagination of the time. Certainly for the handful of geochemists who, whether daringly, or grudgingly like Ourisson, started with the rocks or the oil itself, the two were linked: whatever the source of their curiosity or funding, they found themselves using the same methods, tracking the same molecules, and unearthing clues to both the mystery of petroleum and the long history, if not the ultimate provenance, of life.

Jürgen's collection of papers doesn't begin until the late 1960s, but even in the 1930s the American Petroleum Institute chemists were beginning to note that this one consistent connection to biology, the bending of rays of polarized light, became more pronounced in a certain, relatively plentiful fraction of their oil distillates. It boiled between 300–500°C and was made up exclusively of saturated hydrocarbons with between 27 and 30 carbon atoms in some arrangement of fused five- and six-membered rings. And, of course, somewhere in all or most of the molecules in the mix—for they had yet to separate the individual compounds—were centers of asymmetry with a specific configuration that rotated plane-polarized light consistently, and rather dramatically, to the right. As early as the turn of the last century, German chemists had speculated that these optically active compounds in petroleum might somehow be derived from a family of similarly high-boiling, ring-containing, optically active compounds they'd been extracting from plants and animals. These included the now-infamous 27-carbon alcohol, cholesterol, and a range of similar four-ring alcohols that chemists had found in everything from sheep's wool to algae.

By the mid-1960s, this particular family of compounds was all the rage with organic chemists and biochemists, partly because of the fascinating stereochemistry of their various asymmetric centers, and partly because of the discovery that some of them serve as human hormones. Cortisone, synthesized from a related sterol found in yams, was already working wonders for arthritis sufferers, and Carl Djerassi had just developed the birth control pill. In the process of looking for plant sources of these and similar sterols, natural products chemists were finding all sorts of other interesting compounds, loosely related members of a larger clan of cyclic isoprenoids. Wide-ranging in both structure and function, the common ancestry of the clan, as well as certain family group differences, was evident in the arrangement of rings that form their cores. The biosynthetic pathways to the creation of those rings were discovered in the 1950s and 1960s and affirmed the kinship. All began with the 30-carbon acyclic isoprenoid squalene, and then proceeded in an elegant cascade of enzyme-assisted reactions wherein its isoprene beads looped back on themselves to form the four rings of a sterol or the five rings of a pentacyclic triterpenoid. The sterols required the initial addition of oxygen, so that the ring-forming reactions proceeded from squalene epoxide, but some of the pentacyclic triterpenoids were produced without it.

Once the common ring foundations formed, the construction theme was customized to give a wide variety of compounds, with various carbon side chains suited for one or another biochemical task. Besides their role as human hormones, the four-ring sterols were known to play a more universal, if less romantic, role as components of cell membranes, where their relatively flat, inflexible ring structures lend a degree of structural rigidity to the long, floppy phospholipid chains. In most animals, cholesterol serves this purpose, but natural products chemists were finding all sorts of variations in plants and, once they started looking in the late 1970s, in algae. Most of the pentacyclic triterpenoids they found were in exotic flowering plants, where they seemed to play less essential roles as leaf coatings, protective exudates, poisons, and foul-tasting deterrents to potential

Biosynthesis of sterols and pentacyclic triterpenoids:
a simplified scheme

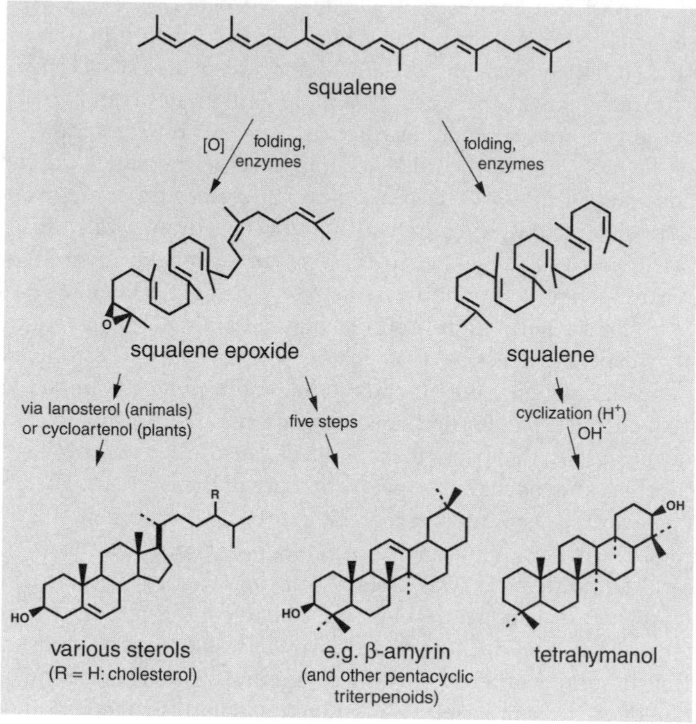

squalene

[O] / folding, enzymes

folding, enzymes

squalene epoxide

squalene

via lanosterol (animals) or cycloartenol (plants)

five steps

cyclization (H⁺) OH⁻

various sterols
(R = H: cholesterol)

e.g. β-amyrin
(and other pentacyclic triterpenoids)

tetrahymanol

grazers—though the exoticism might just have been a by-product of the chemists' penchant for studying unusual tropical plants.

A close look at all of these structures shows that a pronounced optical activity is a familial trait, inherent in the fused ring systems—even the simplest sterol has seven chiral carbons built into its ring system, with asymmetric centers at each ring junction and wherever a methyl group or side chain is attached. With two configurations possible at each chiral center, the same molecule can, in principle, exist as any of 128 possible stereoisomers, and that's not counting any chiral centers it might have in its side chains. But of all those possible stereoisomers, the enzymes of organisms, made up of one-handed amino acids, are likely to make only one. And because the fused ring system freezes certain planes of the molecule in one position and keeps them from flopping about like a simple chain of carbon and hydrogen atoms, the prismlike effect on the rotation of plane-polarized light is particularly pronounced and a solution of that single stereoisomer will exhibit enhanced optical activity.

Even without taking into account their stereochemistry, the steroids and pentacyclic triterpenoids are complicated structures. Determining them precisely and unequivocally was a great challenge for natural products chemists,

Some sterols and triterpenoids found in organisms

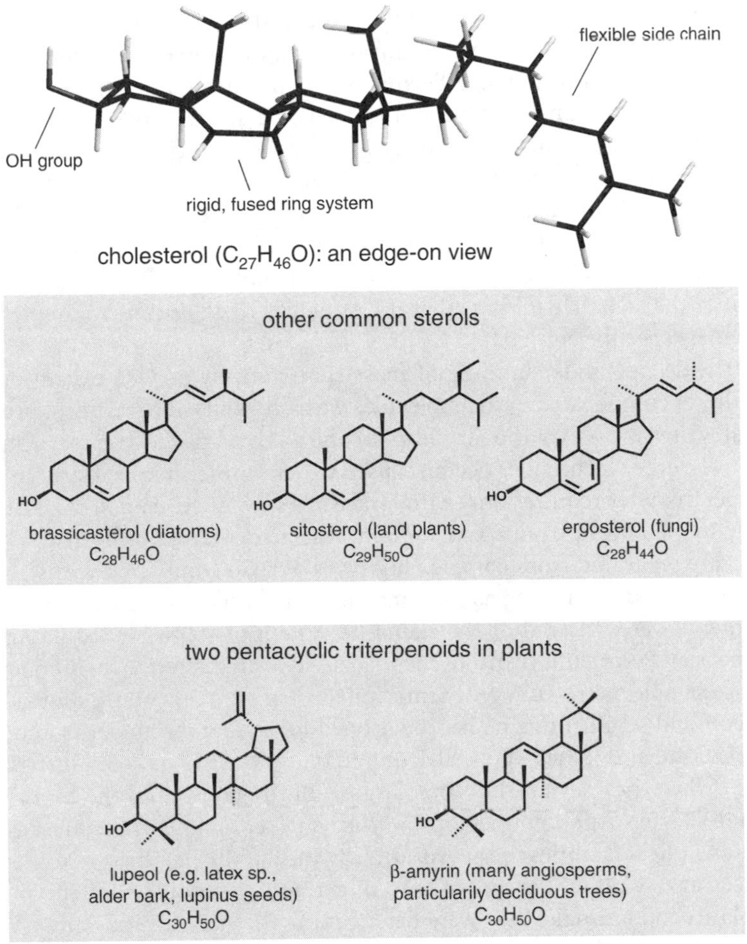
flexible side chain
OH group
rigid, fused ring system
cholesterol ($C_{27}H_{46}O$): an edge-on view
other common sterols
HO
brassicasterol (diatoms)
$C_{28}H_{46}O$
sitosterol (land plants)
$C_{29}H_{50}O$
ergosterol (fungi)
$C_{28}H_{44}O$
two pentacyclic triterpenoids in plants
lupeol (e.g. latex sp.,
alder bark, lupinus seeds)
$C_{30}H_{50}O$
β-amyrin (many angiosperms,
particularily deciduous trees)
$C_{30}H_{50}O$

and they responded by developing new analytical techniques and pushing the limits of the available instrumentation. In the early 1960s, Carl Djerassi went on the offensive with mass spectrometry, which at first hadn't seemed particularly useful for such complex compounds. Its high-energy electron beam left a few molecules intact, so you could easily tell how many carbon and hydrogen atoms a compound had, but most of the molecules broke into what seemed an incomprehensible array of pieces. Rumor has it that Djerassi put everyone in his lab and their wives and uncles to work running mass spectra of every organic compound they could extract, steal, or synthesize, looking for patterns in the way the compounds broke up in the spectrometer. Ourisson, who was among those natural products chemists enamored of the terpenoids, tells me that he still remembers the first time he saw a published mass spectrum of a steroid. "It was a shock," he says. "Just the sheer possibility that something like cholesterol could give a recognizable mass spectrum—that was totally unexpected. It was a revolution!" Ourisson was particularly adept at structure determination using classical chemical means, as well as infrared and ultraviolet absorption spectra, and nuclear magnetic resonance. Seeing the enhanced potential of the mass spectrometer for steroid and triterpenoid analysis, he started coveting one for his lab in Strasbourg, but they were impossibly expensive and beyond the means of most academic chemistry labs. Ourisson might have made do with sending his samples to friends in the pharmaceutical industry, if the geologist who had been pestering him for years to apply his analytical skills to sediment samples hadn't become dean. "I was bribed," Ourisson says, still looking a bit sheepish, some 40 years after the fact. "He told me that if I'd analyze these sediment samples, he'd fund a mass spec for our lab."

If the deeper understanding of mass spectrometry and its expansion into the realm of complicated cyclic molecules was a revolution for natural products chemistry, it was an evolutionary leap for the analysis of sediments, rocks, and oils. Unlike other techniques chemists used to determine chemical structure, the mass spectrometer requires only a tiny trace of a substance. And, it can be linked directly to the output from a GC column, where tiny traces of individual compounds are separated from complex mixtures. For oil company chemists, funding wasn't a problem as long as their management could see how useful the new instruments were. Instead of the elaborate system of stepwise distillation that the American Petroleum Institute chemists used to characterize their Ponca City reference sample, petroleum geochemists started using a procedure similar to the one Geoff and Al Burlingame had developed for Precambrian rocks and lunar samples. Geoff and James Maxwell acquired one of the first GC-MS instruments in 1965, when they set up shop in Glasgow. Burlingame used the huge NASA grant for Calvin's Apollo project in California to set up one of the most sophisticated analytical facilities in the world.... By the late 1960s, they were all hot on Djerassi's analytical tail, except instead of extracting steroids and triterpenoids from plants and animals, they were determining the structures of tiny traces of the long-suspect light-bending compounds in the infinitely more complex mixtures of hydrocarbons in rocks and oils—trying, as Jürgen says, to pin down the "connection to biology" in petroleum, and in the organic-matter–rich sedimentary rocks that purportedly gave rise to it.

Many of the molecular structures that they puzzled out did indeed resemble the optically active cyclic terpenoids that the natural products chemists were collecting from organisms—except that most of these compounds in the oils and shales had been stripped of their double bonds and oxygen-containing functional groups, reduced, as it were, to their bare carbon skeletons. Excited to find what appeared to be molecular skeletons of known biological steroids and triterpenoids, the geochemists named the hydrocarbons accordingly, and sometimes misleadingly—precisely because they were missing some of the key information that distinguished the biological compounds. The sterane they called "ergostane" seemed to bear the carbon skeleton of the sterol ergosterol, which is made by fungi and was quite familiar to chemists. But ergosterol differs from brassicas-terol only by a double bond in the ring structure, which, of course, was missing from the steranes in the rocks, and the relatively large quantities of ergostane that the geochemists found in rocks and oils probably had little to do with fungi and everything to do with a certain very productive group of unicellular algae. Likewise, sitosterol, which is plentiful in land plants, and stigmasterol, which is abundant in land plants and algae, are distinguished by an extra double bond in stigmasterol, so one really can't tell whether the 29-carbon sterane that geochem-ists first called sitostane and later stigmastane derives from land plants or algae or both. Eventually, when the geochemists gave up on trying to distinguish the orientation of the side groups in the fossil molecules, they took to calling them simply 24-methylcholestane and 24-ethylcholestane.

Steranes and pentacyclic triterpanes found in petroleum and rocks

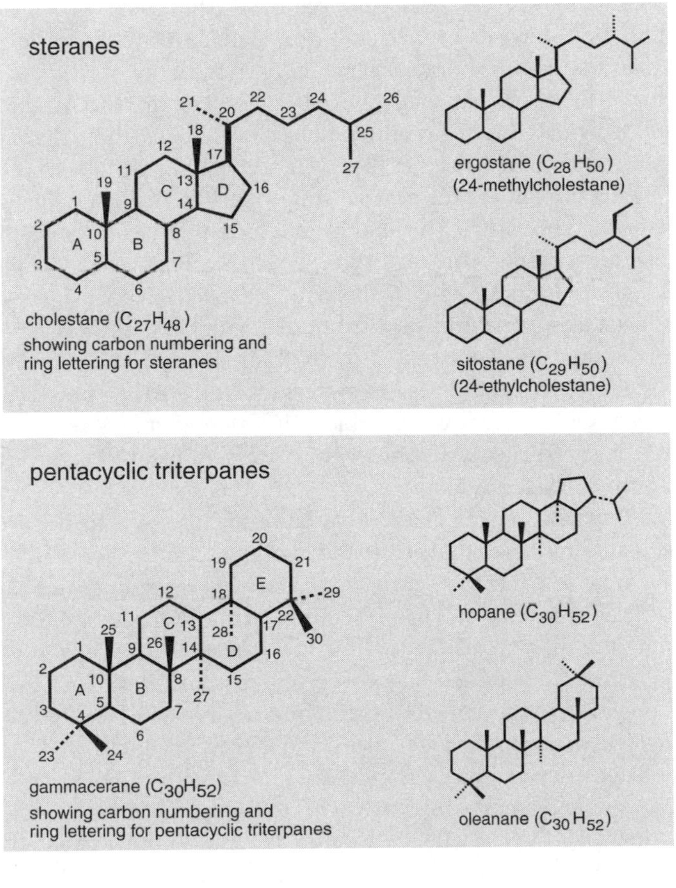

steranes

cholestane ($C_{27}H_{48}$)
showing carbon numbering and ring lettering for steranes

ergostane ($C_{28}H_{50}$)
(24-methylcholestane)

sitostane ($C_{29}H_{50}$)
(24-ethylcholestane)

pentacyclic triterpanes

gammacerane ($C_{30}H_{52}$)
showing carbon numbering and ring lettering for pentacyclic triterpanes

hopane ($C_{30}H_{52}$)

oleanane ($C_{30}H_{52}$)

Nevertheless, as the natural products chemists' catalog of steroids and triterpenoids in organisms grew, a few patterns in distribution that appeared to be indelibly built into the carbon skeletons became apparent: steroids like cholesterol with no branch group attached to the C-24 position of the side chain were present in a large variety of organisms; those with a methyl group at C-24 were common in fungi and algae, and those with an ethyl group—sitosterol and stigmasterol—were prevalent in land plants and certain algae; and some marine phytoplankton had sterols with longer, more complex side chains. Pentacyclic triterpenoids with the β-amyrin carbon skeleton were found only in flowering plants; tetrahymanol was found in marine ciliates—a group of mobile single-celled eukaryotes that feed on sinking detritus in the water—and no pentacyclic triterpenoids whatsoever were found in animals. The triterpenoids in plants were varied and specific enough that they could sometimes be used as a taxonomic tool—much more effectively, in fact, than Geoff's leaf-wax *n*-alkanes. Ourisson was able to distinguish the myriad species of spurge plants from the terpenoids in their milky secretions. He could even track the evolution of the genus as it spread around the world by tallying the increasing numbers and complexity of triterpenoids in species found farther and farther from its point of origin.

The assemblies of steroids and pentacyclic triterpenoids that geochemists were finding in geologic deposits in the 1960s and early 1970s were considerably less revealing than Ourisson's spurge terpenoids. In fact, many of them were downright enigmatic. But reading through the old papers, it's apparent that the scattered chemists who elucidated their structures all had a similar reaction: they had struck an information gold mine. Whether they were trying to help geologists find petroleum, or looking for clues to the past, or simply curious about the fate of organic molecules that had been sitting around in rocks for millions of years, it was clear that enlightenment of some sort was locked into the carbon–carbon bonds of these complex, sturdy molecules. Ted Whitehead at British Petroleum, which was somewhat more open about scientific publishing than most of the American-based oil companies, developed a sophisticated method for isolating and determining the structures of triterpenoids and began a systematic study of their presence in crude oils even before GC-MS became available. His group identified the structures of two pentacyclic triterpenoids that clearly derived from plant terpenoids and proposed that such compounds might be used to distinguish the relative amounts of land-plant and marine organic matter that had gone into a particular crude oil—knowledge that allowed one to predict the quality of refined petroleum it would produce and provided clues to where it had formed and where one might find more. Calvin's Berkeley group was still firmly focused on finding clues to the earliest life-forms, comparing the steroids they identified in the Green River shale with more ambiguous results from the 2.7-billion-year-old Soudan shale. Back in Glasgow, Geoff and James Maxwell were pretty much interested in anything and everything they could get these molecules to tell them, and they were joined, in 1965, by Pierre Albrecht, a like-minded emissary from Ourisson's lab in Strasbourg.

"I think people did not really know what would be the consequences of what they were doing," Albrecht tells me when I ask him and Maxwell about their student days in Strasbourg and Glasgow. "They were just interested that this was an open field and people were gathering information on molecules. You saw peaks in the chromatograms and you wanted to know what they were, and what significance they had in terms of relationships with the living organism."

"People concoct stories afterwards," Maxwell says. "Nowadays people have to focus on what they want to do because they've got to justify the money to someone who's got the whip on their back. But in those days you did what you wanted and you were just desperate to find out something new. Afterwards, in some cases—and it may have been luck, intuition, it may have been something else—it led to something. But most of the time you weren't aware of that."

By the late 1960s the GC in Geoff's lab could separate about 30 acyclic isoprenoids, steroids, and pentacyclic triterpenoids in an extract of an organic-matter–rich rock like the Green River shale. With the mass spectrometer connected, they could scan each peak and get some idea of a compound's size and basic skeleton, even when they couldn't unequivocally determine its precise structure. A sterane like cholestane, for example, with 27 carbon atoms of 12 mass units each, and 48 hydrogens of 1 mass unit, would yield a strong molecular ion of 372 mass units. And one could now predict, to some extent, how a given structure would break up based on the relative strength of the carbon bonds and on the stability of the resulting fragments. A bond between two carbon atoms with side groups and no hydrogens attached would tend to break most easily, and a bond between two methylene groups, with two hydrogens attached, least easily, with the stability of the resulting fragments also playing a role. In the mass spectrum, a small peak just 15 mass units below the molecular ion corresponds to the loss of a CH_3 group. Longer side chains tend to break next to their branches, and a succession of minor peaks usually corresponds to the intact ring structure minus various small pieces of a side chain. How the ring structures themselves break up is often a distinguishing family trait that depends on where the methyl groups and side chains are attached to the rings. Typical of most steranes is an intense base peak at 217 mass units, which corresponds to a stable three-ring fragment composed of the A-, B-, and C-rings, and a less intense peak at 149 mass units that comes from a slightly less stable A–B ring fragment. Cholestane, for example, gives strong peaks at 217 and 149 mass units, and a small peak at 262 mass units that corresponds to the C–D rings with the side chain. The pentacyclic triterpenoids tend to break in the middle of their C-ring, with the sturdy A–B ring fragment giving what is generally the most intense peak, at 191 mass units, and the D–E ring fragment yielding a second strong peak whose mass depends on what methyl groups and side chains are attached. One of the first pentacyclic triterpenoids unequivocally identified in the Green River shale was the symmetrical gammacerane, which has a very simple fragmentation pattern that gives a doubly intense 191 peak, as the mass of the D–E ring fragment is the same as that of the A–B combination.

Even with the expanded analytical power of mass spectrometry, it was hard to determine the precise structures of compounds in the geological samples when there were no mass spectra of reference standards available for comparison. One

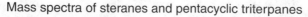

Mass spectra of steranes and pentacyclic triterpanes

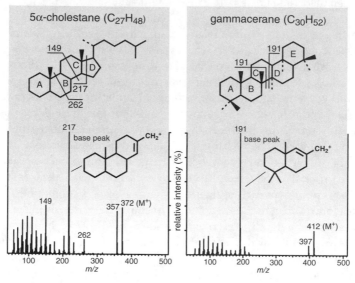

could guess at the structures and attempt to synthesize standards, creating single pure compounds of known structure whose mass spectra could be compared with the unknowns in the rocks—but this was no easy task for even the best synthetic chemist. There was a growing inventory of published mass spectra of steroids and triterpenoids from plants and animals that was helpful, but, of course, the molecules in the rocks were millions of years removed from any biological origin and not likely to be the same as their precursor molecules. And, despite the growing popularity of natural products chemistry as a scientific endeavor, there were still—and are yet—huge gaps in the molecular inventory of organisms. A certain amount of luck and serendipity was involved in determining the molecular structures of organic compounds extracted from rocks and oils, and somehow, in those first analyses, everyone was missing, or disregarding, or simply unable to make sense of the most plentiful and common family of triterpenoids in their rock extracts and crude oils...until Guy Ourisson finally succumbed to his geologist friend's bribe and set his lab to work analyzing the dreaded sediments his friend had been pestering him about for years.

Albrecht says Ourisson handed him the project when he was a first-year graduate student. He was entirely innocent of his mentor's long-standing resistance to analyzing the sort of materials he was being charged to investigate and apparently unaware that Ourisson was embarking on the project not because sedimentary organic matter was a huge, unexplored territory of organic chemistry, or because Calvin's group had convinced him he'd find clues to the origin of life, and least of all because he was interested in petroleum—but because it was a condition for acquiring the lab's most remarkable new analytical instrument.

Ourisson sent him off to Glasgow to learn the new techniques that Geoff's group was developing, and when he returned to Strasbourg, he set up a geochemistry lab with the new mass spectrometer and started analyzing everything from pond sediments to garden soil. "Just to see what we could see," Albrecht says, echoing Geoff's descriptions of much of the Glasgow work. It was more exploration than hypothesis-driven science, and they analyzed everything the geologists could get for them, finally focusing their efforts on samples from the Messel shale, Europe's version of the Green River shale. Like the Great Plains of North America, large areas of Europe had been covered with inland lakes during the Eocene epoch, and the Messel shale in southern Germany had formed from the sediments of one such lake between 45 and 50 million years ago. According to the geologists, it had been relatively undisturbed by the earth's tectonic movements and upheavals during those years and contained a remarkable assembly of well-preserved fossils, as well as large amounts of organic matter for the prospecting chemists. Ourisson says that he wasn't looking for triterpenoid compounds in particular—but they were hard to avoid. And, of course, he had been studying these compounds in plants for more than a decade and was quite well versed in their structural idiosyncrasies.

What turned up, what they saw in the gas chromatograms of cyclic hydrocarbons from every sediment extract Albrecht analyzed, was a cluster of prominent peaks whose mass spectra all contained the strong 191 fragment that Djerassi had recognized as a calling card for pentacyclic triterpenoids, but whose fragmentation pattern did not match that of most of the hundreds of pentacyclic triterpenoids that had been isolated from plants. There was, however, something decidedly familiar about the fragmentation patterns in the sediment compounds. Ourisson had seen it among the triterpenoids he'd isolated from the resins of Southeast Asian trees. And he'd seen it again, years earlier, in a compound one of his students had worked on as a favor to an Italian friend, who'd isolated it from a maidenhair fern. It showed a molecular ion of 412 mass units, indicating a 30-carbon compound, and an intense 191 peak that completely dominated the spectra, as if the ring structure had split into two equal fragments, like gammacerane. But here the reason for the symmetry was less obvious, as a small fragment at 43 mass units below the molecular ion indicated the presence of an isopropyl group—three carbon and seven hydrogen atoms—and a bit of simple math showed that a D–E fragment with an isopropyl group attached would have a mass of 191 only if the E ring were a five-carbon ring, rather than the six-carbon ring of most of the pentacyclic triterpenoids found in plants.

Albrecht was lucky because Ourisson happened to have various closely related compounds stashed away in his lab from his earlier work, and this made it easy to synthesize reference standards and verify some of the unknown structures in the sediments. The first triterpenoid they identified in an extract from the Messel shale in 1968 had the unaltered structure of an alcohol that natural products chemists had isolated from tropical plants. Most of the fossils in the shale were from relatives of Southeast Asian trees and ferns, and the Eocene was a notoriously warm period of geologic history, so this seemed reasonable enough.

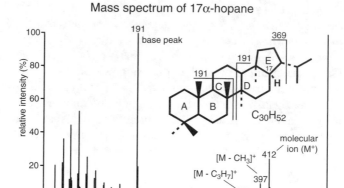

Mass spectrum of 17α-hopane

But then things got confusing. There seemed to be a large variety of compounds with a ring structure that was similar to that of one of Ourisson's compounds, a 30-carbon five-ring affair that people had taken to calling the "hopane" skeleton—not because it inspired particularly positive expectations, but because the genus of Burmese trees where it was first discovered was named after an eighteenth-century English botanist named John Hope. Probably the only reason anyone knew about hopane to begin with was that the British Museum had used the resin from *Hopea* trees as a varnish. What was disconcerting was that compounds with this hopane ring structure were present in every rock or sediment sample one analyzed, no matter when or where it was formed. And yet, as far as anyone knew, such compounds were made *only* by a few exotic species of tropical trees, ferns, and mosses.

Geoff and James Maxwell faced a similar conundrum when they identified hopane as one of the two dominant compounds in the cyclic alkane fraction from the Green River shale. The other compounds they identified seemed to fit with the fossil evidence in the shale. Ergostane could have come from the algal sterol, brassicasterol, and 24-ethylcholestane could have come from stigmasterol or sitosterol made by algae or by plants growing around the lake. Gammacerane was a bit problematic, because there was so much of it, but it could well have derived from the tetrahymanol made by ciliates feeding on the detritus in the lake. But hopane? It could conceivably have come from a compound with the same carbon skeleton that had been identified in some ferns. Bits of fossil leaves in the shale and the presence of long-chain leaf-wax *n*-alkanes in the extracts bore witness to a significant input of organic matter from the land plants in the surrounding watershed, and ferns seemed more plausible than giant Southeast Asian trees—but then, there must have been an awful lot of ferns growing around the ancient lake. And what about all the seemingly related compounds that Whitehead's group was finding in crude oils?

The more they learned, the more samples they analyzed, and the more compounds they identified, the less likely their explanations seemed. This was no modest little family with a single carbon skeleton like what had been isolated from ferns and tropical trees. This was a large, exuberant, extended family of compounds. Their mass spectra had molecular ions at 412, and at 398 and 370 mass units—not just the 30-carbon hopane, but 29- and 27-carbon versions where the 191 ion was less intense and the D–E ring fragments showed up at 177 and 149 mass units, respectively, the isopropyl side chain of hopane replaced by a two-carbon ethyl group or missing altogether. At first it seemed likely that these C_{29} and C_{27} compounds were breakdown products from the 30-carbon hopane, perhaps the result of hungry microorganisms in the sediment chewing on the isopropyl tail, and it made sense that the 28-carbon version was missing because it would require breaking two carbon–carbon bonds simultaneously, which might be too much work for the bacteria. But then, not long after Geoff and his geochemistry group made the move to Bristol, they and the Strasbourg group, working together on extracts of the Messel shale, identified a 31-carbon version of hopane. And it wasn't long before they identified a C_{32} hopane in crude oils, and a C_{33}, a C_{34}, and a C_{35}, and then that, finally, seemed to be the end of it—except, just to complicate things, there appeared to be more than one isomer for each of the homologues in the series.

These "extended" hopanes were present not only in the Messel and Green River shales, both formed during the Eocene in inland lakes, but also in recently formed sediments from the Baltic Sea and from England's Rostherne Mere, in a 150-million-year-old shale from the Atlantic Ocean, and in 25-million-year-old crude oils from Africa and Iran and Texas.... It seemed highly unlikely that bacteria would have *added* carbon atoms to a 30-carbon hopanoid—after all, the bacteria wanted to *eat*, not waste energy adding methyl groups to compounds they couldn't even use. Either some organism could synthesize a whole series of such compounds, or the C_{35} hopane was the parent compound, and the microbes chewed it up to produce the others. The mass spectra of the various homologues showed fragmentation patterns indicating that the "extra" carbons were all tacked onto the side chain of the E ring, as their characteristic D–E ring fragments increased by 14 mass units with the addition of each CH_2 group: from 191 for the C_{30} compound to 205 mass units for the C_{31}, 219 for the C_{32}, and so forth, on up to 261 mass units for the D–E fragment of the C_{35} extended hopane.

It seemed reasonable to assume that a five-carbon side chain attached to a structure that was entirely built of five-carbon isoprene units would itself be an isoprenoid chain, and that its biosynthesis simply started with a longer chain than the usual squalene. No pentacyclic triterpenoid with more than 30 carbon atoms had ever been found in an organism, so the Strasbourg group synthesized this postulated C_{35} hopane in the laboratory and compared it to the one they'd detected in the oils.... But when they ran it on the GC, it didn't even elute from the column at the same time. The extra five carbons, as it turned out, weren't an isoprenoid unit at all, but a straight chain. It was hard to imagine how such a carbon skeleton might be biosynthesized, and yet it clearly derived from a molecule made by some very common organism, or by a lot of organisms, or both. The only universal process when any living organism dies is that it decays, and suspicion was growing in both Strasbourg and Bristol that the extended hopanes might actually come from the bodies of the microorganisms responsible. Around this time in the early 1970s, there were reports of the two C_{30} hopanoids known from ferns, diplopterol and diploptene, in a couple of species of bacteria, which provided some support for this hypothesis—but the extended hopanes remained elusive.

"The bloody things had to come from somewhere!" Geoff says. "But we weren't finding them."

"This was calling for a precursor," Ourisson says. "But none was known! At this time I wrote to about twenty of my friends, telling them, please, if you see anything that resembles hopane plus five carbons as a straight chain, warn me immediately."

"We could have looked and the Bristol people could have looked for ages," Albrecht tells me. "If we had not had the luck that some people, coming from a completely different background, worked on an extract of *Acetobacter*…"

Vinegar? *Acetobacter* is a bacterium known for its role in making vinegar. This is the kind of story that Ourisson relishes, where a bunch of quirky scientific subplots that seem to have nothing to do with each other all weave neatly together at the end. A friend in Zurich responded to his plea for help and told him about a compound that Klaus Biemann's natural products group at MIT had isolated. "They were doing something for Canada's paper industry, looking for bacteria that would produce cellulose—a way to make paper from fermentation, without trees. They were studying one *Acetobacter* species that caused fibers to align in the growth medium, and they isolated a substance that had this property…and it was a 30-carbon hopane skeleton plus five carbon atoms!" The Strasbourg group quickly repeated the work, using their sediment analyses for comparison, and in 1976 they confirmed the complete structure of a compound they called bacteriohopanetetrol.

Albrecht describes it as a "very strange compound," with exactly the same skeleton as their C_{35} hopane, but with hydroxyl groups attached to four of the carbon atoms in its five-carbon tail. There was already substantial evidence that the hydroxyl groups of alcohols could be reduced in the sediments to produce alkanes, so it made chemical sense as a precursor for their extended hopanes, with the hydroxyl groups being removed almost as soon as the organism died. But, well, vinegar? "When we saw that," Albrecht recalls, "we had only two explanations. Perhaps *Acetobacter* is generally active in the subsurface of the sediments. You have many microorganisms, but well, perhaps these particular *Acetobacter* are very widespread and very active and that's why you see these hopanes everywhere. Or, the other hypothesis was to say, well, perhaps it is not only *Acetobacter*—perhaps this 35-carbon hopanoid is widespread in many microorganisms." Then, that same year, a microbiologist in South Dakota who was studying the lipids and membranes of bacteria from extreme environments reported finding the same structure in bacteria he'd isolated from a hot spring in Yellowstone National Park.

Bacteriohopanetetrol (C$_{35}$H$_{62}$O$_4$)

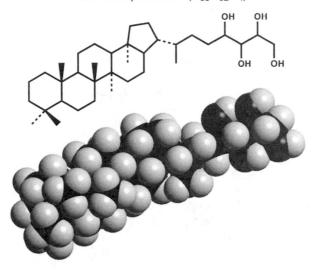

Buoyed by this bit of evidence, Ourisson's student Michel Rohmer began a concerted search for bacteriohopanetetrol—not in the rocks, this time, but in bacteria. Ourisson had no more experience with the fine art of culturing bacteria than he'd had with geology, but he got a friend at the Pasteur Institute to give him a dummy-proof list of bacterial strains: diverse, commercially available, and easy for the unskilled to grow. Rohmer found derivatives of the bacteriohopanetetrol in more than half of them, dozens of strains of bacteria.

The reason the natural products chemists had never found them before was readily apparent. On the one hand, they have a big 30-carbon hopanoid ring structure that is rigid and entirely hydrophobic; on the other, they have a five-carbon tail that is coated with water-loving hydroxyl groups or other polar derivatives. Chemists would extract an organism with water and solvent, collect the water layer containing the hydrophilic compounds, and the solvent layer with the hydrophobic ones, and throw away the thin, cloudy film that collected between the two. But it is just such a murky neither-nor layer where a molecule with an identity crisis like bacteriohopanetetrol is likely to be lurking. "Poor old molecule doesn't know which way to go!" Geoff says. "It's like a soap, with all those hydroxyl groups. That was why we were missing these key compounds—the damn things were being dumped down the sink!"

Rohmer, however, developed a very careful technique for extracting them and, in the ensuing years, isolated dozens of different structural variations of bacterial hopanoids, with various hydrophilic attachments from sugars to amino acids, in more than a hundred strains of bacteria. The reason for their ubiquity has become clear: these are not just excretion products and luxury compounds that allow an organism a wider range or enhance its competitive advantage, like the pentacyclic triterpenoids from higher plants. These bacterial hopanoids, rather, are an integral part of the bacterial cell membrane. Years after their discovery, while he was trying to solve the puzzle of how, exactly, the bacteria make them, Rohmer discovered quite unexpectedly that even the biosynthesis of the basic isoprene building blocks, hitherto considered a universal process in *all* organisms, proceeds by a completely different pathway in many bacteria. He and Ourisson have hypothesized that the bacteriohopanoids are to bacteria what the sterols are to eukaryotes—that the flat, rigid five-ring hopanoid structure serves as reinforcement in the cell membrane, much like the four-ring steroid structure in animal and plant cells—and suggested that hopanoids may have been evolutionary precursors to steroids.

No one, least of all Ourisson, had suspected that something as complex and messy as a sediment or oil would afford such general and far-reaching results. Certainly they had never imagined that their explorations would lead to discovery of a whole new family of biological molecules, or to the promotion of terpenoids from their supporting actor status in life's cast of players, to a role as a vital main character—from secondary metabolites, along with vitamins, hormones, and tree resins, to an essential component of the bacterial cell membrane. Such ubiquity among common organisms, combined with a sturdy carbon skeleton, gives rise to hopanoids in the organic matter of most sedimentary deposits. They may be a small fraction of a small fraction, some one-tenth of one percent of the sediments, but given the millions of cubic kilometers of sediments on Earth, it is enough to place hopanoids among the most abundant organic substances on the planet, as Ourisson was quite fond of pointing out. The identification of hopanes in crude oil and discovery of their biological precursor molecules in bacteria affirmed, once again, that oil derived from the accumulated debris of living things, rather than from abiotic synthetic processes within the earth. It also showed that the hydrocarbons in petroleum were the stripped-down remains not just of molecules made by plants and algae, but, likewise, of those synthesized by the sediment bacteria that feed on their detritus. Petroleum geochemists, however, weren't so much interested in the source or biochemistry of the hopane skeleton as they were in its long-term fate and the yet-to-be-deciphered history that its structure seemed to have recorded not of living things, but of rocks and oils.

With the connection between biological molecules and petroleum hydrocarbons established, the question of the ultimate provenance of petroleum was fast becoming a moot point, superseded by the question of precisely how and under what conditions the former became the latter. One could try to simulate the process in the laboratory by heating the biological molecules with clays and sediment materials to see how the molecules changed, like Geoff did when he heated cholesterol to see if it generated cholestane, or as geochemists at Shell's Dutch facility did when they heated a carboxylic acid to see if they would produce *n*-alkanes like the ones they found in oil. The case of cholesterol was relatively straightforward: it was reduced to cholestane, as predicted, though precisely how was unclear. But contrary to the Mobil Oil group's original hypothesis that the preponderance of *n*-alkanes in crude oil came from gradual removal of the carboxyl group in fatty acids, which so neatly explained the distributions of alkanes and acids in some ancient sediments and oils, production of *n*-alkanes from carboxylic acids was minimal or nonexistent in the Dutch experiments. That the hydrocarbons in oil came, ultimately, from biological molecules was clear, but the path that connected them was not necessarily a direct one. If one heated a relatively young, organic-matter–rich rock like the Messel shale or Green River shale in the laboratory, the amount of oil produced was some 20 to 30 times *greater* than the amount of organic matter that could be extracted: it had to be coming from the 80–90% of the organic matter in the rock that was insoluble. By the mid-1960s, when Albrecht started work on his doctoral thesis, there was already abundant evidence that crude oil was generated not from the soluble organic matter, what petroleum geochemists called bitumen, but via the dreaded kerogen.

Attempts to characterize the organic conglomerate that somehow—with a capital S—formed in the sediments not long after they were deposited had met with limited success. Petroleum geologists had defined several classes of kerogen based on color and general appearance under a microscope, but to a chemist this was like looking at three blocks of silver, gold, and tin and saying they were the same because they were all metallic and rectangular. The chemists were equally unsatisfied with their own efforts, which only told them that when they broke up the molecular conglomerate with blasts of excessive heat or strong chemicals, it produced some molecules with the same structures and biological imprints as those that could be extracted directly from the bitumen in the rock, on the one hand, and from petroleum on the other. What else the kerogen contained, or how it was all bound together in the solid matrix and how, when, and where it generated oil, remained a mystery.

Of course, the best way to get an idea of the requisite conditions for oil formation was to drill deep into the earth and take samples from places where oil had formed in the past, was forming at present, and would presumably form in the future. But only oil companies had the wherewithal to drill so deeply, and they were not keen to provide samples or share data acquired by their own scientists. Even the biennial conferences on organic geochemistry that started up in the early 1960s and brought academic and industry scientists together, purportedly to exchange ideas and knowledge, couldn't break through this culture of industrial secrecy, antithetical though it was to the scientific process. For the geologists charged with finding new reserves of oil in the late 1960s and 1970s, the question of how and under what conditions petroleum formed took on a new imperative. Exploration geologists had relied on the observation and comparison of physical features, along with a good measure of intuition and luck, in their hunt for oil. But once the shallow reservoirs had all been discovered and the easiest targets drilled, petroleum companies started moving offshore, where it was expensive to drill exploratory wells. They could no longer afford the prospector's one-in-ten success rate. It wasn't enough to observe that oil seemed to form in the organic-matter–rich shales of large sedimentary basins and then migrate through cracks and sandstone into rock traps with a particular morphology. One could identify all the requisite geologic structures, and still not find oil—or vice versa.

The Paris Basin in France was a good case in point. It seemed to have all the geological requisites for oil production, and the French worked hard to find the payoff in the 1950s and early 1960s, peppering the region with exploratory wells. It seemed that small quantities of oil had indeed formed in deeply buried layers of shale, but in most areas it had been expelled and migrated through fractures in the rock, gradually moving upward and dispersing without ever encountering an appropriate trap. Commercially significant oil reservoirs were not discovered until the 1980s, when deeper drilling and new exploration strategies revealed large accumulations trapped *beneath* the shale that was the focus of the earlier attempts, having formed, apparently, in still older, more deeply buried layers. In the meantime, the French failure to strike gold in the Paris Basin in the 1950s and 1960s turned out to be a great boon for the scientific community. Having given up hope of finding commercial reserves in the vicinity, the oil companies that had drilled exploratory wells had nothing to lose and everything to gain from open scientific studies. There happened to be a very effective bridge between the French academic community and industry: the Institut Français du Pétrole, spawned just after World War I and developed in the wake of World War II, was based on the premise that knowledge of the industrial world's most essential commodity was in the national interest. Funded by oil industry taxes, it provided support for research collaborations between industry and academic scientists. The IFP's Bernard Tissot was able to obtain samples from the oil companies' failed exploratory wells—not a random sample here or there, but an entire sequence from various horizons deep within an oil-generating shale.

While most petroleum geologists were still observing and comparing the geologic forms where oil collected, Tissot started attacking the problem from the other end, trying to understand the chemistry and physics of oil generation in organic-matter–rich sediments: how deeply did they have to be buried, for how long, and at what temperature, to produce what? Petroleum geochemists at Shell were working along similar lines in Southern California, as was a group with the Soviet Union's Academy of Sciences, but the Paris Basin was an ideal site for such studies because the history of its formation was relatively straightforward. Some 180 million years ago, the place where the Gauls would eventually build their illustrious city was a shallow, highly productive sea, with the remains of microscopic marine organisms collecting in a thick layer of sediments at its bottom. Stretched by tectonic movements, this organic-matter–rich layer subsided in the middle and was slowly buried by some 2,500 meters of sediments, this time from a less productive sea. As the organic-matter–rich layer sank ever deeper into the earth's crust, it was exposed to progressively higher geothermal temperatures: the oil companies could provide samples from various depths in this shale, and the geologists could even say how long they had been buried there and what temperatures they'd been exposed to. Tissot's analyses showed clearly that the asphaltenes and extractable hydrocarbons in the shale began to increase in quantity at 60°C and continued to increase with depth and temperature up to 100°C, and that the kerogen decreased in proportion—solid evidence that most of the oil came out of the kerogen. For chemists this was a little like saying that you dumped a bunch of algal remains into a black box, baked it, and got petroleum out the other end, but for exploration geologists it was a boon: Tissot was able to quantify and model the process whereby burial depth, temperature, and time conspired to produce oil from an organic-matter–rich shale.

Albrecht and Ourisson took the work a step further in 1968 when they acquired a suite of samples from an exploratory drilling operation in the Douala Basin in West Africa—another lost gamble for French oil companies and, likewise, a fortuitous success for geochemists trying to figure out how oil formed. The exploratory holes went down 4,200 meters, into a layer of shale that had formed between 90 and 70 million years ago and contained little lenses of oil at various depths. There were no big accumulations of oil in the vicinity, but the Douala Basin provided an even better sequence for discerning the effects of temperature than the Paris Basin had. The makeup of the sediments had been more or less constant over the 20 million years of deposition, and they had accumulated rapidly, so there was little variation in age or source material over a range of temperature and depth of burial within the shale. What Albrecht saw, when he analyzed the hydrocarbons in the entire sequence, was that there was a narrow window of opportunity for petroleum generation: the rock had to get hot, but not *too* hot. As in the Paris Basin, the total quantity of extractable hydrocarbons increased at the expense of the insoluble kerogen. But then, at about 2,200 meters depth and 90°C, it reached a maximum and began to decrease rapidly until, at temperatures above 100°C, there were only small traces of extractable hydrocarbons in the shale.

"The classic hydrocarbon generation profile," Maxwell says, over lunch with Pierre Albrecht and me. "That's actually in Pierre's thesis. That's the first mention. The whole classic curve."

This was before the Strasbourg scientists got the mass spectrometer set up and started on their hopane saga, so they couldn't yet do any complex structural analysis of triterpenoids. But gas chromatography readily revealed that the *distribution*, as well as the quantity, of *n*-alkanes changed with depth in the shale. The nicely resolved, limited distributions at the top of the sequence gradually disappeared as they went deeper, progressively blurring into a hump of inseparable hydrocarbons. At the top of the sequence, the distribution of *n*-alkanes resembled that of marine algae, limited to short chains of 15 to 22 carbon atoms, with the 17-carbon alkane dominating. Deeper in the sediment, as the total quantity of hydrocarbons increased, longer chain *n*-alkanes appeared and the distribution broadened and evened out to include

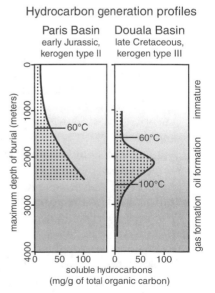

Hydrocarbon generation profiles

Paris Basin
early Jurassic,
kerogen type II

Douala Basin
late Cretaceous,
kerogen type III

maximum depth of burial (meters)

soluble hydrocarbons
(mg/g of total organic carbon)

n-Alkane distributions at various
depths in Douala Basin shale

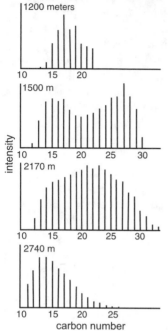

a range of compounds from 13 to 30 carbon atoms long. But in the middle of the oil window where the total hydrocarbon generation reached a maximum, the amounts of long-chain compounds began to decrease again, and the distribution shifted toward ever shorter chains.

The picture that had begun to emerge at the beginning of the decade—when Keith Kvenvolden and geochemists at Mobil noted differences in n-alkane distributions between young unheated rocks and crude oil, and Geoff and Warren Meinschein noted similar differences between the Green River shale and ancient Precambrian rocks—was somewhat clearer in this commercially disappointing West African formation. As the kerogen was heated, long-chain n-alkanes were released, and then, as the temperature rose above 90°C, these newly freed long-chain compounds began to crack, breaking into progressively smaller chains. Cracking occurred at random intervals in the carbon chains, producing a mix of n-alkanes with both odd and even numbers of carbon atoms and obliterating the biological pattern. Eventually, if things were hot enough long enough, the only hydrocarbons left were small molecules such as methane, ethane, and propane, all gases that either escaped from the sediment entirely or formed accumulations of natural gas. Here, finally, was an explanation for the central paradox of petroleum formation, how the plethora of hydrocarbons in crude oil arises from the molecules of life, which include few hydrocarbons: long after the onslaught of bacteria and loss of functional groups that mark the shift from biochemically produced molecule to geologic biomarker, the latter continues to change as it's incorporated into the kerogen, heated in the kerogen, or released from the kerogen.

"We were lucky," Albrecht says. "Because the Douala Basin samples covered the whole oil window. Going from the kerogen to petroleum, and then to cracking going to gas…"

It was not only the amount and type of organic matter in a rock that determined whether it produced oil. It was also its *maturity*. And this depended not only on age or depth of burial, but more specifically on how long it had spent in the oil window at temperatures between 60°C and 100°C. In other words, maturity depended on all the conditions of the rock's past: how deeply and rapidly it had been buried, how much the temperature increased with depth, and if and when it had been uplifted by tectonic movements. Industry scientists were developing tools to gauge if a rock was ripe for oil production, even before all this was understood, in the early 1960s. The idea was that they might develop empirical tools for exploration that didn't require untangling the precise chemical effects of maturity, sources of organic matter—whether from an open ocean, coastal, or inland lake ecosystem—and conditions in the sediments when the rock first formed. The different types of kerogen—as classified by color, physical characteristics, and the ratios of hydrogen or oxygen to carbon—were roughly correlated with the extent of thermal alteration it had undergone. Having noted that the biological signature of *n*-alkanes disappeared in ancient rocks and crude oil, the petroleum geochemists started using what they called the "Carbon Preference Index," or CPI, which was basically the ratio of *n*-alkanes with an odd number of carbon atoms to those with an even number. One could sample a rock layer, either at the surface or deep in an exploratory drill hole, and eliminate it as a possible source rock for petroleum if it had a high CPI, or use the kerogen as a basis for guessing whether or not it had ever been heated enough to generate oil. Inspired by gas chromatography in much the same way Geoff had been when he first analyzed the leaf waxes and noted the patterns of *n*-alkane peaks, industry scientists had started using the patterns of peaks in the GC traces as fingerprints to distinguish crude oils. Such a chemical fingerprint was the idiosyncratic product of a rock's singular origin and history, which meant it could be used to distinguish the oils without understanding the processes that had formed them—or so the reasoning went. The fingerprints of an oil in a sandstone trap could be compared with those of a source rock where it was suspected to have formed, or oils found in different traps in an area of exploration could be compared to see if they'd migrated from a common source rock.

Ourisson recalls the first meeting at the Institut Français du Pétrole that he and Albrecht went to in 1973, before they had published much about the hopanes. "I gave the introductory speech, and described this pattern of hopanes...you had the C_{27}—that's just the skeleton of five rings—no C_{28}, then C_{29}, C_{30}, and then C_{31} two peaks, C_{32} two peaks, up to C_{35}. So this was the pattern, to have one peak missing, C_{28}, and after C_{30} to have doublets....And, okay, that went through without any comment. But then we had a series of lectures by people in the petroleum industry showing their own chromatograms. And the moment they showed a chromatogram everybody gasped, because there was this same pattern. But no one had noticed it before. It became comical, to hear people gasp at first sight."

At the time, the Strasbourg scientists were still trying to find a biological source for the extended hopanes, but they already suspected that the whole series of structures came from one precursor molecule and were connected by some sort of degradation process. The doublets represented two stereoisomers each for the C_{31} through C_{35} compounds, and life's one-handed enzymes were unlikely to have created more than one version of the same carbon skeleton. Immature rocks such as the Messel shale contained another distribution of hopanes altogether, one that lacked the larger homologues and exhibited different stereoisomers for the C_{27} through C_{31} ones. Indeed, Albrecht says, this was one of the reasons it had taken so long to figure out the structures and their sources, and why Ted Whitehead's early work at British Petroleum, where he'd used X-ray crystallography to determine the precise stereochemistry of some of the compounds in oils,

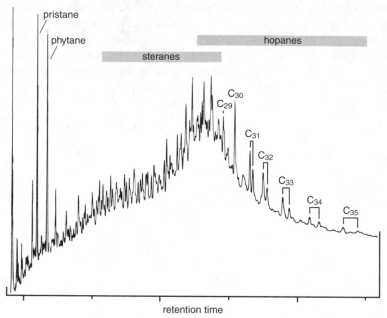

Gas chromatogram showing extended hopanes, 1975
deeply buried early Jurassic shale, Paris Basin

retention time

had been so critical. What immediately caught the attention of the petroleum geochemists was that, like the *n*-alkanes, the distributions of hopane homologues in crude oils were consistently different from those in immature rocks, but the patterns were much more interesting and potentially informative. While the academics in Strasbourg and Bristol were trying to figure out where the compounds came from and how they were transformed, the industry scientists immediately started making use of the patterns in their fingerprint analyses.

By the time Jürgen started out in petroleum geochemistry in 1975, fingerprint analyses for crude oil identification had become standard procedure in the industry, and the more fingers there were to print the better. "It was like Christmas," he says, "every time someone came up with a new compound class." But despite industry scientists' enthusiasm, the fingerprints didn't actually reveal what they wanted them to reveal. One could match up oils that had formed in exactly the same place at the same time and then migrated in different directions into separate traps in a basin. But one could not, in fact, connect a crude oil found in a trap with the source rock where it had formed, tens of kilometers away. Jürgen says that people weren't thinking about how maturity might affect the fingerprints, how a layer of rock sampled in one place might have generated an oil with an entirely different fingerprint if it were more deeply buried somewhere else in the basin. The whole point of a fingerprint, after all, is that it is specific to its material of origin and doesn't change with age or conditions. But though the distribution of steranes and *n*-alkanes might reflect an origin in a specific plant and animal community, the hopanoids are produced by a huge variety of sedimentary microorganisms that differ little from place to place and, as it turns out, from epoch to epoch. And *n*-alkane, sterane, and hopane fingerprints did change continually over the entire history of a sedimentary rock—this was abundantly clear in the results from the Strasbourg and Bristol groups and, ironically, in the industry's own research results.

Only by unraveling the effects of maturity from the imprint of the source material and the effects of migration were geochemists able to develop really useful tools for petroleum exploration, and that required a more expansive interpretation of the biomarker concept. If it was the perseverance of a hydrocarbon's skeleton that carried life's most promising message from the past, it was ultimately the changes in those skeletons—processes that, like petroleum, had a trajectory in geologic time—that held the clues the petroleum geologists needed. The residual imprint of life's one-handed enzymes, first observed in the deflection of polarized light by extracts of oil, had helped chemists to track down the steroids and triterpenoids and convinced most of them that oil was indeed formed from the remains of living organisms. But it was the gradual *loss* of that imprint that would provide the dynamic variables of time and temperature, without which Nature could not perform her alchemy—and petroleum geochemists could not predict the whereabouts of its product.

Maxwell's early attempts to determine the stereochemistry of isoprenoid acids in sediments and rocks had confirmed that the original stereochemistry of the phytyl tail of chlorophyll was unchanged by passage through the food chain, and that it persisted unchanged for millions of years. The asymmetric centers in the middle of the sturdy isoprenoid chain were not affected by cleavage from the chlorin part of chlorophyll, or by chemical and bacterial action on the hydroxyl group at the end of phytol or on its double bond. By the early 1970s, Maxwell and his students had developed GC techniques for separating and distinguishing the stereoisomers of pristane and phytane and could analyze older and more mature rocks and sediments. Here they found that the stereochemistry of the isoprenoid carbon skeleton did change over time, that the hydrogens at the two chiral centers in the middle of the pristane chain and the three in phytane effectively switched sides as the sediments were buried and heated, and that mature shales and crude oil contained a near-equal mixture of stereoisomers. It appeared that these chiral isoprenoids underwent a reversible interconversion something like that between L and D amino acids. But how?

In an amino acid, the negatively charged planar intermediate produced by removal of a proton from the chiral carbon is stabilized by the neighboring nitrogen atom and carboxylic acid group, lowering the energy required to generate it. But an alkane like pristane or phytane has no such energy-lowering mechanism, and pulling a hydrogen off seems an almost impossible task. The musings in the 1967 paper by Calvin's group, which had so inspired Maxwell in his student days, were based on the assumption that pristane and phytane would retain their original configuration from chlorophyll for billions of years and provide evidence of the first photosynthetic organisms. "They just didn't know," Maxwell says. "I didn't know." But a decade later it was abundantly clear that, somehow, phytane and pristane became isomerized over geologic time. Maxwell proposed that the hydrogen came off with both electrons, as a hydride anion, and generated a positively charged planar intermediate, but it was just speculation, based on organic chemists' reports of laboratory experiments with alkanes and extremely strong acid catalysts. That the process is much slower and more difficult than the isomerization of amino acids, requiring higher temperatures and an excess of time, was evident. The original analyses of pristanic acid in the Green River shale had indicated that no isomerization had occurred in more than 50 million years, and the new techniques now showed that only 20% of the phytane and pristane had isomerized. And in the slightly younger and decidedly less mature Messel shale, which had never been heated above 40°C, *all* of the pristane and phytane retained the chlorophyll-derived configuration.

Stereoisomers of the structurally more complex steranes and hopanes were, in some senses, easier to separate than the identical and near-identical twins of pristane and phytane that Maxwell started with, and even in the early GC-MS analyses it was evident that the singular biological stereoisomers were somehow giving way to their twins and siblings in sediments and oils. One noted it in the doublets that appeared among the hopane peaks in the gas chromatograms from crude oils, and in the six peaks that could, with some difficulty, be identified as C_{27}, C_{28}, and C_{29} steranes—two for each carbon skeleton—in the chromatograms from immature rocks, and the unresolvable hump of steranes that appeared in those of mature source rocks and oils. It would take geochemists until the end of the 1970s and the development of more refined GC columns to separate the compounds in that hump of steranes. It contained many, but certainly not all, of the stereoisomers from the biologically generated carbon skeletons. The 27-carbon cholestane molecule has eight chiral centers, but two of those cannot be changed without breaking a carbon–carbon bond and tearing the whole molecule apart. Of the 64 stereoisomers that might be produced if the hydrogens at the remaining six chiral centers were to switch sides, some were blatantly unstable and unlikely to form under any circumstances. But that still left a dizzying array of possible structures to choose from when trying to figure out what steranes were in fact present in the geologic deposits. The stereoisomers of the extended hopanes presented similar difficulties, but here there was only one biologically generated design, and its homologues were slightly easier to separate.

Ourisson's work from a decade earlier, when he had been looking for steroids with hormonal activity and designing ways to synthesize variations on the theme of testosterone—never imagining that he'd use the knowledge to identify compounds in petroleum—served the Strasbourg group well. Not only was Strasbourg the only place in the world where there was a stash of purified hopanoids and similar compounds for comparison, but Ourisson already had an intimate knowledge of their stereochemistry. When he and Albrecht began trying to make sense of the hopanes in sediments and oils in the late 1960s, they knew which configuration of the ring structure was most likely to be chemically

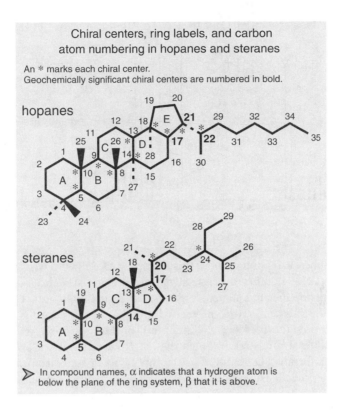

Chiral centers, ring labels, and carbon atom numbering in hopanes and steranes

An * marks each chiral center.
Geochemically significant chiral centers are numbered in bold.

> In compound names, α indicates that a hydrogen atom is below the plane of the ring system, β that it is above.

stable, and which one organisms were likely to make—the least stable, as is commonly the case, with the two hydrogens at C-17 and C-21 in the E ring in the β position, sticking up from the plane of the ring system on the same side as the methyl group at C-10. They found that the mass spectrum of this 17β,21β configuration was easily recognized for most of the sediment hopanoids, as its D-E ring fragment—at 177, or 205, 219, 233, 247, or 261 mass units, depending on the homologue—gave a more intense signal than the A–B fragment at 191 mass units.

The ββ configuration was the one that was most plentiful in contemporary sediments and in the Messel shale, the same configuration that was eventually identified for bacteriohopanetetrol…but it was nowhere apparent in crude oils. That the hydrogen atom at C-17 had somehow switched sides, like in pristane and phytane, was readily apparent: even in the immature Messel shale, each of the ββ compounds in the series was preceded by a baby peak that the mass spectra showed to be its 17α,21β version. In the more mature Green River shale, these αβ compounds had grown to dominate and the ββ peaks had shrunk to babies. And in crude oils the original, biological ββ isomer had disappeared completely, and some of the 17α compounds appeared as doublets in the chromatograms.

Mass spectra of 17β,21β- vs. 17α,21β-hopanes (C₃₁H₅₄)

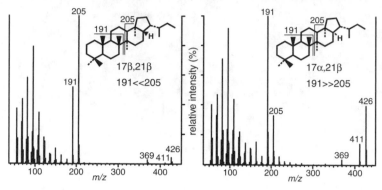

From a chemical point of view, this seemed to fit well with what was known about the relative stability of these structures. A look at a three-dimensional model of the hopane ring structure shows that in the biological, 17β,21β configuration, the side chain and the methyl group at C-18 are in uncomfortable proximity. Switch the hydrogen at either the 17 or the 21 position to the α configuration, and the strain is relieved, but the 17α,21β configuration takes the prize for most comfortable. Differences in stability between stereoisomers of amino acids, or even acyclic isoprenoid chains, are relatively minor, because the molecules have some freedom to twist around their bonds, relieving any interference between attached groups. If a hydrogen at a chiral carbon in pristane or phytane could be removed and replaced, it would be relatively unimportant which side it ended up on, and an equilibrium mixture with the two isomers approaching a 50:50 ratio would eventually prevail. But if 17β,21β-hopane were to convert in a similar manner to a more comfortable, lower energy configuration, it would be unlikely to change back and the ββ isomer would eventually disappear altogether. Likewise, the 17β,21α isomer, more stable than the ββ but less stable than the αβ isomer, would eventually give way to the latter, which was consistent with its low abundance in mature rocks and oils. The chiral center at C-22 in the extended hopane side chain, on the other hand, resembles the chiral centers in acyclic compounds and seemed to behave accordingly: immature sediments contained only the biological configuration, whereas mature sediments and crude oils contained what appeared to be an equilibrium mixture of isomers. According to the *R/S* chemical convention used to designate the arrangement of atoms at chiral centers in hydrocarbon chains, the biological 22*R* isomer gave way to a 60:40 mix of 22*S* and 22*R*, which showed up in the gas chromatograms as doublets of 17α,21β compounds.

That the progress of these apparent isomerizations in rocks and oils might provide a measure of their maturity was noted by both the Strasbourg and Bristol groups almost as soon as they were able to identify the stereoisomers. And one industry geochemist was paying particularly close attention when they noted it.

3-dimensional models of hopane stereoisomers

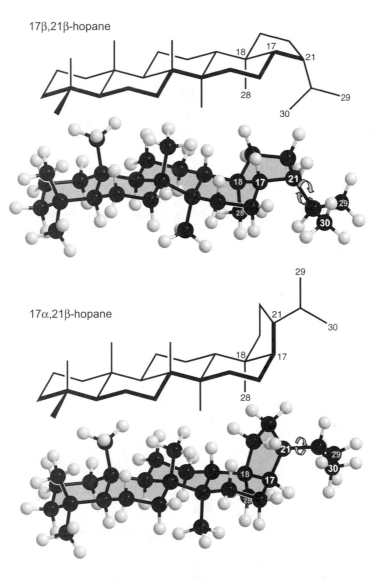

17β,21β-hopane

17α,21β-hopane

"I remember that Seifert actually chased us," Geoff says when I ask him about Wolfgang Seifert, who headed the organic geochemistry division of Chevron's northern California research facility and was pivotal in the development of biomarker tools for petroleum exploration in the 1970s and early 1980s. "I mean he physically chased us when we were at the Gordon Conference. He was an incredibly active climber and walker, very physical, like a Schwarzenegger type." Jürgen and I are both laughing at this description, which Jürgen seems to find quite

Hopane stereoisomers in an immature rock and a crude oil

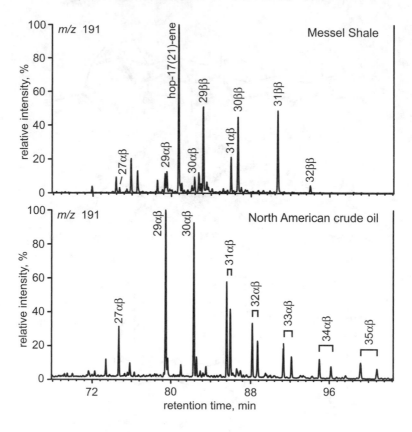

apt. "No, I'm serious," Geoff protests. "He actually trailed James Maxwell and I around the campus, and followed us on the afternoon hikes, asking questions. He was one of the few from the oil side who would actually talk about their work."

Seifert's group at Chevron included an analytical chemist who was apparently working on ways to separate and identify the stereoisomers of the steranes in oil even before Seifert got involved. "Emilio Gallegos," Jürgen says, smiling as we flip through his collection of papers from the 1970s, "a nice quiet guy, not this Terminator type." In the late 1960s, new types of GC columns were becoming available that had a vastly increased power of separation. Here the liquid phase was coated directly inside the walls of a narrow glass or metal capillary, which could be dozens of meters long, coiled up to fit in the GC oven. As soon as Geoff and his students in Bristol started experimenting with capillary columns on their GC-MS, they noted that there were at least two stereoisomers of cholestane in the Green River shale. By 1971, Gallegos, also using GC-MS with a capillary column, had clearly separated and identified six steroid peaks that corresponded to the 5α and 5β stereoisomers of the C_{27}, C_{28}, and C_{29} steranes. The presumed precursors for these steranes—cholesterol, and algal and plant sterols like brassicasterol and sitosterol—have a double bond between C-5 and C-6 of the B-ring, and it made chemical sense that hydrogen would add to either side of the ring system when it was reduced, producing a mixture of α and β stereoisomers at the newly formed chiral center according to their stability, with the relatively streamlined α isomer dominating. When the Bristol group tried to simulate the effects of time and burial on cholesterol, mixing it with ground-up shale and heating it at 200°C for a month and a half, they obtained just such a mixture of 5α- and 5β-cholestane. Separating, identifying, and explaining the mix of steranes in the hump that appeared in the chromatograms from extracts of mature oils was a more difficult task, but by the end of the 1970s the combination of capillary GC columns and new advances in GC-MS instrumentation allowed most of the peaks to be resolved. This time, the analytical advances didn't have to be appropriated from the food industry or from biomedical researchers, but were the result of concerted efforts by organic geochemists themselves, now fueled by petroleum industry money.

Mike Moldowan, whom Seifert recruited for his Chevron research team in 1974, says that Gallegos invented a real-time printout for the MS, where you could monitor a few ions at a time instead of just one. The new setup generated a monumental amount of data, and the commercial instrument designers decided to develop a computerized data system to handle it. "We let them use our lab," Moldowan explains, "and they developed a computerized data system for GC-MS, right there at Chevron." Now one could use the mass spectrometer as a detector for the compounds of choice in a GC trace, creating what came to be known as a "mass fragmentogram." If one set it to detect the 191 ion, one obtained a clear fingerprint of all the hopanes, or one could easily pick out the various extended hopanes by setting it to the masses of their different D–E ring fragments at 205 mass units, 219, and so forth. And if one wanted to obtain a clear sterane fingerprint, unhindered by overlapping hopanes, one could set the mass spectrometer to detect the 217 ion.

Such a system, combined with the best new capillary GC columns, allowed Seifert and Moldowan to compare the fingerprints of individual classes of compounds across the entire spectrum of steroids and triterpenoids in crude oils and source rocks. Rather than blindly apply it to exploration problems, they used it to garner a deeper understanding of how the fossils of biologically generated organic compounds changed over time in the geosphere, a basic inquiry that united them with the academic community and might have put them at odds with the corporate culture of research, if it hadn't been for Seifert's leadership. "In the 1970s," Jürgen tells me, "basic geochemical studies had no direct application to oil exploration. It took someone with Seifert's energy and PR skills to convince the management to support it." A forceful personality by all accounts, Seifert managed to engage his lab in a good measure of basic research *and* publish the results. Moldowan says that though Chevron's management didn't give them a green light to publish everything, they did have quite a bit more freedom than most industry scientists. "Where we had trouble was identifying samples . . . They wouldn't let us do that. But eventually we got stuff out." Though the oil companies partially sponsored government research institutes like France's Institut Français du Pétrole or Germany's Institute of Petroleum and Organic Geochemistry, or

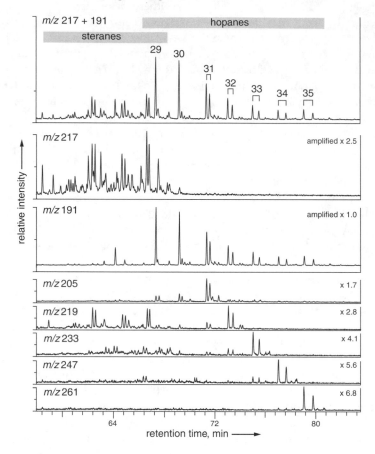

Mass fragmentograms for biomarkers in crude oil

had programs to fund university researchers like British Petroleum's no-strings-attached grants to the Bristol lab, support for basic research within the companies generally required a direct economic payoff. At least that's one version of the story. The other is that the industry scientists were actually doing all sorts of exciting things that they weren't allowed to divulge. Ourisson says he always suspected that the scientists at Shell's research facility in the Netherlands already knew everything his group discovered. "I heard that Shell had a huge amount of data," he says cheerfully. "But they were not publishing. Only a few were publishing." Geoff worried about the same thing when he first started with geological samples back in 1963. "They were so secretive that I thought, well maybe it has all been done. Maybe we're just idiots!" At the International Meetings on Organic Geochemistry and Gordon Research Conferences where industry geochemists and academics came face to face every year, the former would sit quietly in the back, eagerly soaking up whatever hot new ideas and information the academics had to offer, while the academics fumed because they suspected the industry scientists of harboring precisely the information and samples they needed to answer the questions at hand. That the two came from different work cultures was visibly apparent—stylish suits and big expense accounts on the one hand, frugal budgets and sports shirts on the other. "We used to have big rows at the Gordon Conference," Geoff says, "because the people from Esso and Shell would just give these very general talks, and then listen to us and not say anything." Except for Wolfgang Seifert. "Seifert was different. He would talk about the molecular approach and fingerprinting. Whether the others were doing very much, we couldn't find out. But Seifert was doing a lot of work at Chevron. And also dying to know what the academics were doing."

Seifert had been one of Treibs's students in Munich, weaned on the nascent concept that the structure of organic molecules could provide key information about the history and sources of oils and rocks. In the 1960s he'd worked in the refining part of the oil industry, identifying all the different kinds of carboxylic acids in crude oils—including some in the steroid family. The acids were a minor constituent of the oil but caused problems during the refining process, and the idea was simply to identify the various types in different oils and adapt refineries to deal with them. Seifert, however, wanted to know where the different acids came from to begin with. Ironically, given that his mentor had provided the first unequivocal evidence for the plant origin of petroleum, one of Seifert's first biomarker publications at Chevron in 1973 was rather spectacularly construed as evidence for an origin in the remains of animals. He determined the structures of steroid acids that bore some resemblance to the bile acids of invertebrates and amphibians, and hypothesized that zooplankton made a small contribution to the predominantly plant-derived ingredients of petroleum. The paper itself wasn't particularly sensational, but rumor has it that it was responsible for a popular Chevron TV commercial that showed a dinosaur strolling across the plains, falling over dead, and then melting into little droplets of oil. Perhaps Seifert's paper simply caught the fancy of an imaginative marketing director, but no one who knew Seifert puts it past him to have used the study for a little PR of his own, playing up its importance with the company managers and promoting it as something new and exciting.

"Seifert was really impossible to say no to," Moldowan says. "He made demands, threatened to quit. And management thought we were contributing so much to their goals." He himself was still a postdoc in Carl Djerassi's lab at Stanford when the so-called dinosaur paper came out. Like Jürgen, who was then studying with Djerassi's counterpart and collaborator at the University of Cologne, Moldowan was one of a small contingent of upcoming analytical and natural products chemists who were educated in the most cutting-edge new techniques in mass spectrometry. "Seifert called Djerassi," he recalls, "and started talking about this exciting new field, asked if Djerassi had any natural products chemists who'd be interested in getting out of chemistry into something a little more risky. I was in Djerassi's office at the time, and kinda raised my hand. Looked interesting to me—natural products chemistry with a real economic payoff at the end."

Jürgen's impulse to apply for a job with Germany's new, government-sponsored Institute of Petroleum and Organic Geochemistry at the Jülich Research Centre a year later was even more prosaic. He had just taken an academic post, developing new analytical techniques and determining structures of natural products sent in by chemists and botanists from all parts of the globe, but as a young researcher in a well-established lab, he didn't have much independence or room to develop. He was thinking about looking for a job in industry when he saw Dietrich Welte's newspaper ad for an analytical chemist and applied just to get some practice interviewing—though when Welte greeted him in a suit that was 20 years out of fashion, if not 20 years old, he wondered if it was quite the sort of practice he needed for a company interview. If the truth be told, Jürgen was as uninterested in petroleum as Ourisson had been in the 1960s and had no intention of taking the job. "I didn't see what use I could be," he says, "surrounded by a bunch of geologists who didn't care about chemistry or biology." He laughs now, recalling his own ignorance. He knew nothing about geochemistry, didn't realize that chemists at British Petroleum were pivotal in designing new capillary columns for GC separations or that analytical chemists at Chevron were developing a computerized data analysis system for mass spectrometers, and he had no idea that Dietrich Welte and Bernard Tissot were in the midst of writing the pivotal text on petroleum formation. But when Welte showed him what the new group's chemist was doing, analyzing seemingly impenetrable geological samples on the mass spectrometer, identifying steranes that resembled the sterols in organisms in an extract of sediment from a petroleum bore hole, Jürgen's response was similar to Mike Moldowan's: interesting. And feasible. When Welte told him the perks—the money, the technicians, the engineers, and, most important, a good dose of independence—he took the job. It was, as both he and Moldowan would soon see, one of the most exciting times to be working in the field.

In several benchmark papers in 1978 and 1979 Seifert and Moldowan elevated the use of chemical fingerprinting to a higher art. They looked at the hopane fingerprints of crude oils from a range of depths, ages, and geographical locations within a single geological basin, and determined which of the various homologues were reflecting the oils' maturity. They found, empirically, that the ratio of hopanes with longer side chains to those with shorter chains decreased with increasing maturity, presumably because the hopanes with shorter chain lengths came from more extensive cleavage of carbon–carbon bonds in the hopane side chain. These ratios, taken together with the overall steroid and triterpenoid fingerprints, allowed them to distinguish oils of differing maturity that had been generated from the same layer of source rock at different times during the subsidence and burial of the sediments. But, to complicate matters, their comparisons showed that the distance an oil migrated after its formation also had an effect on its steroid and triterpenoid composition, and this needed to be taken into account when making correlations. It was as if the rocks were acting as a sort of chromatography column through which some components of the oil moved faster than others. Acyclic alkanes generally moved faster through the pores of the rocks than did the cyclic triterpenoids, and the relative amounts of the two compound classes could give some indication of the distance an oil had traveled from its source rock—except that the maturity of the oil also affected these quantities, because the cyclic compounds apparently broke down faster than the acyclic compounds.

As the full spectrum of sterane structures in crude oils began to reveal itself in the late 1970s, it was apparent not only that organisms contribute a huge variety of sterols to the sediments, but that a given sterol could yield more different steranes than chemists had predicted. Mike Moldowan rather fondly remembers a 1979 paper where he and Seifert puzzled out the structures of dozens of compounds, all variations on a single theme, like Bach's Inventions for chemists. The Bristol and Strasbourg groups were finding an impressive variety of sterols in algae, zooplankton, and microbial mats, and an even more dazzling collection of structures in contemporary surface sediments, where the sterols were joined by various intermediate structures that seemed to punctuate the passage into the geological realm. They were trying to understand the relationships between what they found in organisms and what they found in the sediments, and between geological and chemical processes, whereas the Chevron chemists were mostly interested in the latter, and like others affiliated with industry, they ultimately needed tools that would help geologists find commercially viable accumulations of oil. But the challenge of the day for all of these chemists—including those at British Petroleum and Shell U.K., and the purported know-it-alls at Shell Netherlands, and Shell defector Pieter Schenck's organic geochemistry group at the University of Delft, and Jürgen starting out among the geologists in Welte's Jülich group— was still to sort out the incredible mix of complex triterpenoids in the rocks and oils and determine their precise chemical structures. "You've got to know what you're dealing with," Maxwell says, "before you can do anything. There were hundreds of compounds. People were slapping structures around. I've made the mistake myself, using these molecular parameters when you aren't really sure about the structure. Pierre Albrecht was one of the people who made sure that he, and other people, got the structures of the compounds right. And this fellow George Ryback, an organic chemist who worked for Shell in the U.K. Those two were always right."

It was Maxwell's own student Andrew Mackenzie, initially trained as a geologist, who correctly identified two GC peaks that Seifert and Moldowan got wrong in their 1979 sterane paper. Seifert was famous for his competitiveness in everything from skiing and ping-pong to structure determination, and it was with some relish that the Bristol group wrote up the finding and swore everyone to secrecy until the paper went to press...but Seifert happened to visit the lab in the meantime and, true to form, teased the information out of someone, publishing his own correction almost simultaneously with Mackenzie's paper. By this time, the work of the three groups had converged, with Chevron providing the European labs a bit of research money, and Seifert making annual visits to Bristol and Strasbourg to pick everyone's brains. As Geoff recalls, Seifert was the one who coined the term "biomarker" from their original "biological marker," perhaps because it sounded catchier, especially in an industry that was more interested in rocks than biology. The term caught on, even as the work with petroleum biomarkers moved farther and farther from the original concept of a biological marker—from understanding the history of life or the biological origin of hydrocarbons in petroleum, to chronicling the history of rocks and the fate of the hydrocarbons themselves.

From the assembly of sterane structures identified in recent sediments, it was apparent that the first round of bacterial and chemical breakdown, while producing a mixture of α and β configurations at C-5 where the double bond had been, preserved biological configurations at the C-14, C-17, and C-20 positions. Maxwell's Bristol group had embarked on an ambitious set of studies with the Paris Basin shales, examining how the distributions of sterane, hopane, and acyclic isoprenoid structures changed with burial depth and temperature. They

3-dimensional models of sterane stereoisomers

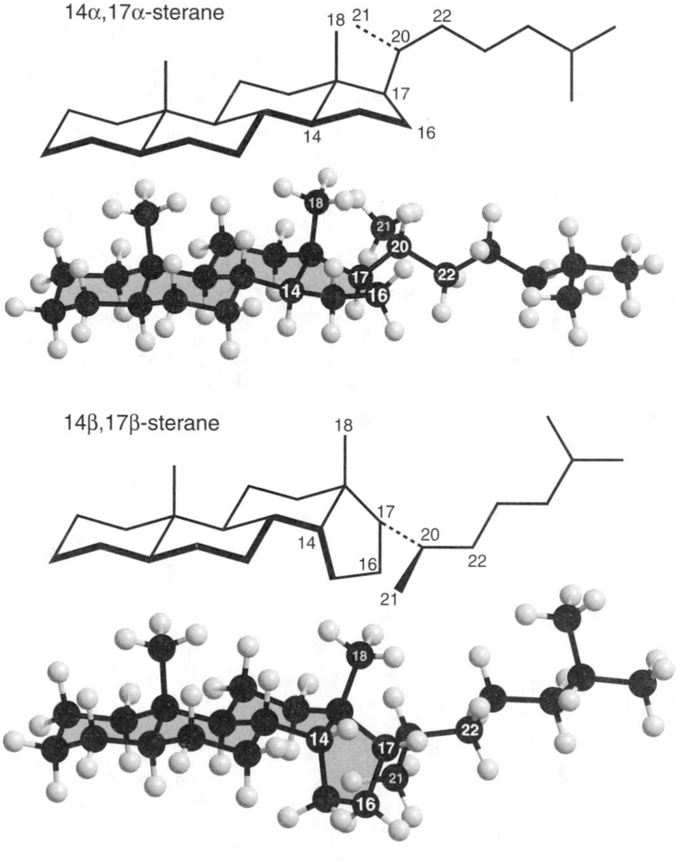

14α,17α-sterane

14β,17β-sterane

found that the sterane ring structure behaved much like that of the hopanes, relaxing into a more comfortable configuration than the flat 14α,17α one that life had chosen. Unlike the hopanes, however, the steranes required quite a bit of heat for this process to get going, and the transformation was not entirely one-sided: the more comfortable 14β,17β configuration didn't begin to appear until the rocks had been heated almost to oil-generating temperatures, and even the most mature oils still contained some 25% of the original 14α,17α isomer.

Just as it was beginning to seem that one might make some sense of all these isomers and their relationships to biological sterols and to the history of the rocks, a whole new group of steranes was identified that defied the emerging logic. Jürgen says he'd first got wind of them in 1977, when a visiting petroleum geochemist from the U.S. told him about some chromatographic peaks that were mystifying him. Their mass spectra included the telltale 217 ion of a sturdy A–B–C ring fragment from steranes, but they had an intense ion at 259 mass units that the steranes didn't exhibit—an intact four-ring fragment that would have been unstable—and they eluted from the GC column earlier than steranes. Then, later that same year at the International Meeting on Organic Geochemistry, Alexander Petrov and his Moscow-based research group reported the structures, a variation on the sterane ring structure that was nonexistent among known biosynthesized sterols. Indeed, it appeared that the ring structures of many of the known steranes had been broken apart and rearranged, with the methyl groups at the C-10 and C-13 ring junctions shifted to the C-5 and C-14 positions, a design that made for a stable four-ring ion when the branch at C-17 was broken off.

It was the first time most of the Russian geochemists participated in the IMOG, which was held in Moscow that year, and they had been rather isolated from the knowledge that had accumulated in the West since the war. They tended to read only Russian journals, and many of them still adhered to an outdated but nationally accepted theory that the vast petroleum reserves in Siberia had been produced by abiotic chemical processes within the earth rather than by the gradual transformation of deposits of organic matter. But Petrov and his group clearly had no doubts that they were dealing with ancient relics of biologically produced molecules from debris that had been buried and heated over millions of years, and they postulated a viable chemical mechanism for the formation of the rearranged steranes, or diasteranes, from known biological sterols. Pierre Albrecht was among

C_{27} sterane C_{27} diasterane

both configurations

the few westerners at the meeting and happened to be reporting on the same family of compounds. The Strasbourg group had identified the structures of unsaturated versions in extracts of slightly acidic contemporary sediments and had done laboratory heating experiments that supported the Russian group's hypothesized scheme for their formation. Apparently, the rearranged compounds formed in the sediments during the initial steps in the transformation of sterols to steranes. The Strasbourg group's experiments showed that the rearrangement of the methyl groups on the ring structure occurred only in the presence of an acid catalyst, and that this could be provided by clay minerals in the sediments. The original, biological configurations at the various chiral centers were lost during this process, which produced a mix of stereoisomers and seemed to occur in all of the sterols regardless of side chain. By the time of the Gordon Research Conference in New Hampshire the following year, these rearranged steroid structures were common knowledge, and it was fast becoming apparent that they, indeed, occurred only in sediments with a high clay mineral content. Because clay minerals generally come from the erosion of landmasses, this meant that rearranged steroids were absent from many ocean sediments, as well as from the sediments of saline lakes that were cut off from river input. Petroleum geochemists were quick to put this information to use: the ratio of diasteranes to steranes in an oil could provide clues as to the nature and location of the source rock that generated it, specifically, whether it came from a shale formed in a near-shore, clay-rich marine or freshwater lake environment, or from carbonate marine rocks or evaporites.

Another set of steroid compounds that started attracting the attention of geochemists in the late 1970s was even farther removed from the sterols produced by organisms than the diasteranes. Not only are the methyl groups switched around, but the rings have lost their interesting sterol geometry and taken on the flat pancake form of the aromatic ring.

Like the diasteranes, the steroids with an aromatic ring appear in immature sediments. In this case, the initial loss of hydrogens and formation of double bonds in one of the rings might well be due to microbial action. But unlike the diasteranes, which, once formed, seem to persist largely unchanged through the oil generation process, these monoaromatic steroids appeared to be further transformed at oil generation temperatures. Seifert and Moldowan found that the length of their side chains tended to get shorter with increasing maturity, as

sterols   →→ multiple steps →→   5α & 5β C-ring monoaromatic steroids   →→ aromatization →→   triaromatic steroids if the aromatic ring somehow facilitated pieces breaking off. But Tissot had noted something that would turn out to be even more useful in the Paris Basin shales: as the shale matured, the monoaromatics disappeared and triaromatic steroids appeared. In the late 1970s, the young Andrew Mackenzie—whom Maxwell pronounces the brightest of his students and proudly claims as a "fellow Scot"—did a systematic study of the process for his doctoral thesis and showed that aromatization of the monoaromatics proceeded slowly and steadily, from the first signs of heat until well into the oil window, when all of the monoaromatics disappeared and only triaromatics were left.

"How molecules transform," Moldowan told me in answer to my question about what he'd found exciting in petroleum geochemistry. "That's interesting stuff for an organic chemist." Indeed. But it wasn't just the transformations that were exciting, any more than it was just the structures: it was their trajectory in geologic time, what they could tell us about the history of the earth, on the one hand, and what the laboratory of rock and time revealed about the molecules, on the other. By the early 1980s, the Bristol and Strasbourg groups had tracked an entire parade of molecular transformations that began at different temperatures and times in the burial history of the basin and proceeded at different rates. Pristane seemed to lead the parade, with the pure 6R,10S isomer inherited from organisms gradually giving way to an unchanging equilibrium mixture. Fast on its heels came the relaxation of the hopane ring structure from its biological 17β,21β configuration to the 17α,21β one. As one went deeper in a sedimentary

Transformation of biomarkers with increasing maturity

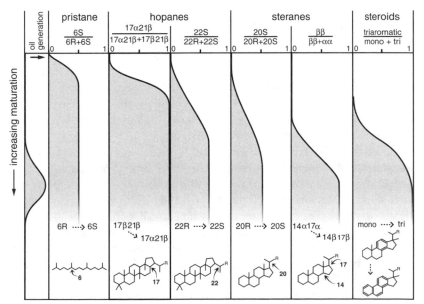

sequence, into older and more mature sediments, where temperatures rose above 80°C, the nonbiological *S* configurations began to appear at the branch carbons of the hopanes and steranes. But though the nonbiological 22*S* hopanes and 20*S* steranes made their first appearance almost at the same time, still well before oil production began, isomerization at the C-22 of hopanes was much faster than at the C-20 of the steranes: the hopanes reached equilibrium about the same time the kerogen in a good petroleum source rock began to release hydrocarbons, whereas the 20*R* and 20*S* steranes didn't attain equilibrium until well into the oil window, when hydrocarbon release from the kerogen was near its maximum. Quite a bit of heat was required before the ring structure of the steranes could relax from its 14α,17α configuration, but once it got going it was fast to convert, and by the middle of the oil window an unchanging mixture that was 75% 14β,17β had appeared. The triaromatic steroids, however, won the prize for endurance: they made their first appearance in the marginally mature shale, during its late adolescence, so to speak, and continued accumulating all the way through the oil window, not reaching their maximum until gas generation was already underway.

Clearly, this procession of transformations was reflecting the time and temperature history of the rocks, and Seifert and Moldowan had already started using the ratio of mono- to triaromatic steroids and the relative quantities of the hopane isomers to compare maturities of different source rocks and crude oils. But could one put numbers to that? Could they do more than compare maturities of different source rocks and oils, and actually determine the temperature a rock had been exposed to, when and for how long? As Mackenzie started work on a postdoctoral project, he and Maxwell began to discuss the possibilities. The rates of the reactions would depend on the amount of precursor and on the temperature at any given time, with the extent of the temperature dependence specific to each reaction. Equations describing these relationships and techniques for determining the specific temperature dependence of a reaction—its kinetic constants—date back to the nineteenth century and had been in wide use among organic chemists since the 1940s. If one could determine the kinetic constants for several of the transformations in the rocks and measure the quantities of products and precursors at different depths in a sedimentary basin—corresponding to different times in the progression of the reactions—then one could, theoretically, determine how the temperature had changed over time in the basin, and in which layers of rock conditions might have been right for oil generation. But in order to do this, they first needed to understand the characteristics of the reactions and follow their progress at various temperatures in the laboratory—both problematic.

Mackenzie heated an immature shale at different temperatures for different lengths of time and then extracted the bitumen and measured the amounts of aromatic steroids and of the various sterane, pristane, and hopane isomers. The relative rates of the various isomerizations were in keeping with what had been observed in shales of different maturities, and the conversion of monoaromatic to triaromatic steroids was more sensitive to changes in temperature than were the isomerizations, which was likewise consistent with what they had been seeing in the Paris Basin sequence. Seifert and Moldowan had attempted to analyze the components of the kerogen itself, using very brief blasts of excessive heat to release the bound molecules without subjecting them to further thermal transformation or break-down. When they compared the isomer ratios of the hopanes and steranes from the kerogen with those in the bitumen they extracted from the same rock, they found that the isomerizations had apparently proceeded at significantly slower rates in the kerogen than they had in the bitumen. And when Mackenzie heated his imma-ture shales for extended periods, progress toward an equilibrium mix of sterane or hopane isomers seemed to slow or even reverse, as if the less extensively isomerized compounds from the kerogen were now being released into the bitumen.

Normally, a chemist would be able to use such experiments to understand the reaction mechanisms of the aromatization and isomerizations and determine the desired kinetic equations and constants. But determining a reaction mechanism within an irregular amorphous macromolecular substance like kerogen seemed virtually impossible. And even in the bitumen, the only way to get these reactions to produce measurable quantities of products in anything less than a million years was to crank the temperatures up way beyond those experienced during sediment subsidence and burial. Wouldn't such high temperatures open up new reaction pathways, with different kinetic constants from those actually occurring in the rocks? These were the questions and quandaries Mackenzie and Maxwell were tossing around in the Bristol lab in 1981, when Geoff went to a Royal Society event and got to talking to Cambridge geologist Dan McKenzie.

McKenzie was one of the discoverers of plate tectonics, interested in under-standing the movements and properties of the earth's mantle. This is a realm where organic molecules just don't figure, and though he and Geoff had been elected to the Royal Society at the same time, they'd never had much to discuss in the way of common research interests. The sedimentary basins where petroleum is generated are, however, formed by tectonic movements, and in 1981 McKenzie happened to be working on a quantitative model of the stretching and subsid-ence that created them and the changes in temperatures during such processes. This time, when he and Geoff met and started talking about their research, they immediately realized that they were holding the opposite ends to the same string, and Geoff, amused by the prospect of a "Mackenzie and McKenzie" collaboration, invited the geologist to Bristol to see what Maxwell's star student had going.

Andrew Mackenzie needed a laboratory in the rocks, a place where his isomerizations had been running for millions of years at known temperatures. And Dan McKenzie had a mathematical model that could reproduce the temper-ature gradients in a sedimentary basin with a simple geologic history, where the temperatures had increased at a constant rate over an extended period of time. If they could get samples from at least two basins with such histories, they could use McKenzie's model temperatures, measure the product to precursor ratios for at least two of the biomarker reactions Andrew had picked out, and try to estimate the kinetic constants for the reactions. With these in hand, they could eventually turn the process around: calculate the actual temperature history in a sedimen-tary basin with a complex unknown history from the biomarker ratios in the rocks, and then plug these temperatures into a geologist's mathematical model to reproduce a complete depositional and petroleum generation history of the basin, including the precise timing of tectonic events such as stretching and uplift.

Simon Brassell, who was one of Geoff's students at the time, recalls being particularly impressed by Dan McKenzie's visit and his eloquent explanation of his basin model. The whole lab group was gathered in Geoff's office to discuss what basins would make the best kinetics laboratory and how Andrew could get hold of samples. In order to span the full range of temperatures and times for their reactions, they needed one basin where temperatures had risen gradually over a long period of time, and they needed one that had been heated faster and to higher temperatures. The first was easy—the North Sea offered a perfect example of a cold basin with the requisite simple geologic history, and samples were readily available because oil companies were required to file cores from wells they drilled there with the U.K. National Core Repository. Brassell says they were just plain lucky with their second basin: they had a visiting Hungarian scientist in the lab at the time, and he had a series of samples from Hungary's Pannonian Basin, which was a perfect example of a hot basin.

At first it appeared to be a grand wedding of chemical theory, geology, and empirical pragmatism. I still remember reading the first Mackenzie and McKenzie publication when I was a grad student and being completely enthralled, despite my prejudices about anything that was useful to the petroleum industry. They used a simplified isomerization scheme and rate equations that didn't include all the possible sterane interconversions—something of an anathema to chemists, including those in the Bristol lab itself—but the very concept that one might use molecular reaction kinetics to document the massive movements of the earth's lithosphere was so magnificent that the geochemical approximations could easily be forgiven...if they worked.

In order to estimate the kinetic constants, Mackenzie and McKenzie had to base their equations on the premise that the isomerizations in the side chains of the steranes and hopanes were simple reversible reactions that proceeded to equilibrium, like the isomerization of amino acids with one chiral center, and that any breakdown of hopanes or steranes during the millions of years of heating would occur equally to the two relevant isomers and not affect their ratios. Likewise, they had to assume that the aromatization proceeded directly from monoaromatic steroids to triaromatics, and that there were no new additions of monoaromatics, other sources of triaromatics, or preferential destruction of either. And, finally, they had to focus on the bitumen, which they could analyze with some precision, and ignore the kerogen, which they couldn't. Their measurements were relative amounts, determined by comparing the peak areas of the various isomers in the gas chromatograms and mass spectra, because at the time it was difficult to determine absolute quantities of these compounds even in the bitumen, let alone know how much might have been released from kerogen as it matured. If, as appeared to be the case, these reactions proceeded at different rates in the kerogen and in the bitumen, then one could only hope that the quantities of steranes and hopanes released from kerogen at any given time in the long subsidence and heating process were small compared to those in the bitumen. The apparent decrease in isomerization rates due to release of the less extensively isomerized compounds from the kerogen during their laboratory heating experiments would, then, be due to the artificially high temperatures they'd had to use.

These were all pretty big assumptions, as Andrew Mackenzie and the Bristol crew were well aware, but, after all, they were trying to reenact chemistry that had happened deep within the earth, over millions of years. When Mackenzie moved on to Welte's lab and teamed up with Jürgen and a couple of Canadian geologists to reconstruct the temperature and burial histories of potential petroleum source rocks in Alberta and Nova Scotia, the excellent agreement between the geological observations and thermal histories determined from biomarker data seemed to validate the assumptions. In fact, the reconstructions worked so well and were so useful to exploration geologists that the kinetic method, as it came to be known, was immediately embraced by the industry and incorporated into exploration strategies, launching what one might call the halcyon years of petroleum geochemistry, from 1983 until the mid-1990s. Biomarker methods came into wide use, Andrew Mackenzie took a job with British Petroleum and quickly worked his way into a top management position, and industry geochemistry sections enjoyed greater research freedom and generally flourished.

Jürgen thinks that the euphoria over this grand success in the industry tended to obscure the fact that the chemistry behind the kinetic measurements was not rigorous, that it was, in effect, a geochemist's provisional working approximation of the real chemistry that was going on in the rocks. Indications that the isomerization reactions were not as simple as assumed were noted in the early papers from Maxwell's group, but then somehow dropped from the discussion once the actual basin studies began. The lack of insight into the chemical reaction mechanisms, however, continued to nag at chemists. Some of the values estimated for the kinetic constants, which ultimately depend on the reaction mechanisms and the intermediate structures formed, just didn't jive with anything known from laboratory experiments. And the constants weren't as constant as they should have been, but rather seemed to vary from one sediment type and location to another. Clay mineral catalysts are known to facilitate reactions with charged intermediates like those proposed for the isomerizations, and yet laboratory heating experiments using such catalysts failed to drive the reactions in solutions of steranes or pristane. Seifert had proposed a mechanism wherein the hydrogen atoms are extracted by free radicals in the sediment matrix and a planar free radical intermediate is formed, and the members of Maxwell's group had some success driving the isomerization of pristane when they heated it with elemental sulfur, a catalyst known to encourage the production of free radicals. But, Jürgen says, the rates of isomerization in different rocks showed no correlation whatsoever with the presence or absence of sulfur, and the high temperatures and large amount of sulfur used in the laboratory experiments were likely to drive reactions that didn't occur in nature. By the late 1980s, he was beginning to suspect that the changes in carbon skeleton stereochemistry they observed in nature might not be due to interconversion of free steranes and hopanes in the bitumen at all, but rather to some unknown process that was occurring either in the kerogen or during release of the compounds from the kerogen.

In a series of hydrous pyrolysis experiments designed to simulate the maturation of shale and production of petroleum, Jürgen's student Roger Marzi heated rock samples under high pressure in the presence of water, for various lengths of time, from a few hours to weeks. The kinetic constants that Marzi obtained were quite different from those estimated by Mackenzie and McKenzie, but Marzi's most surprising result harked back to the little-heeded caveats the Bristol and Chevron groups had posed almost a decade earlier. The sterane 20S/20R isomer ratios measured in the bitumen increased to the expected equilibrium value of about 1. But rather than persisting at equilibrium like a simple reversible interconversion between two stereoisomers, the ratios then dropped back to smaller values. Even the aromatization of monoaromatic to triaromatic steroids, which should have been a one-way street, appeared to reverse itself in some of the experiments. Around this same time, a group of researchers headed by Ian Kaplan at UCLA reported a similar result for isomerization at the C-22 position in hopanes. Like Seifert and Maxwell in the 1970s, they took this as evidence that isomerization had occurred more slowly in the bound molecules of the solid kerogen than in the free ones of the solvent-extractable bitumen. Jürgen, however, suspected that the story was even more complicated. In his lab they had found that the absolute quantity of steranes in the bitumen increased dramatically during initial heating, and then decreased just as dramatically, both during simulation experiments and in rocks of increasing maturity. It seemed that the apparent isomerization and attainment of equilibrium with increasing maturity that they'd observed in so many shale and crude oil samples had more to do with the release of the compounds from the kerogen and their subsequent decomposition than with reversible isomerization reactions in the bitumen—which would refute one of the kinetic method's core assumptions.

Researchers at the University of Newcastle in the United Kingdom—Geoff's son Tim Eglinton among them—were coming to a similar conclusion. The Newcastle research group was founded in 1966 by Geoff's old partner in crime from Glasgow, Archie Douglas, and known for its assaults on "the kerogen problem." Anyone in the Bristol lab who showed a more than passing interest in the intractable stuff would surely end up there. Paul Farrimond, who had been a student of Geoff's, and Geoffrey Abbott, who worked with Maxwell for years, both joined the Newcastle faculty in the late 1980s. It was only fitting that Tim, who had inherited his father's passion for chemistry and rocks, should do his Ph.D. under Douglas, and that it should include an attempt to understand the nature of a substance that had, to date, pretty much eluded Geoff and everyone else. Tim Eglinton's early experiments quantified, for the first time, the amounts of specific steranes and hopanes released during artificial maturation of kerogen by hydrous pyrolysis and showed that these amounts were often enough to swamp the steranes and hopanes in the rock's preexisting bitumen. In another hydrous pyrolysis experiment, Abbott heated the kerogen together with a solution of synthetic 20R sterane that had been labeled with deuterium, the heavy isotope of hydrogen. When he ran the solution on the GC-MS after periods of heating, he found that the unbound deuterium-labeled sterane in the solution had remained unchanged, whereas the unlabeled steranes released from the kerogen were progressively more isomerized. To what degree this relatively high-temperature laboratory result could be translated to the natural, low temperature maturation of a shale over millions of years remained to be seen, but it certainly refuted earlier interpretations that isomerization proceeds more *quickly* in the bitumen than in the kerogen.

Evidence was mounting that the assumptions behind the successful kinetic method were not just the obligatory oversimplifications of a more complex reality but, rather, categorically flawed. In 1998 Farrimond published a study that in some ways emulated the Paris and Douala Basin studies of earlier decades. Farrimond and his colleagues analyzed a sequence of progressively more deeply buried and mature samples from a single core in the Barents Sea, beginning at 1,300 meters burial depth and continuing down to 3,000 meters. But now, instead of assessing the amounts of bulk hydrocarbons like Bernard Tissot had done, or comparing *n*-alkane patterns like Albrecht, or determining the ratios of various isoprenoid stereoisomers like Maxwell, they determined the absolute quantities of specific sterane and hopane stereoisomers in the bitumen.

The Farrimond study is enlightening: the concentrations of both the biological *and* the nonbiological stereoisomers increase significantly with increasing burial depth, even before the oil window is reached. And neither the amounts nor the timing of increases of the nonbiological 22*S* hopane and 20*S* sterane isomers in the bitumen are mirrored by decreases of the 22*R* and 20*R* isomers.

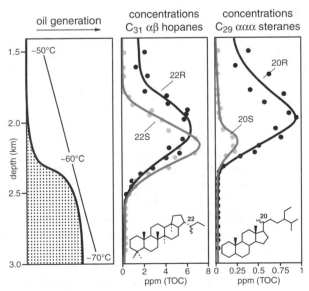

Changes in amounts of hopane and sterane stereoisomers with burial depth
Barents Sea core, Eocene claystone, 1998

The *S* isomers in the bitumen must have been generated by something besides, or in addition to, direct isomerization in the bitumen—they must, somehow, have come from the kerogen. All of the stereoisomers clearly begin to decompose with increasing depth in the core, that is, with increasing temperature, and the authors note that the biological isomers begin to decompose at lower temperatures than their nonbiological counterparts, while the latter pick up speed at higher temperatures—which might explain the apparent retreat from equilibrium that was observed in the bitumen.

Like Jürgen, and Abbott and others in Newcastle, Farrimond and his colleagues concluded that the apparent isomerization of these isoprenoids occurs either during their release from the kerogen or within the kerogen, that they are apparently decomposing in the bitumen at variable rates, and that any interpretation of the rates of these processes needs to consider both kerogen and bitumen in concert. Finally, if we are to understand Nature's alchemy—the transformations of organic matter that has been buried and heated deep within the earth for millions of years—we have no recourse but to tip up the lid of the black box and peer into the chemically murky interior of kerogen.

Though there is yet no consensus on the true story of kerogen formation, significant evidence indicates that it begins early in the sedimentation and rock-forming process, as part of the low-temperature chemical and microbial transformation of organic matter that organic geochemists refer to as diagenesis. This initial process appears to involve the condensation and polymerization of lipids, fatty acids, sugars, amino acids, alcohols, and so forth released during cell death and decay into a solid, amorphous macromolecular structure that includes undigested bits of cell wall and membrane and is sometimes rather loosely referred to as "protokerogen." There are many variations on the theme, depending on the organisms, the productivity, and the amount of oxygen and sulfur in the environment where the protokerogen forms. Over time, reactions of an unknown nature, which probably involve free radical mechanisms, transform this initial protokerogen into the insoluble residue we find in rocks, known as kerogen.

The Newcastle group is not the only one to have become obsessed with the kerogen problem over the years. It was an account of the mystery of kerogen during a seminar at Delft University that convinced organic chemist Jan de Leeuw that trying to characterize the earth's largest store of organic matter was a more worthwhile activity than the monotonous routines of chemical synthesis he'd begun his career with. He joined petroleum geochemist Pieter Schenck's lab at Delft, but concentrated his efforts more on the formation and makeup of kerogen than on its production of petroleum. De Leeuw suspected that the bulk of the kerogen might consist of biological macromolecules that were resistant to breakdown by bacteria, like the lignin in plants. In rocks with a large amount of organic matter, whole pieces of cell wall and the outer casings of plant spores and pollen could be seen under a microscope, and plant cell wall components such as lignin, cutin, and tannin could all be detected using chemical analysis. But the largest contribution of organic matter to most sediments comes from unicellular algae, and it wasn't until the late 1980s that de Leeuw and his cohorts found a

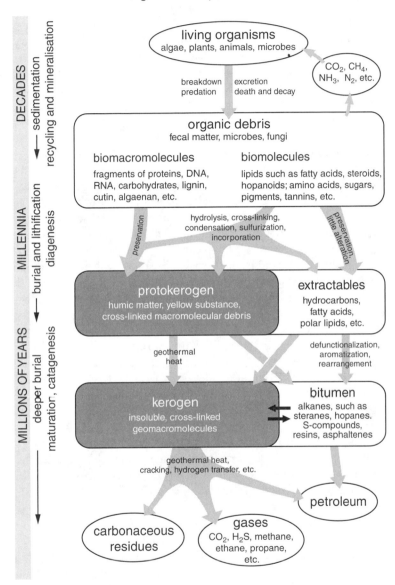

Kerogen and biomarkers: the journey from organisms to petroleum

DECADES — sedimentation — recycling and mineralisation

MILLENNIA — burial and lithification — diagenesis

MILLIONS OF YEARS — deeper burial — maturation, catagenesis

living organisms
algae, plants, animals, microbes

CO_2, CH_4, NH_3, N_2, etc.

breakdown predation | excretion death and decay

organic debris
fecal matter, microbes, fungi

biomacromolecules
fragments of proteins, DNA, RNA, carbohydrates, lignin, cutin, algaenan, etc.

biomolecules
lipids such as fatty acids, steroids, hopanoids; amino acids, sugars, pigments, tannins, etc.

preservation

hydrolysis, cross-linking, condensation, sulfurization, incorporation

preservation, little alteration

protokerogen
humic matter, yellow substance, cross-linked macromolecular debris

extractables
hydrocarbons, fatty acids, polar lipids, etc.

geothermal heat

defunctionalization, aromatization, rearrangement

kerogen
insoluble, cross-linked geomacromolecules

bitumen
alkanes, such as steranes, hopanes. S-compounds, resins, asphaltenes

geothermal heat, cracking, hydrogen transfer, etc.

petroleum

carbonaceous residues

gases
CO_2, H_2S, methane, ethane, propane, etc.

similarly resistant macromolecule in the cell wall of such organisms. They called it algaenan and spent a good part of the next decade trying, with mixed success, to figure out its structure, its distribution among algae, and its role in kerogen formation. It is a three-dimensional matrix comprised of unbranched chains with 22 to 34 carbon atoms, cross-linked to each other by mid-chain ester and ether bonds. Artificial maturation by hydrous pyrolysis results in the production of a plethora of *n*-alkanes resembling those in oils, and there is some speculation that a relative abundance of algaenan-forming algae may have been responsible for the formation of certain types of petroleum source rocks. But attempts to distinguish an unequivocal algaenan component in kerogen have given ambiguous results, and knowledge of its prevalence and distribution in living organisms is still incomplete. It appears to be present in many freshwater green algae and generally lacking in most marine algae, but only a small percentage of the abundant algal groups have, to date, been analyzed. To what extent biological macromolecules such as algaenan actually comprise the macromolecular material in the protokerogen, and to what extent it formed from chemical condensation and polymerization in the sediments, remains a topic of debate.

There is now substantial evidence that fatty acids, alcohols, aldehydes, isoprenoid ethers, hopanoid acids, sterols, and other polar structures are incorporated into the protokerogen during early diagenesis, before they've been fully reduced and stripped of their functional groups. Artificial maturation experiments using hydrous pyrolysis and deuterium-labeled water—something Thomas Hoering first tried in the mid-1980s—can show how the compounds released into the bitumen were bound in the kerogen, because the deuterium then marks the carbon atoms where a bond was cleaved. The Newcastle group found that the biomarkers produced in such experiments—extended hopanes, steranes, and acyclic isoprenoids—were attached to the kerogen at the points where the original biological molecules had reactive polar groups. Hopanoids, for example, are linked by one or more sites in their side chains, which were originally coated with hydroxyl and other polar groups, and a simple sterane at the C-3 position, where the hydroxyl group of the original sterol was. Experiments using specific chemical reagents and carefully controlled heating to cleave specific types of bonds likewise reveal a full array of lipid molecules that have been bound into the kerogen not only by oxygen–carbon bonds and carbon–carbon bonds but also, to a varying extent, by sulfur–carbon and sulfur–sulfur bonds. Both de Leeuw's and Albrecht's groups hypothesized that unsaturated biological molecules have undergone a sort of pseudo-polymerization process in which they become cross-linked by such bonds, often at points of unsaturation in the original biological molecules.

The presence of organic sulfur compounds in oils, coal, and kerogen was first noted in the late nineteenth century, and in the 1970s people began seeing hints of such compounds in the bitumen of recent marine sediments as well. But it wasn't until de Leeuw and his student Jaap Sinninghe Damsté began determining the structures of these compounds in the mid-1980s that their ubiquity and role in diagenesis became apparent. Until it occurred to someone to use a detector that would specifically detect sulfur, they were just lost in the crowd of unidentified compounds—though, of course, if one took a closer look at their mass spectra it was readily apparent that they contained atoms other than carbon, hydrogen, or oxygen. They turned up in all fractions of the bitumen, and indeed, the structures that Sinninghe Damsté determined paralleled those of the hydrocarbons that had been identified—unbranched carbon chains, isoprenoids, steranes, extended hopanes, and so on—except they had cyclic sulfide

Some organic sulfur compounds and possible precursors

groups of one sort or another, usually in positions where the related biological molecules sported a double bond or other reactive functional group. The Strasbourg group suggested that these compounds formed when hydrogen sulfide and other inorganic reduced sulfur compounds, often plentiful in sediment pore water, reacted with organic compounds released from the decaying organic matter, and Sinninghe Damsté demonstrated this in laboratory experiments where hydrogen sulfide reacted with double bonds and oxygen-containing functional groups to form organic sulfur structures like those found in the sediments. Whatever the specific reactions actually involved in their formation, it is now abundantly clear that, particularly in marine environments where sulfur is available in excess, organic sulfur compounds can comprise a major component of the kerogen and be as plentiful in the bitumen as the hydrocarbons, particularly in the asphaltene fraction. They can be freed from the otherwise difficult-to-analyze asphaltenes by using a so-called desulfurizing agent such as Raney nickel, which breaks the sulfur bonds and releases the compounds as soluble hydrocarbons. These molecules had been sequestered and protected from microbial attack in the asphaltenes early in the diagenetic process, often with their original functional groups "labeled," so to speak, with sulfur, and they comprise a treasure trove of biomarker information.

Do any of the new insights into the structure of kerogen and its role in the journey of biomarkers from organism to oil allow us a better understanding of the true chemistry of that journey? At least we can finally explain the paradoxical presence of reactive carboxylic acids, which generally disappear from the extractable organic matter within a few thousand years of sediment deposition, in petroleum and ancient rocks: sequestered and protected from bacterial consumption during formation of the protokerogen in the surface sediments, they are then released at oil generation temperatures millions of years later. The nature of the reactions that produce the mixes of triterpenoid stereoisomers in aging rocks and oils, however, remains elusive. It's clear that the measured quantities are affected, if not determined, by release of compounds from the kerogen, and that they do not, in fact, proceed according to the simplified scheme of reversible isomerization reactions that Mackenzie and McKenzie used in developing their kinetic method. Many petroleum geochemists nevertheless maintain that the kinetic method accurately reproduces the temperature history of a rock or oil and, combined with geologists' models of basin formation, has saved the industry millions of dollars over the past two decades. This success may stem more from the fortuitous numerical adjustment of the so-called kinetic constants to fit a given basin—what some geochemists now refer to as "pseudokinetics"—than from the tenets of physical chemistry or reaction kinetics that gave it birth. There is little doubt that the sterane and triterpenoid stereoisomer and aromatic steroid ratios are determined by the time and temperature history of the rocks where they've resided—their maturity. But for all that we've learned about the process—that it occurs, to what extent, what it looks like in which molecules, and even the approximate temperatures and timescales required—*how* it occurs still poses as much of a mystery as it did when Maxwell first noted the mix of stereoisomers in pristanic acid extracted from the Green River shale, and the question that he and Mackenzie tossed around in the Bristol lab in 1980 remains unanswered: can the process be rigorously quantified?

Geoff Abbott and his Newcastle group have proposed that isomerization of hopanoids and steroids *accompanies* their release from the kerogen, as part of the bond cleavage that releases them from the macromolecular matrix, with the stereoisomers being generated according to their relative stabilities—a 60/40 mix of 22S and 22R hopanes, for example—and the concentrations of the stereoisomers in the expelled bitumen at any given time is determined both by their rate of release from the bitumen and by preferential decomposition of the less stable isomer. They designed a quantitative model of this process and found that it accurately reproduced the isomer ratios obtained during the artificial maturation of a shale by hydrous pyrolysis.

For their part, the members of the Delft group created a kinetic model based on the reversible interconversion between stereoisomers via a positively charged planar intermediate, as originally proposed by Maxwell and McKenzie, but taking into account interconversions between *all* of the possible stereoisomers of the hopanoid. They used the standard kinetic equations for isomerization at all of the chiral centers with a hydrogen attached, and instead of trying to determine the kinetic constants empirically, they calculated them on the basis of the molecular mechanics of their postulated intermediate...and found that they could also reproduce hopanoid isomer ratios in bitumen expelled during artificial maturation experiments.

If we consider the substantial evidence from Abbott's and Farrimond's studies that no isomerization occurs in the free molecules in the bitumen, then success of the Dutch model would imply that isomerization at C-22 of hopanoids occurs *within* the kerogen matrix itself and their molecular mechanics calculations are valid for the bound molecules. It also implies that the stereoisomer ratios in the bitumen are unchanged by release of molecules from the kerogen or by their degradation. And yet we know that the isomer ratios in the bitumen must necessarily reflect the timing of additions from the kerogen and subtractions due to decomposition, and that these are neither negligible nor independent of stereoisomer stability. The Newcastle group's model, on the other hand, implies that the compounds bound in the kerogen isomerize only upon release—and yet recent work indicates that some degree of conversion does occur *within* the kerogen itself. In Newcastle and at the University of Strathclyde in Glasgow, researchers have been experimenting with a technique known as hydropyrolysis—pyrolysis in the presence of pressurized hydrogen gas and a catalyst—which makes it possible to cleave the bonds in kerogen at lower temperatures, without affecting their stereochemistry: extended hopanes released from several shale samples of increasing maturity include small but significant proportions of 22S compounds. Add to these findings results from hydrous pyrolysis experiments that indicate that kerogen-bound polar compounds such as hopanoic acids, which are released from the matrix at even lower temperatures than their alkane counterparts, are partially converted to their corresponding alkanes *during* release...and it seems unlikely that eloquent chemical solutions like the Delft or Newcastle models will ever suffice to explain the gradual loss of life's one-handedness in the geologic realm.

The ultimate provenance of petroleum, a riveting mystery for generations of chemists, was long in solving, and long since solved. But the machinations of Nature's alchemy, hidden within the enigmatic structure of kerogen, have eluded understanding and, until recently, largely repelled chemists' curiosity. Not until the last decades of the twentieth century did geochemists develop the wherewithal to open the black box of kerogen and begin, systematically, to examine its decidedly unsystematic contents. And only now, in the first decade of the twenty-first century, do they find themselves on the cusp of sorting out some of its messy secrets...even as we search for the last holdings of its coveted product.

5

Deep Sea Mud

Biomarker Clues to Ancient Climates

But nothing, from mushrooms to a scientific dependence, can be
discovered without looking and trying.
—Dmitry Ivanovich Mendeleev, 1834–1907, creator of the first periodic table
From *Principles of Chemistry*, Vol. 2 (1905)

It is a capital mistake to theorize before one has data.
—Sherlock Holmes, fictional detective created by Sir Arthur Conan Doyle in
the late nineteenth century
From the short story "A Scandal in Bohemia" (1891)

Though the concept of the biomarker emerged from attempts to infer the provenance of petroleum and the incidence of life on the young earth—for all the successes and disappointments of the early studies on Precambrian rocks, lunar dust, and oil shales—it was in the sediments of the deep sea that biomarkers really came into their own. The Deep Sea Drilling Project (DSDP) was initiated in the 1960s by a consortium of American oceanographic research institutions, but institutions in Russia, the United Kingdom, France, and Germany were quick to sign on. In what began as an effort to understand the makeup and dynamics of the earth's crust and mantle, the DSDP's special research ship traveled the world's oceans, drilling thousands of meters into the seafloor to retrieve sediment cores that soon became coveted objects of study for geologists, oceanographers, biologists, paleontologists, and geochemists around the world.

When Geoff's group started analyzing the DSDP sediments in the early 1970s, most of the organic chemists involved with the program were from the oil industry and formed part of the drill ship's safety program, monitoring the cores as they were brought on deck to ensure that dangerous accumulations of gas or liquid hydrocarbons weren't being penetrated. But Geoff saw the DSDP as the perfect opportunity to wean his Bristol lab of its dependence on NASA's Apollo program—a chance to bring his full attention back to Earth and its still largely unexplored realm of fossil molecules. The British Natural Environment Research Council had earmarked a large pot of funding for work on the cores, which would be unencumbered by the narrow commercial goals and secrecy that surrounded the limited offerings from oil-company bore holes. Geoff's budding Organic Geochemistry Unit would be aligned with a multidisciplinary community of scientists who were all studying the same cores, working cooperatively, and publishing freely. And, unlike the lunar samples, ocean sediments were rife with interesting organic compounds, including many entirely unforeseen structures. Most of the cores consisted of sediments that had been laid down and buried sequentially without ever being subjected to the tectonic turmoil of stretching and subsidence, and the overlying kilometers of cold water had kept their temperatures relatively low. The result was that many of these organic structures were phenomenally well preserved, with vestiges of some of their biological functional groups persisting for millions, sometimes even tens of millions, of years. Here was an even better sequential record of diagenesis and the progressive fossilization of biological organic molecules than either the Messel or Green River shales could provide. Eventually, as the DSDP ship set out on ever more expansive drilling expeditions, its cores would also provide a record of almost every imaginable marine environment to have graced the earth in the last 150 million years of its history.

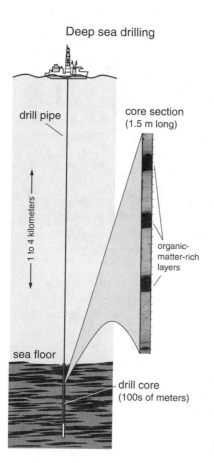

Deep sea drilling

drill pipe

core section
(1.5 m long)

1 to 4 kilometers

organic-
matter-rich
layers

sea floor

drill core
(100s of meters)

The grand majority of the organic matter formed in the ocean comes from microscopic algae, zooplankton, and microbes, and yet knowledge of the chemistry of such organisms was limited. Many of the compounds the geochemists found in surface sediments had never before been identified in organisms. When Geoff and his cohorts first started analyzing the DSDP cores, their biggest task was to identify as many of these new compounds as they could, an endeavor that the geologists who were running the show viewed with some disdain. The geologists had just had huge success with a more hypothesis-driven breed of science, having found evidence for the theory of seafloor spreading and continental drift during the drill ship's first transect of the mid-Atlantic ridge, precisely where they had predicted it. Geoff recalls that one of them accused him of "just stamp collecting," an insult that gets him riled up to this day. After all, geologists had spent more than a century obsessively identifying and classifying minerals, rock types, and fossils, whereas the organic geochemists had only recently begun their surveys—though there's no denying that they were obsessed with their molecular finds. Geochemists in the Strasbourg and Bristol labs, and in Jan de Leeuw's group in the Netherlands, and even Jürgen working in

Dietrich Welte's petroleum-oriented group in Jülich, all gleefully indulged their passion and made a fine art of determining the precise structures of the compounds they found in marine sediments, as well as lake sediments, microbial mats, and any other promising detritus they could get their hands on. The natural history of molecules was still in its infancy, and there were vast, virgin territories yet to be explored. But the grand attraction of the sediment cores from the drilling project was that they provided both a chemical chronicle of the last 150 million years of earth history *and* a lesson in how to read it. Despite the geologists' disdain, the geochemists with their molecule collecting were in no way divorced from the DSDP's lofty goals of understanding the history of the earth and its climate and life—they just thought it might be a good idea if they learned how to read before attempting to write a magnum opus.

The layers of sediment in the DSDP cores were dated by interdisciplinary teams of scientists using everything from the slow decay of naturally occurring radioactive elements in minerals to the changing assemblages of microfossils. Thousands of meters and hundreds of millions of years of ocean sediment can accumulate without ever reaching more than 50–60°C, temperatures that are generally too low to break down the insoluble kerogen and release hydrocarbons into the bitumen. It was thus possible to observe the various intermediate compounds in the low-temperature stepwise conversion from reactive biological compounds to stable molecular fossils by analyzing the soluble organic matter of the sediments, without having to consider the largely unfathomable processes occurring within the solid matrix of the kerogen.

Most of the organic material that forms in the surface waters of the ocean is recycled there; of the 1–2% that does reach the seafloor, more than 90% is usually burned for energy by bacteria in the top few centimeters of the sediments, and much of the remainder is bound up in the insoluble protokerogen. By the mid-1970s, a number of organic chemists at Woods Hole Oceanographic Institution in Massachusetts were following Max Blumer's lead, tracking the fate of specific organic compounds from marine organisms to the sediments and trying to determine what happened in the water column—what compounds were recycled and what survived to reach the sediments. In keeping with what the Bristol, Strasbourg, and Dutch groups were finding in lake sediments, the tiny proportion of the organic matter generated by phytoplankton in the surface waters that persisted in the sediments and eluded incorporation into the insoluble kerogen consisted mostly of pigments and the lipid components of cell membranes: chlorophyll and carotenoids, fatty acids and alcohols, acyclic isoprenoids, sterols, and hopanoids. In the DSDP cores, the loss of oxygen-containing functional groups and reduction of these biological molecules—decarboxylation of acids, dehydration of alcohols, hydrogenation of double bonds—to their bare carbon and hydrogen skeletons proceeded so slowly that one could work from the surface sediments downward into ever more ancient sediments and infer the progress, step by step. These long, coherent sequences of immature marine sediments contained actual evidence of chemical and microbial transformations that geochemists had hitherto only been able to theorize about.

The Bristol group immediately set about trying to follow the convoluted fate of the sterols produced by organisms. And James Maxwell, of course, looked for the intermediates in the hypothesized transformation of chlorophyll—phytol being oxidized or reduced to pristenic acid and phytenes, and eventually to pristane and phytane, and the chlorin hearts that were gradually aromatized and transformed to the blood-red porphyrins found in oil and ancient rocks. Jürgen says he got involved with the DSDP in the 1970s because Germany, as a new member of the international consortium, had been asked to assign an organic geochemist for one of the cruises and a colleague in the Jülich lab was drafted for the task. The colleague, a geologist and organic petrographer, found the cores so interesting that he put in requests for samples from future cruises and convinced Jürgen to analyze them on the GC-MS—but then quit the lab to pursue other endeavors. Jürgen was left with what seemed to be a relentless supply of DSDP samples to analyze and report on, a task that at first seemed rather pointless. But it wasn't long before he was thoroughly addicted and clamoring for more; the samples were a treasure trove of new natural products and structure identification challenges, not to mention a virtual connect-the-dots picture of diagenesis. Despite a lack of interest on the part of the petroleum companies that the Jülich institute was supposed to cater to, he convinced Welte to continue the work with the DSDP cores.

As the sediments aged, there were systematic trends in the relative abundances and nature of the compound classes. Short-chain and unsaturated fatty acids in the soluble organic matter disappeared first, either consumed by bacteria or incorporated into the kerogen; likewise, most alcohols and ketones disappeared rapidly, giving way to alkenes and alkanes. The sterols made by organisms lost their hydroxyl groups in a dehydration reaction, and the resulting sterenes, with two double bonds, underwent a complex array of rearrangements and then could be reduced to steranes or, in certain kinds of acidic sediments, to diasteranes, or somehow oxidized to monoaromatic steroids. The A- and B-ring monoaromatics appeared in early diagenesis, but then disappeared, their fate unknown, whereas the C-ring monoaromatics showed up in late diagenesis and survived for hundreds of millions of years, and the triaromatics didn't appear at all in most of the relatively low-temperature marine sediments. In keeping with Guy Ourisson's and Pierre Albrecht's petroleum studies and their observations from modern lake sediments, bacteriohopanetetrol produced by the bacteria in the surface sediments underwent analogous defunctionalization, this time with accompanying degradation of the side chain containing the hydroxyl groups and stereochemical isomerization at the 17 and 21 positions of the pentacyclic structures. And, as Jaap Sinninghe Damsté and de Leeuw would discover in the 1980s, at some point quite early in this process, the double bonds in alkenes and sterols and carbonyl groups in ketones reacted with dissolved sulfide compounds in the sediment pore waters to produce a plethora of organic sulfur compounds.

How fast and to what extent these transformations occurred seemed to depend strongly on conditions in the water column and sediments at the time of deposition. Within a hundred thousand years in a typical deep-sea sediment—pale gray

Diagenesis of steroids and hopanoids

or white limestone and mudstones containing less than 1% organic matter—the fatty acids had been radically depleted and transformed, most of the alcohols had lost their hydroxyl groups, and alkenes and reduced alkanes ruled the day. But in some sediments where rapid burial and a low oxygen concentration had conspired to preserve the organic matter from decomposition—places where upwelling of deep water resupplied nutrients at the surface and productivity was high, or along continental shelves where rivers had deposited a lot of refractory organic matter from land plants—sterols, saturated fatty acids, and long-chain alcohols could persist for more than 10 million years.

Whereas kerogen still stood like an impenetrable knowledge blockade between the molecular fossils in mature rocks or petroleum and their sources, the emerging understanding of diagenesis made it possible, at least in principle, to link many of the compounds found in ancient ocean sediments with the specific organisms or groups of organisms they came from. In some cases, biogenic lipids replete with their double bonds and oxygen-containing functional groups could be found in ocean sediments that were laid down millions of years ago, which was almost like finding a dinosaur bone with some organs or pieces of muscle still attached. Even so, it was no easy task to link a compound found in the sediments, even recent sediments, to the organism or group of organisms that made it—and ultimately, it was this connection that would allow the molecule collectors to read a history of life that was invisible to the geologists, climatologists, biologists, and paleontologists. Despite their collector's mania, these were, after all, chemists with a sense of adventure, refugees from the tedium of synthetic chemistry and veterans of the search for natural products who were intent on exploring the interface between organisms and their environment through geologic time.

They used the same basic extraction and separation techniques that Geoff and Al Burlingame had started using in Melvin Calvin's lab in Berkeley, procedures that had been refined to perfection for the first Apollo samples: extract all the soluble organic compounds from the sediment into some solvent mixture such as methanol/chloroform, pour the mixture onto a chromatography column or thin-layer plate to separate them into groups of different polarity—nonpolar hydrocarbons, then aromatics, esters, ketones, alcohols, and so forth—and run each fraction on the GC-MS to separate and, with any luck, determine the structures of some of the hundreds of compounds it contained. When Sinninghe Damsté and de Leeuw began to realize how much information was hidden away and protected in the sulfur compounds, they added desulfurization and other steps that allowed them to separate and analyze the full array of sulfur components. But the basic extraction and separation scheme used for the DSDP sediments in the 1970s and 1980s was not hugely different from that developed in the 1960s and, for that matter, the one used by organic geochemists to this day.

In the 1970s, the so-called molecule collectors looked for suites of compounds that might be biosynthetically related or linked through diagenesis: acids, alcohols and ketones with similar chain lengths, or sterols, sterenes, and steranes with the same side chains. They focused on the most plentiful compounds partly because these were the easiest to analyze, and partly because they promised to come from some prominent member of the ecosystem. And they focused on the most unusual structures, not only because they were the most fun to figure out, but also because they were likely to contain more specific information about the organism or process that made them. Occasionally, they got lucky and the association between compound and organism was readily apparent.

Though it was the mystery of kerogen that had initially inspired de Leeuw, the soluble organic matter in freshly deposited surface sediments obviously held some clues about what had gone into its making and, as it turned out, offered more immediate gratification than the recalcitrant kerogen. Here, at least, one could have the satisfaction of proving a new chemical structure, not to mention lots of publications and successful Ph.D. theses. It wasn't long before de Leeuw had hooked up with Burlingame's state-of-the-art analytical laboratory, gained access to samples from DSDP cores, and earned his Dutch group a reputation as a

Molecular analysis of immature marine sediments
(one example: alcohols)

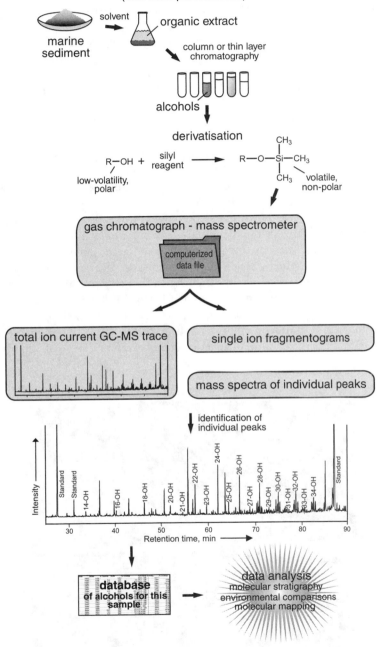

marine sediment

solvent → organic extract

column or thin layer chromatography

alcohols

derivatisation

$$R{-}OH + \text{silyl reagent} \longrightarrow R{-}O{-}\underset{\underset{CH_3}{|}}{\overset{\overset{CH_3}{|}}{Si}}{-}CH_3$$

low-volatility, polar

volatile, non-polar

gas chromatograph - mass spectrometer

computerized data file

total ion current GC-MS trace

single ion fragmentograms

mass spectra of individual peaks

identification of individual peaks

Standard Standard 14-OH 16-OH 18-OH 20-OH 21-OH 22-OH 23-OH 24-OH 25-OH 26-OH 27-OH 28-OH 29-OH 30-OH 31-OH 32-OH 33-OH 34-OH Standard

Intensity

30   40   50   60   70   80   90
Retention time, min

database of alcohols for this sample

data analysis
molecular stratigraphy
environmental comparisons
molecular mapping

bastion of first-class molecule collectors. One of its first big successes came, however, not from the long deep-sea cores, but from a 5,000-year-old layer of surface sediments in the Black Sea. Here the group found massive amounts of a sterol with a "peculiar" structure, similar to one that had puzzled the Strasbourg group when they identified it in the Messel shale a few years earlier. The Black Sea sediments, however, were also loaded with the distinctive fossils of dinoflagellates, a large, diverse genus of single-celled algae, and the Dutch group suspected this was the source of the sterol. In 1978, around the same time they determined its structure, a group of Rhode Island natural products chemists who were studying the toxin-producing algae responsible for the deadly "red tides" that periodically occur in coastal waters began isolating sterols from dinoflagellates and confirmed de Leeuw's suspicions: the most prevalent of the sterols, which they christened dinosterol, was none other than the Black Sea sterol, a hitherto unknown structure, with a methyl group attached to the A-ring, and an extra branch attached to its side chain.

Dinoflagellates are known to paleontologists from the cysts they form during dormant periods, which had allowed them to distinguish hundreds of living and extinct species. These are made of a resistant organic polymer that is readily fossilized and can maintain its distinctive shape and decorations for millions of years, but the organisms only make them when they go dormant, and many species of dinoflagellates don't make them at all. Most of them do, however, make dinosterol. Once it had been identified, geochemists were able to detect it or its diagenetic successors in sediment cores dating back hundreds of millions of years. Sterols with the same or similar skeletons have been found in small amounts in

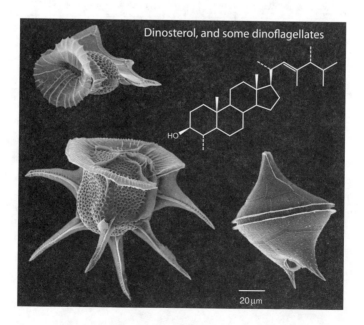

Dinosterol, and some dinoflagellates

20 µm

a few other microorganisms, but 25 years' worth of research has confirmed that concentrations of dinosterol in sediments generally depend on dinoflagellate populations. It can be used to track the periodic bursts of dinoflagellate productivity in surface waters, providing valuable insight into how the ecology of a lake or coastal marine environment has changed over time. Petroleum geochemists use the presence of dinosterane in petroleum source rocks to provide information about the environment where they were first deposited, and environmental scientists measure the content of dinosterol in the sediments of lakes and estuaries as a way of gauging nutrient inputs and tracking the effects of fertilizer use and runoff in a region.

Few of the thousands of compounds identified in sediments during the 1970s and 1980s were as readily linked to a specific class of organisms as dinosterol. Many were constituents of a broad range of organisms and too general to provide much information. Others, like the extended hopanoids, had never been identified in organisms at all, and it took a fortuitous mix of experience, insight, and luck to find their biological parents. A series of long-chain alcohols that the Dutch group found in its Black Sea cores in the 1970s would remain orphans for more than 20 years. These alkane diols constitute a homologous series of simple *n*-alkyl chains with 28–32 carbon atoms, slightly glorified by two hydroxyl groups, one on the first carbon, and a second in the middle of the chain at C-14 or -15. They turned up regularly in surface and ancient marine sediments from all over the world, but neither they nor any plausible precursor molecules could be found in organisms—not that most natural products chemists were really looking for such compounds.

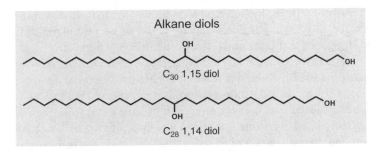

A series of somewhat more distinctive compounds, also noted in the sediments in the mid-1970s, were so difficult to separate from the mixtures of branched and cyclic hydrocarbons that it was years before anyone determined the precise structure of their carbon skeletons. Dubbed "highly branched isoprenoids" or HBIs, they are composed of a saturated or unsaturated isoprenoid chain with another branching from its middle in an odd T-shape. James Maxwell and one of his students first determined the precise structure of a 20-carbon HBI alkane in a sediment extract in 1981, but despite the near ubiquity of such compounds in marine and lake sediments—and, as it turns out, some pharmaceutical potential—another 10 years went by before a source of HBIs was identified in marine organisms.

Highly branched isoprenoid (HBI)

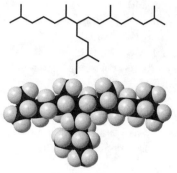

C$_{25}$ HBI alkane

Another series of compounds that the Dutch group identified in sediments from an Atlantic DSDP core in 1978 took only a few years to trace, but this required a complex coincidence of international encounters and interdisciplinary collusion. Indeed, if it hadn't been for a few worried Scottish fishermen with a lemonade bottle in the North Sea, the most useful and powerful biomarker to date might well have languished among the molecular orphans with the alkane diols for years to come.

In the spring of 1975, amid concerns about the local fishing industry, British Petroleum installed the first of its Forties oil field production rigs off the coast of Scotland. A few days later, when a small commercial fishing boat sailed into an expanse of eerily opaque white water in the normally dark-blue North Sea, the fishermen were convinced that it was pollution from the oil production rig. They filled a lemonade bottle with the milky water, which extended for hundreds of kilometers, and when they returned to shore they delivered the sample to the Aberdeen Marine Laboratory. From there, it made its way around the area's research labs, and eventually, when no one could detect any petroleum hydrocarbons, it ended up on biochemist John Sargent's desk.

Sargent was a specialist in the lipids of marine organisms, particularly in the fats and long-chain esters, or wax esters, used for energy storage in zooplankton. When he analyzed the milky water, he found it was loaded with wax esters from copepods, the tiny shrimplike crustaceans that feed on microscopic algae and bacteria. He figured that there must have been a copepod population explosion in the area and their waxes had made the water look milky. He'd observed a similar phenomenon in British Columbia a few years before, when the die-out of a big population of copepods had left the beaches literally covered with sticky wax. But a colleague suggested that the wax esters might have come from the sediments instead, accumulated over time from the usual cycle of growth, death, and sedimentation, and then stirred up when the oil platform was installed. Sargent didn't really think this was the case, but then there was no denying that the installation had churned up massive amounts of sediment. And when he thought about it, he had no idea what sort of organic compounds accumulated in the sediments. Copepod waxes? Perhaps. Curious, he requested a North Sea sediment core from the British Geological Survey.

When the core arrived, Sargent says he and his coworkers didn't quite know what to do with it. As biochemists, they were used to working with extracts from organisms, not meter-long tubes of mud. He says they treated it like an oversized chromatography column, turning it on end and pouring solvent through it for several days. It wasn't the most efficient extraction technique in the world, but it was effective enough to extract the most plentiful lipids, and these could then be separated by thin-layer chromatography. The compounds in the sediments were more varied and less abundant than what the biochemists were used to extracting from single organisms, and only two bands on the thin-layer plate could be clearly distinguished from the smear of slightly polar, high-molecular-weight compounds that appeared where the wax esters should be—but neither band matched that of the wax esters they'd found in the milky water, nor, for that matter, did it correspond to any of the long-chain waxy lipids they were familiar with. At this point, Sargent was convinced that his initial assessment was correct in that the milky water had nothing to do with the installation of the oil platform and disturbance of the sediments. But now he was curious about the unknown lipids they'd found in the sediment extract.

He scraped the bands off the thin-layer plate, extracted the compounds from the chromatographic solid phase, and turned them over to a colleague who had a GC-MS. But his colleague couldn't get the compounds to pass through the GC column and had to analyze the two mixtures from the thin-layer bands by introducing them directly into the mass spectrometer. This was a little like throwing several picture puzzles into one box and trying to reconstruct each of the pictures, and all the colleague could tell Sargent was that they were very long carbon chains with carbonyl groups, which fit with their position on the thin-layer plate, and that the carbonyl groups appeared to be near the ends of the chains. Meanwhile, unbeknownst to the Scottish biochemists, de Leeuw was in Berkeley with Burlingame, analyzing DSDP sediments from the Walvis Ridge off the Atlantic coast of southwest Africa and wondering about a group of compounds that behaved like very large, heavy ketones in thin-layer separations and produced broad humps tailing off the ends of the gas chromatograms.

While the Dutch group set about trying to isolate the ketone-like compounds and decipher their structures, Sargent, who had always been concerned with the biochemical systems and ecology of live animals and had never given a lot of thought to their remains on the seafloor, was now ruminating about sediments and petroleum. He says Aberdeen was then in its boomtown days, and everyone was thinking and talking about oil, which somehow brought to mind an amusing lecture he'd heard at a Marine Biological Association meeting in Plymouth several years earlier, in 1971. It was about coccolithophores, a group of unicellular algae that were then known more for the tiny calcium carbonate shells they left as testament in marine sediments than for their living presence in the water. The shells of the most common species of the group, *Emiliania huxleyi*, comprised the bulk of many marine sediments and carbonate rocks, and were quite beautiful when viewed through an electron microscope. According to the speaker that day, "Emily," as the Plymouth biologists fondly called the algae, contained a large oil globule that made it float near the surface of the water. The speaker innocently speculated that this oil might also accumulate in sediments and even have something to do with the formation of petroleum, and a rather rancorous but humorous argument erupted about how an oil that made coccolithophores float could sink into the sediments. So now Sargent found himself wondering if the unknown lipids he'd found in such quantity in the North Sea sediments

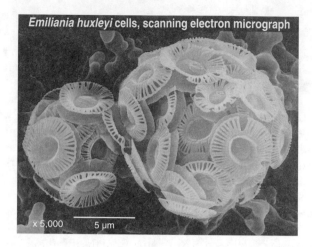

Emiliania huxleyi cells, scanning electron micrograph

x 5,000   5 μm

might come from this same oil. Just for fun, he wrote to Plymouth and requested a starting culture of Emily.

The algae grown from the Plymouth culture, as it turned out, produced the exact same long-chain ketones his group had isolated from the North Sea sediment extract—an amazingly fortuitous result, given the string of free association that inspired him to request the culture. Their precise chemical structures remained a mystery, but they were clearly the same compounds, and so plentiful that Sargent figured they must be concentrated in some sort of droplet inside the cell, perhaps even the oil globules he'd heard described at the Marine Biological Association meeting. He would have liked to know more about the compounds and their biochemical or physiological roles, but he could find no reference to such oil globules in the literature and he was feeling pressured to get on with the work that his laboratory had been funded for. With no intention of publishing the results from the study, they stashed the sediment and algal extracts in the deep freeze and went back to work on their live zooplankton—until some six months later, when biochemist John Sargent happened to meet organic geochemist Geoff Eglinton.

In the fall of 1976, the DSDP ship made a port call in Aberdeen, and Geoff, as a British participant in DSDP advisory committees, was invited to tour the ship and join the reception and press conference, which was held at Aberdeen's Institute of Marine Biochemistry, where Sargent worked. Sargent had nothing to do with the DSDP, but he wandered over to the reception just to see what was going on and get some free coffee. The two got to talking, and Sargent told Geoff about his foray into the sediments and the strange algal lipids he'd isolated. Geoff was immediately intrigued, and of course, when he heard that the Scottish team hadn't figured out the chemical structures, the challenge was more than he could resist. "We can crack it!" he exclaimed enthusiastically, if a bit too optimistically. He headed back to Bristol with several vials of the mystery compounds from Emily, but then they languished in the freezer there for another year before anyone in the lab had the time, and the interest and particular skills, to analyze them.

Australian John Volkman came to the Bristol lab on a postdoctoral fellowship that was part of Geoff's and James Maxwell's plan to end the lab's reliance on chance discoveries. They wanted to embark on a more concerted, tailor-designed investigation of lipids in marine plankton and bacteria. Volkman was fresh out of grad school, but he had already demonstrated a talent for instrumentation and an interest in marine natural products. Almost immediately upon his arrival from Australia, Geoff sent him off to Switzerland to learn the most cutting-edge techniques in capillary gas chromatography from the royal family of GC, the Grob family—a father-mother-son team of world-class analytical chemists. When Volkman started trying to analyze the Emily extracts and, like the Aberdeen biochemist, couldn't get the long-chain ketones through the lab's commercial GC columns, he was well prepared to make his own capillary columns. He used a material that could withstand the high temperatures needed to vaporize the unwieldy compounds, but they still got stuck in the injection system and he finally resorted to dipping the end of the column directly into his extracts. The reward was a chromatogram with eight clearly separated peaks, four from each band scraped off the thin-layer separation plates.

While Volkman was trying to figure out the structures of the Emily ketones, other Bristol postdocs and students were busy working on new DSDP cores from the Japan Trench, a deep ravine in the ocean floor immediately southeast of Japan. They were analyzing organic-matter–rich sediments that had been deposited during a period of high productivity, extracting, separating, and identifying as many of the lipids as they could. Simon Brassell, who was still a student at the time, was responsible for the steroid and ketone fractions. Using a standard, commercially available GC column, he was finding ketones that seemed to mirror the *n*-alkanes in chain length and distribution, and he suspected they'd been produced by bacterial oxidation of the alkanes in the top layers of sediment shortly after deposition. One day he forgot to turn the chart recorder off and by the time he remembered, paper was spilling out across the room. As he gathered it up, he noticed a massive, ill-defined hump as if something had bled off the column at the end of the chromatogram, long after the column had reached its maximum temperature and his ketones and everything else had come through. At first he was afraid that the material inside the column was breaking down, but a blank run produced only the expected long, straight line of nothingness. He purposely left his next sample running for an extended period, and the hump showed up again, now followed by several others. At that point, he decided to try running his Japan Trench extract on John Volkman's new system and, lo and behold, the humps were replaced by 10 clearly separated peaks, eight of which matched those of Volkman's Emily ketones.

By this time, Volkman had identified eight different long-chain ketones, or "alkenones" as they came to be known, all variations on a theme: long unbranched carbon chains with several double bonds and a carbonyl group near the end, just as Sargent had deduced. They had from 37 to 39 carbon atoms and two or three double bonds, and the carbonyl was either at the C-2 position or the C-3 position. He was still trying to determine exactly where in the carbon chains the double bonds were positioned when Jan de Leeuw visited Bristol for a thesis defense. De Leeuw had a reputation for being a whiz at mass spectra interpretation, so Volkman showed him the alkenone mass spectra, in the hope that de Leeuw might be able to figure them out. To Volkman's surprise, de Leeuw didn't have to figure out anything—he recognized the spectra immediately and drew out the compounds, double bonds and all. It was the same compound his student Jaap Boon had found in a DSDP core from the Walvis Ridge. "What sediment is this from?" he asked Volkman, and then it was de Leeuw's turn to be surprised: Volkman told him the compounds hadn't come from a sediment but, rather, from an extract of the common marine coccolithophore *Emiliania huxleyi*.

With the realization that they'd been investigating the same compounds, the Dutch and Bristol groups began to swap information and samples. At the next international meeting of organic geochemists in the fall of 1979, they presented their papers together, detailing the structures of the alkenones and reporting their presence in a variety of marine sediments and in the coccolithophore *Emiliania huxleyi*. Volkman had followed Sargent's lead and collaborated with the biologists in Plymouth, requesting cultures of a number of other algae and bacteria, but he found no alkenones in any of them, not even in other species of

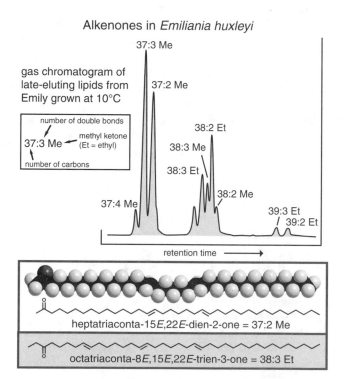

Alkenones in *Emiliania huxleyi*

gas chromatogram of late-eluting lipids from Emily grown at 10°C

number of double bonds
methyl ketone (Et = ethyl)
37:3 Me
number of carbons

37:3 Me
37:2 Me
38:2 Et
38:3 Me
38:3 Et
37:4 Me
38:2 Me
39:3 Et
39:2 Et

retention time ⟶

heptatriaconta-15*E*,22*E*-dien-2-one = 37:2 Me

octatriaconta-8*E*,15*E*,22*E*-trien-3-one = 38:3 Et

coccolithophores. He also analyzed some of the original milky water from the North Sea lemonade bottle, about which there was still some confusion, due to yet another coincidence that Sargent was initially unaware of: *Emiliania huxleyi* is often associated with just such a milky water phenomenon. When nutrients are high and their numbers swell, the billions of microscopic calcium carbonate shells make the water look as if it's full of powdered chalk and create spectacular expanses of ghostly pale, turquoise-colored water that seafarers have been commenting on for centuries. It would have seemed quite logical for Sargent's Emily quest to have begun with finding the ketones in the milky water the fishermen brought back, and in fact this is the myth that was and is still recounted in Bristol and beyond—that discovery of the alkenones began with a coccolithophore bloom in the North Sea. But the truth of the matter is that the water in the lemonade bottle was completely devoid of ketones and chock-full of copepod wax esters, as Volkman's analyses confirmed. Sargent's quest was more whimsical than logical. When he saw the alkenone structures he was curious anew about the biochemistry of the compounds—how they were synthesized and used, and if they were in fact concentrated in oil globules as he suspected. But he was too busy with his zooplankton to follow up, and the project was left to the geochemists…who had started finding alkenones in many of their DSDP cores, and noticing some very intriguing patterns.

Simon Brassell had been working with DSDP sediments from the moment he'd started as Geoff's research student three years before. For his doctoral thesis he had tracked the early stages of the complex web of reactions that converted the sterols in organisms to the steranes in petroleum and mature rocks that Andrew Mackenzie was working on with Maxwell. When he first noted that the distribution of the various alkenones in the Japan Trench sediments differed from what Volkman had found in his Emily cultures, he thought it was due to similar diagenetic reactions in the sediments. The algae contained mostly alkenones with three double bonds, whereas the sediments had a higher proportion of those with only two, so it seemed that the double bonds were gradually being reduced, just as the double bonds in sterols were reduced in the upper few centimeters of the sediments. Of course, there was also the possibility that some other, yet unidentified species made the same compounds in slightly different proportions. Other species must have made alkenones in the past, because he found them in segments of sediment core that were several millions of years old and, according to the fossil record of coccoliths, Emily hadn't appeared on the scene until about 250,000 years ago.

While Brassell was working on the Japan Trench sediments, Geoff attended his first meeting as a member of the DSDP scientific advisory committee on paleoenvironments, a weeklong brainstorming retreat with a group of some fifteen oceanographers, geologists, and paleontologists. They were supposed to define the most pressing questions about conditions on Earth over the past 300 million years and recommend drilling sites in the world's oceans, but the meeting was as memorable for its parasitic worms and nocturnal mosquito fests as it was for any daytime brainstorming. The meeting place was chosen such that all the international participants would have the shortest number of miles to travel, which somehow put them in a rather rustic, so-called conference center in Barbados. Geoff spent the hot, sleepless nights in conversation with the Austrian geologist he'd been assigned to room with: Michael Sarnthein, from the University of Kiel, heard about organic chemistry and molecular fossils, while Geoff learned all about the ice ages that geologists had spent the past century trying to understand.

Naturalists had recognized signs of long-extinct glaciers and extremely cold climates in now temperate regions of the northern hemisphere at the beginning of the nineteenth century, but their attempts to explain the phenomenon all failed to account for later observations that there had been not one period, but multiple periods of glaciation that had alternated with periods of warmer climate over the past few million years. The most plausible explanation for such oscillations between glacial and interglacial climates was presented in 1930 by the Serbian mathematician Milutin Milankovitch, who suggested that it might be due to cyclical changes in the earth's spatial orientation with respect to the sun. Milankovitch had determined that the slowly oscillating shape of the earth's elliptical orbit around the sun, the shifting tilt of its axis of spin relative to the plane of its orbit, and the wobble in its spin would affect the distribution of solar radiation and intensity of seasons in cycles of 100,000, 41,000, and 23,000 years, respectively. The variations in solar radiation were minuscule, but he reasoned that they might be enough to trigger feedback effects. If, for example, the amount of sunlight decreased just enough that some patches of snow survived the summer, then the snowfields would build up over the years, reflecting away more sunlight as they expanded and amplifying the small cooling from the orbital changes, until the snowfield had grown, over the centuries, into a thick sheet of ice that covered half a continent. Based on these variations, Milankovitch estimated the climate fluctuations over the past 450,000 years and came up with cycles of cold and warm periods, the so-called Milankovitch cycles.

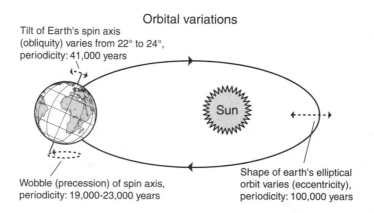

Orbital variations

Tilt of Earth's spin axis (obliquity) varies from 22° to 24°, periodicity: 41,000 years

Wobble (precession) of spin axis, periodicity: 19,000–23,000 years

Shape of earth's elliptical orbit varies (eccentricity), periodicity: 100,000 years

Geologists were unable to determine the timing of the glacial periods from the rocks, and Milankovitch's astronomic theory was just the hypothetical musing of a star-gazing mathematician until the 1970s, when the DSDP made marine sediments from around the world available for study in overlapping sequences of tens of millions of years, and chemists and paleontologists developed reliable means for dating them and estimating paleotemperatures. Paleontologists had spent nearly a century classifying the tiny fossils of unicellular zooplankton—the silica shells of radiolaria, and the calcium carbonate shells of foraminifera—and could distinguish tens of thousands of extant and extinct species. As they learned which species thrived where and at what temperatures, they developed statistical treatments of the relative abundances of microfossils that not only allowed them to correlate layers of sediments deposited during the same time period in different regions, but also provided some indication of the climate at the time. These techniques became increasingly difficult when the sediments were dominated by extinct fossil species, and were not very dependable for sediments that were more than 30,000 years old. Chemists soon learned, however, that the calcium carbonate shells of the foraminifera had an index of seawater temperatures built into the atoms of the mineral itself, and this was less dependent on the precise identification of species. When the shells precipitate from the dissolved carbonate in seawater, they preferentially incorporate carbonate that contains oxygen-18, the rare heavy isotope of oxygen, which has an extra two neutrons in its nucleus compared to oxygen-16. The lower the temperature, the stronger this preference for ^{18}O: when researchers carefully picked out the tiny, millimeter-sized foram shells from the rest of the sediment and measured the relative amounts of ^{18}O and ^{16}O—a difficult undertaking in the 1950s, but an elementary task for a mass spectrometer in the 1970s—they could, theoretically, determine the temperature of the sea at the time the shells had formed.

The story that began to emerge from such painstaking analyses of marine sediment cores in the late 1950s was of a climate that had shifted back and forth between cold glacial and warmer interglacial periods dozens of times during the past few million years. And by the mid-1970s, Milankovitch's theory had descended from the stars and settled, with dramatic certainty, in the solid realm of geologic evidence: researchers were able to combine ^{18}O and microfossil species distribution data for DSDP cores spanning the past 450,000 years and, using established mathematical methods to determine their cyclical components, show that the cooling and warming periods alternated with frequencies that combined components of Milankovitch's three orbital cycles.

Sarnthein explained to Geoff how he and others had been gathering data from DSDP cores in an attempt to both verify and understand these results. Though it was clear that the orbital cycles were in some way responsible for triggering the glacial–interglacial climate shifts, it wasn't at all clear *how*. As Milankovitch himself had noted, such minuscule changes in solar radiation would have to set in motion other processes that magnified their effect. But it wasn't clear what the timing of ice formation had been with respect to other climate variables. Indeed, there wasn't even a clear picture of how the whole climate system had operated during the last ice age—what the wind patterns had been, the differences between land and ocean temperatures, patterns of heat-transporting ocean currents and deep ocean circulation....Even the overall latitudinal temperature differential was unknown. First and foremost, they needed maps of sea surface temperatures dating back a few million years. Sarnthein said he and others had been trying to obtain enough microfossil species distribution and ^{18}O data to construct such maps, but both techniques were incredibly labor intensive. Plus, in some areas of the ocean the calcium carbonate shells had dissolved before they reached the seafloor, so there were few fossils to be found and the technique was rendered completely useless. Another big problem, he said, was that the ^{18}O to ^{16}O ratios in fossil foram shells depended not only on temperature, but also on the ^{18}O to ^{16}O ratio in the ocean water, and this had not been constant. Evaporation and cloud formation preferentially removed H_2O with the lighter ^{16}O isotope; if rain and river runoff didn't return all the evaporated water to the ocean, the water would become enriched in ^{18}O, and this was what happened during an ice age when most of the rain that fell over landmasses was tied up in glaciers. The oxygen isotope ratios in the fossil forams thus gave an ambiguous measure of ocean temperature *and* ice volume.

What Geoff learned from Sarnthein during that Barbados "retreat" in 1979 was that oceanographers, paleontologists, and geologists—everyone who wanted to know anything about the climates, oceans, and ecology of the past, and even those who were trying to predict the future—were all desperate for an independent sea surface paleothermometer to fill in the gaps and clarify the intriguing but ambiguous message they were getting from the ^{18}O data. A few months later, when Brassell came bursting into his office suggesting that the alkenone distributions in the sediments he'd been analyzing might be the result not of diagenesis but of the different temperatures in the water where the algae had been growing, Geoff realized that just such a paleothermometer might already be under design in his own laboratory. The oceanographers would call it a temperature "proxy" because it was a way to determine temperatures through a third party, so to speak, by measuring some other variable that acted in response to temperature. He looked at the chart of alkenone distributions Brassell showed him with a mix of jubilation and scientific cynicism—it was such a fantastic possibility, he thought, that it couldn't possibly be true...or could it?

Brassell had been doing a preliminary write-up of analyses of a DSDP core from a site near the Middle America Trench, off the Pacific coast of southern Mexico, comparing the lipid distributions to those in his Japan Trench sediments from the same period, which extended from the present back to about five million years ago. Having finished his thesis on the diagenesis of sterols, he was on the lookout for anything that might link the various lipids the Bristol team was finding not only to their source organisms, but also to specific environments or processes. He noticed that the distribution of alkenones in the Middle America Trench sediments was, like that in his Japan Trench sediments, inconsistent with the distribution in John Volkman's Emily cultures. But the cumulative alkenone data from all the cores the group had analyzed revealed none of the gradual,

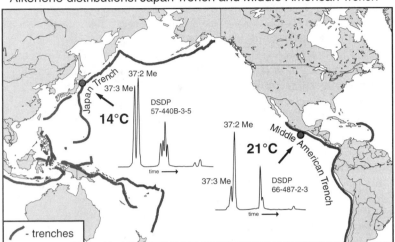

Alkenone distributions: Japan Trench and Middle American Trench

systematic changes with age that one would expect if the difference was due to dia-genetic alteration. When he noted that the alkenone distributions in the Middle America Trench and Japan Trench sediments of the same age also differed dra-matically—that at the tropical site alkenones with two double bonds were much more plentiful than those with three double bonds, whereas in the Japan Trench sediments they were only slightly more plentiful—the most dramatic difference in the two environments came immediately to mind: in the modern ocean, at least, there's a cold current running from the polar region south along the coast of Japan, whereas the water off the coast of southern Mexico is as warm as a baby's bath.

Was it possible that the algae adapted to water temperature by changing the degree of unsaturation in their alkenones, making more double bonds if they lived in cold water, fewer if they lived in warm? For his part, Volkman says this possibility simply didn't occur to him, despite having recently collaborated on a project where certain bacteria were found to vary the degree of unsaturation in their fatty acids as a means of maintaining the consistency of their cell mem-branes at different temperatures. He'd been working with biochemists at the time, not thinking about paleoenvironments or biomarkers. And when he did start thinking about them, the fatty acids were not in the running. Their car-boxylic acid groups, relatively short carbon chains, and closely spaced double bonds made them too reactive and easily transformed; they were synthesized by all types of organisms, including bacteria in the sediments, and the species-specific distributions of chain length and unsaturation that could help to distin-guish organisms were quickly lost during diagenesis in the sediments. But the alkenones were another matter altogether. They showed every promise of being perfect biomarkers: their excessively long carbon chains made them insoluble in water, immobile in the sediments, and presumably difficult to degrade; their double bonds were spaced far apart and relatively unreactive; they were unusual, apparently limited to a few species of algae; and *Emiliania huxleyi*, their main producer in modern times, was plentiful and cosmopolitan, present in all the world's oceans except the Southern Ocean around Antarctica. That such a perfect biomarker might preserve a record of sea surface temperature seemed too good to be true…and so Geoff and Simon Brassell immediately sat down and listed all the reasons it might not work. Then they wrote a proposal to fund a graduate student project so it could be tested—outlining all the key experiments, but carefully avoiding direct reference to the suspected temperature dependence of the alkenones, just in case. They didn't want to set some reviewer off on the same path and risk being scooped.

The first thing, of course, was to see if Emily really did manufacture different combinations of alkenones when grown at different temperatures, or if the differences observed in the sediments were the result of something else entirely. And then, too, were the alkenones really as sturdy and resistant to microbial attack as they looked? What happened to them before they got to the sediments? *How* did they even get there? Emily's cells were so small that they were hardly visible under a traditional light microscope, just a thousandth the size of the copepods and foraminifera that fed on them. Calculations indicated that if the dead cells were left floating in the water column, it would take them more than a hundred years to sink the thousands of meters to the seafloor. Most phytoplankton, however, were eaten by zooplankton, whose millimeter-sized fecal pellets could sink in a matter of days and made up much of the organic component of the sediments. But wouldn't a voyage through a zooplankton intestine have some effect on the structures of the alkenones?

The Plymouth marine biologists and biochemists had been studying the fecal pellets of Emily's most likely grazer, the copepod, for years, and Geoff sent Fred Prahl, one of the Bristol postdocs, to work with them. Together, they grew cultures of Emily, fed them to copepods, collected the fecal pellets, and extracted and analyzed the lipids—and found that the alkenones did indeed pass undigested and unchanged through the zooplankton intestines. The Plymouth group also started growing some cultures of Emily at different temperatures, and Ian Marlowe, the student funded by the paleothermometer proposal, was charged with analyzing their lipids. His first graph of temperature versus the ratio of diunsaturated to triunsaturated alkenones only had three points—at 20°, 15°, and 5°C—but it was enough to show a distinct trend toward more triunsaturated compounds with decreasing temperature. In addition to a series of related 37- to 39-carbon esters and simple alkenes that showed similar trends, Marlowe identified alkenones with four double bonds in the low-temperature cultures, and it became clear that the overall degree of unsaturation in the algae's lipids increased systematically with decreasing temperature.

The Bristol researchers moved into high gear. They continued the collaboration in Plymouth, searching for alkenones in other species of algae and exploring factors other than temperature that might influence the distributions—the growth phase of the algae, the nutrient levels in the water, microbial activity,

Average degree of unsaturation versus algal growth temperature
for long-chain unsaturated lipids in *Emiliania huxleyi,* 1982

or chemical breakdown in the sediments. They searched for a single index that would reflect the temperature dependence of the overall degree of unsaturation in the family of long-chain compounds, and finally focused on the 37-carbon alkenones, which were the most plentiful and consistently present, and included only methyl ketones. They went on a massive data-gathering campaign that would eventually allow them to calibrate their unsaturation index against actual water temperatures, using the alkenone distributions not only in laboratory-grown cultures of algae, but also in natural populations of phytoplankton and marine sediments. And, finally, Geoff cut to the chase and contacted Michael Sarnthein.

Where could they obtain a sediment core to test their developing temperature proxy against the ^{18}O record of the ice ages, he asked Sarnthein—one that spanned the last million years of glacial–interglacial climate cycles, had an uninterrupted record of forams for ^{18}O analyses, and contained sufficient quantities of alkenones for their measurements? Sarnthein immediately recommended a site in the Kane Gap, off the coast of northwest Africa, where the German research vessel *Meteor* was scheduled to go in 1983. The *Meteor* wasn't a big drill ship like the DSDP's, but it could retrieve short gravity cores of 10 or 15 meters, which should serve their purposes, and Sarnthein already had a place secured as a principal researcher. He could easily take Marlowe along to collect the samples for organic analysis.

In the fall of 1984, four years after Brassell burst into Geoff's office with his too-good-to-be-true proposition, members of the Bristol group met in London with Sarnthein to exchange and discuss their Kane Gap results. Geoff says he couldn't make heads or tails of their alkenone data. They had plotted their unsaturation index based on the C_{37} compounds against depth in the core, but it just

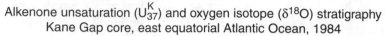

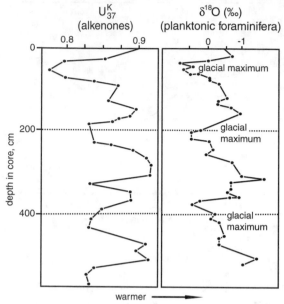

Alkenone unsaturation (U^K_{37}) and oxygen isotope ($\delta^{18}O$) stratigraphy Kane Gap core, east equatorial Atlantic Ocean, 1984

looked like a bunch of irregular squiggles. He apologized for its inelegance as he handed it to Sarnthein, but Sarnthein took one look at the graph and exclaimed, "That's it! There they are! But it's easier to follow if you flip it over." Stable isotope compositions were usually given as "delta" values, he explained. $\delta^{18}O$ was the amount that the proportion of ^{18}O in a sample deviated from that in an ocean water standard and became more negative with increasing temperature, so he was accustomed to looking at what amounted to inverted temperature curves. He turned the Bristol group's unsaturation index graph around and proceeded to explicate each of the little squiggles in terms of the waxing and waning glaciers that he and others had been seeing in the $\delta^{18}O$ data from sediment cores around the world.

In Bristol they were still working to calibrate their unsaturation index with seawater temperature, so they couldn't put precise sea surface temperatures on their graph. In the months that followed the meeting with Sarnthein, they finally settled on the unsaturation index that gave the best calibration, would increase with temperature and give values that conveniently varied between 0 and 1. They named it U^K_{37}—U for unsaturation, K for ketone (with a nod to national vanity), and 37 because they'd limited it to the 37-carbon compounds—and defined it as:

$$U^K_{37} = (C_{37:2} - C_{37:4}) / (C_{37:2} + C_{37:3} + C_{37:4}),$$

where $C_{37:2}$ was the concentration of compounds with 37 carbons and 2 double bonds, and so forth. When Geoff displayed Sarnthein's $\delta^{18}O$ data, plotted right-side up, along with U_{37}^K plotted upside down, at the next international organic geochemistry meeting, the excitement in the room was palpable: the ice-age climate oscillations were as apparent in the biomarker data as they were in the foram data, which, of course, were registering the combined effects of changing temperatures and changing ice volume. It was the 12th International Meeting on Organic Geochemistry, held in Jülich in the fall of 1985, and it was packed: what had been a scattered handful of maverick chemists and geologists when Geoff extracted his first hunk of Green River shale two and a half decades before was now a community of organic geochemists some 400 members strong. Purely by chance, Geoff had been scheduled to present the first paper of the conference, and for the rest of the week people would stop him in the hall to remark on the Bristol-Kiel cooperative and offer congratulations and encouragement: the alkenones had set the community abuzz. There was a sense that organic geochemistry had crossed a new threshold and was ready to take on some of the most interesting and difficult questions in the earth sciences.

Complete results from the study appeared in a 1986 paper in *Nature* under the audacious title "Molecular Stratigraphy: A New Tool for Climatic Assessment." As intended, the paper captured the attention of the oceanographic community, including the once-disdainful marine geologists and paleoceanographers. For their part, the marine chemists at Woods Hole had been interested in the alkenones since John Volkman had visited them in 1980, well before anyone had any idea that they might be useful as a paleothermometer. Max Blumer had died in 1977, but his legacy of marine organic chemistry at Woods Hole lived on in the work of John Farrington, Bob Gagosian, and Cindy Lee, who had been analyzing everything from hydrocarbons and sterols to the cycling of amino acids in the water. When Volkman visited, Stu Wakeham had recently joined their ranks and was developing new and improved methods of collecting suspended and sinking detritus from different depths. The group immediately added the alkenones to its repertoire and found it to be the most stable of the compounds made by marine organisms that they had come across. Not long after the U_{37}^K paleothermometer was unveiled, Fred Prahl, who had worked on the coccolithophore feeding experiments in Plymouth, also did a stint at Woods Hole, teaming up with Wakeham to determine U_{37}^K values from the alkenones in sinking and suspended particles. Plotting these against temperatures in the overlying surface water from a range of environments produced a surprisingly straight line that was remarkably close to the one determined from the alkenones of Emily grown at different temperatures in the laboratory. This served to calibrate the new paleothermometer, though efforts to extend, refine, and validate that calibration would continue for another decade.

Geoff had started the Bristol group collecting data for a definitive calibration of U_{37}^K as soon as he'd seen Marlowe's first graphs. But he'd reasoned that such a calibration should be based on a broad survey of marine sediments—where seasonal temperature changes, biological effects, and diagenetic loss of the alkenones would all be empirically averaged out—and it took years to acquire the requisite surface sediment samples from all the world's oceans. In the meantime,

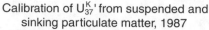

Calibration of $U^{K'}_{37}$ from suspended and
sinking particulate matter, 1987

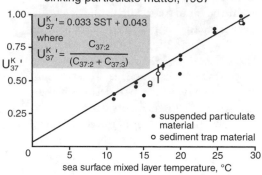

$$U^{K'}_{37} = 0.033 \ SST + 0.043$$

where

$$U^{K'}_{37} = \frac{C_{37:2}}{(C_{37:2} + C_{37:3})}$$

• suspended particulate material
○ sediment trap material

sea surface mixed layer temperature, °C

the group tried to determine if any extant organisms other than *Emiliania huxleyi* made alkenones, and what the source of the compounds in the sediments might have been before Emily came on the scene. Marlowe did a study of cores that extended back into the middle of the Eocene epoch to about 45 million years ago, comparing alkenone distributions with the microfossil records in the same sediments. Paul Farrimond, another student of Geoff's who was working on a completely different project, forgot to turn the chart recorder off after a GC run and discovered quite by accident that alkenones were present in the 100-million-year-old Cretaceous sediments he was analyzing. The hunt for alkenones in living algae eventually revealed their presence in a handful of closely related species, all, like the coccolithophores, members of the Haptophyta division, and, with the exception of Emily, uncommon in contemporary oceans, or limited to coastal regions. Later, in the mid-1990s, Volkman and his colleagues in Australia would find alkenones in *Gephyrocapsa oceanica*, which can be as plentiful as Emily in tropical or marginal seas. This may well have been a main source of alkenones in the past, as *Gephyrocapsa* coccoliths comprise the dominant form of haptophyte fossil for the three million years or so before Emily took over the show.

Volkman says there was a period after he left Bristol when he just couldn't get away from the alkenones, even when it seemed most unlikely he would encounter them. By the mid-1980s, he was back in Australia with a post at the Marine Research Division of the Commonwealth Scientific and Industrial Organization, and he and one of his students were studying the ecology and history of a large saline lake in Antarctica, analyzing lipids and pigments in its sediments. The Southern Ocean is one of the few regions of the world where Emily doesn't live, so though the lake was quite saline and had been connected to the sea in the past, they were surprised to discover that the sediments deposited at various times in its history were loaded with alkenones—so loaded, Volkman says, that he and his student were able to obtain their infrared spectra. The compounds with four double bonds predominated, as might be expected in such a cold region, but the species responsible couldn't be identified, and when they tried to get a record of the lake's temperature from the U^{K}_{37} proxy, it gave unrealistic values—highlighting the necessity of defining limits to the proxy's use in areas and periods of time when Emily was not the main alkenone producer. More importantly, the infrared spectra revealed a hitherto unnoted quirk in alkenone molecular structures, which had, to date, still not been confirmed by synthesis: of the two geometric forms possible for double bonds, *cis* and *trans*, the alkenone double bonds seemed to be

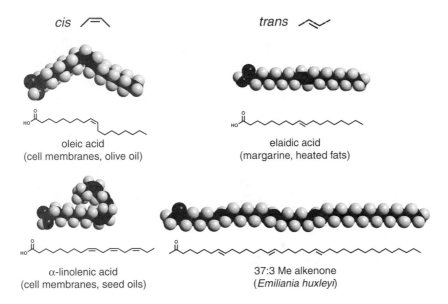

cis and trans double bonds in fatty acids and alkenones

cis

trans

oleic acid
(cell membranes, olive oil)

elaidic acid
(margarine, heated fats)

α-linolenic acid
(cell membranes, seed oils)

37:3 Me alkenone
(*Emiliania huxleyi*)

trans, a very unusual configuration for straight-chain biological lipids and con-fined to a few special cases, such as in carotenoid pigments. Maxwell had a post-doc in Bristol trying to synthesize the *cis* alkenones in the laboratory, so Volkman immediately sent him a note. Sure enough, when they changed tactics and accom-plished the difficult syntheses, they duplicated the compounds made by algae and found in the sediments, replete with all-*trans* double bonds—but how the algae themselves accomplish this synthesis remains a mystery to this day.

Until very recently, it wasn't even clear where in the cell the alkenones reside. The Bristol group speculated that they were bound in the cell membrane, and that the change in saturation was a means of regulating its viscosity or fluidity at a wide range of temperatures, as had been observed for fatty acids in some bacte-rial membranes. This speculation has been repeated as if it were documented fact in just about every alkenone paleothermometer reference since 1986, but there were, in fact, few investigations into its verity. Sargent and Maxwell had wanted to set up a graduate student project to study the biochemistry of the alkenones, but there just wasn't funding for such basic research in the 1980s and 1990s. For his part, Sargent says he never subscribed to the idea that the alkenones were membrane-bound lipids. His Scottish brogue makes everything sound like a question, but there's nothing questioning in his opinion on the subject. "That's just rubbish," he says bluntly.

Sargent claims the alkenones are too long to fit into the membrane, for one thing. And whereas *cis* double bonds in fatty acids give them a kink that fits loosely into the flexible membrane, the *trans* double bonds make the molecule into a straight, rodlike structure that seems an unlikely choice for a membrane lipid. Furthermore, Sargent says, there is no way that the large amounts of alkenones found in Emily cells could be bound to the cell membrane. There must be globs of them inside the cell, enclosed in little vesicles of some sort—quite possibly the same oil globules he'd heard described by the Marine Biological Association biologist back in 1971. Indeed. Not long after I talked to Sargent in 2004, more than two decades after the discovery of the alkenones, biochemists began making concerted attempts to ascertain their position in the cell and found that they are not, in fact, bound in the cell membrane as most of the geochemists had been assuming. Just as Sargent expected, the entire series of alkenones and related compounds appears to be contained in vesicles of oil *inside* the cell, and there is now significant evidence that they are used for energy storage much like more conventional fat molecules are used in other algae. But why, then, is the degree of unsaturation temperature dependent? We seem to have returned to first base with this question.

Certainly understanding the biochemistry of the alkenones would help in their use as a paleotemperature proxy, which is ultimately dependent on the consistency of that biochemistry during the evolution and diversification of alkenone-producing species. The sources of the alkenones found in sediments that predate Emily's emergence remain a matter of speculation, and laboratory culture studies with the handful of extant alkenone-producing species indicate that the temperature dependence of alkenone unsaturation can vary significantly among species, and even between strains of algae obtained from two different geographical regions. There are also concerns that the temperature dependence might be nonlinear at lower temperatures or that triunsaturated alkenones degrade more quickly than diunsaturated.... And confusion reigns about whether the alkenone signal represents average annual temperatures, or is skewed toward spring and summer when the algae are at their peak, or whether, on the contrary, it is biased toward the low-nutrient periods in the surface water when the algae produce more alkenones to store energy they can't use....But amid all these ifs, buts, and maybes—and now, without even a working hypothesis to explain the correlation between water temperature and the degree of unsaturation to begin with—compilations of average annual sea surface temperatures and U_{37}^K values from surface sediments around the world yield a consistently linear relationship, and after years of challenges and recalibrations, the equation that is still used to assign average annual sea surface temperatures to measurements of U_{37}^K in ancient sediments is almost identical to the one first derived from sediment trap data at Woods Hole in 1987. And not long after its introduction, it started to yield new—and surprising—information about past climates, information that has played a part in a virtual paradigm shift in ideas about how the earth's climate system operates.

By the end of the 1980s, the idea that climate change during the past hundred million years was driven primarily by plate tectonics and the earth's orbital geometry had become a paradigm, supported by a large body of evidence from ocean

Worldwide calibration of $U_{37}^{K'}$ in surface sediments (1998)

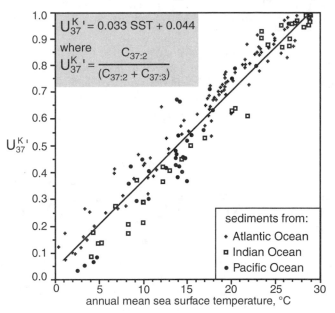

sediments. The picture that had emerged was of a climate system that changed very gradually, over millions of years, in response to factors associated with the slow movement of the earth's tectonic plates and concomitant uplift, subsidence, and volcanism—shifting continents, changing topography and ocean currents, and changing concentrations of atmospheric CO_2. The earth's orbital cycles caused average global temperatures to oscillate, with frequencies of 40,000 and 100,000 years, around the mean determined by this gradually changing climate system, which over the past 60 million years had become progressively more sensitive, with the oscillations growing in amplitude.

Abundant geologic evidence indicated that as long as the earth was free of ice, as it appears to have been throughout the long Cretaceous period and perhaps even back to the beginning of the Mesozoic era some 248 million years ago, the climate's response to orbital variations was limited. But when ice began to form in Antarctica about 33 million years ago, after a prolonged period of gradual cooling in the Eocene, subtle 40,000-year oscillations in global climate began to appear in the geologic record, increasing in amplitude as the amount of ice increased. Presumably, conditions on Earth were now such that the small variations in the angle and intensity of incoming solar radiation were triggering the feedback effects imagined by Milankovitch: the tendency of the ice to reflect heat back into space, or alterations in ocean circulation due to the changing salinity and density of ocean surface waters as ice melted or formed. When ice began to form in the northern hemisphere a little more than three million years ago, the oscillations increased dramatically in amplitude, first at 40,000-year and later at 100,000-year intervals, and eventually gave way to the bimodal, glacial–interglacial cycles that have characterized the earth's climate for the past million years.

Throughout the development and reign of human beings, according to this paradigm, the climate had alternated between long, steady periods of cold and warm, separated by transition periods where glacial ice advanced or retreated, sea levels fell or rose accordingly, and temperate sea surface temperatures decreased or increased by as much as 6°C over a relatively short, 10,000-year period. So went the most popular version of the story. But despite the clear association of the Pleistocene ice ages with the Milankovitch cycles, and despite evidence of feedback effects in the earth's internal climate system that could have been triggered by small variations in the distribution and magnitude of solar heating, no one had come up with a plausible combination of feedbacks that produced *large* enough effects to account for the switch from ice age to interglacial period, or vice versa.

Most paleoclimate analyses required tediously picking through each sample to find well-preserved fossil shells, which were sometimes few and far between. And since some species of foraminifera were planktonic and some lived on the seafloor, the only way to distinguish the temperatures at the sea surface, which was in contact with the atmosphere, from the much colder and less variable deep sea temperatures was to pick out a chosen species one by one. Alkenone analyses, however, could be performed rapidly on small samples of bulk sediments. In coastal zones or other areas with lots of nutrients and high primary productivity in the surface waters, the sediments accumulated fast enough that one could obtain such samples at relatively short time intervals and, potentially, obtain a more detailed record of climate change during the Pleistocene than had to date been possible.

In 1991, scientists from the Bristol-Kiel alliance obtained just such a record from alkenone analyses of sediments in a core taken off the northwest coast of Africa, where they were able to sample at one- or two-centimeter intervals, about every 75–200 years, over part of the past 650,000 years. The last three ice age cycles were readily apparent in the U_{37}^{K} data, with glacial to interglacial transitions beginning 340,000, 250,000, 150,000, and 25,000 years ago. But superimposed on this now-familiar climate curve were rapid fluctuations in sea surface temperature, sudden leaps and dips of several degrees that occurred within 300 to 1,000 years—fluctuations that clearly weren't triggered by Milankovitch's orbital cycles, and less so by tectonic factors. So surprising were the results, Geoff says, that when they submitted their paper to *Nature*, the reviewers objected, saying that temperature changes could not possibly be *that* fast, and when the paper did appear, duly fortified with more data, it was received with some skepticism. Hypothesis-loving geologists were loathe to sully their neatly oscillating oxygen isotope curves and eloquent, if incomplete, astronomical explanation with such baffling patterns of temperature swings—especially when they came from an unorthodox method.

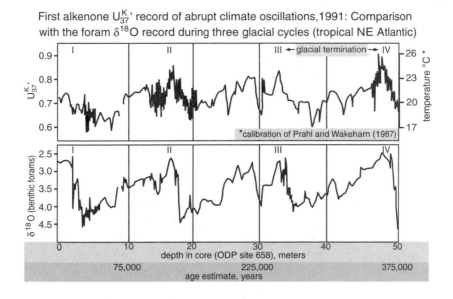

First alkenone $U_{37}^{K'}$ record of abrupt climate oscillations, 1991: Comparison with the foram $\delta^{18}O$ record during three glacial cycles (tropical NE Atlantic)

There had, in fact, been some scattered evidence of abrupt climate changes in the North Atlantic during the last glacial period, but it had hitherto been dismissed as a regional phenomenon. The alkenone evidence, however, came from a tropical region and indicated that the changes were global in nature...and fast on its heels came verification and clarification from other methods and cores. Analyses of the ^{18}O content of ice samples from cores drilled deep into the ice sheets covering Antarctica and Greenland—ice that had accumulated in annual layers during tens of thousands of years—showed a similar pattern of rapid change. And as the Ocean Drilling Program (the DSDP's successor in the 1980s and 1990s) retrieved more sediment cores where high-resolution studies were possible, further alkenone and improved ^{18}O analyses of foram shells confirmed that abrupt climate oscillations had indeed affected the entire globe.

It was fast becoming clear that the long glacial and interglacial periods of the Milankovitch climate cycles were not as stable, and the transitions between them were not as decisive, as hypothesized. Rather, the earth's climate had flickered back and forth between warm and cold modes every 2,000 years or so, with sea surface temperatures dropping some 4°C within a few hundred years, even in tropical regions. Joan Grimalt, who participated in the initial development of the U_{37}^{K} index when he was a postdoc in Bristol, went on to refine its use and application in paleoceanographic studies of the Mediterranean and North Atlantic. One recent study by his Barcelona-based group shows that the climate flickers occurred throughout the past 250,000 years and that they were generally more frequent during glacial than during interglacial periods, and most pronounced during the transitions from interglacial to glacial. Indeed, after a relatively long

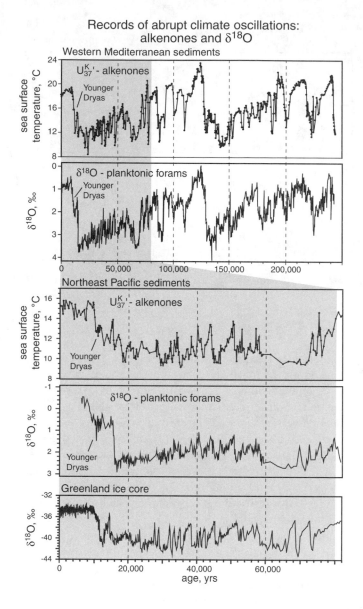

Records of abrupt climate oscillations:
alkenones and $\delta^{18}O$

period of interglacial stability like the current one, the temperature swings could be so dramatic as to flip the climate from interglacial to glacial and back again in the space of a few hundred years.

Still more detailed investigations of Holocene sediments, and more and better ice core measurements from Greenland, have brought the paleoclimate record into historical time, where the only abrupt climate change was much more subdued than those that punctuated the last three glacial cycles—but nevertheless

From paleoclimates to historical climates: an alkenone ($U_{37}^{K'}$) record of sea surface temperatures off the north coast of Iceland

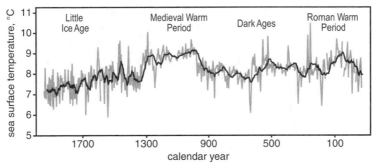

wreaked havoc on the developing civilizations of Northern Europe. Around the turn of the fourteenth century, a 300-year period of mild weather and stability known as the Medieval Warm Period—a time when grapes were grown in England and the Vikings thrived in Greenland—ended abruptly, and the climate shifted gears. Within a few decades, Greenland's coastal settlements were buried under its advancing ice cap, and the seas around Iceland were clogged with ice for months at a time, while the Thames River was freezing over in the winter. Dutch artists immortalized the era with romantic winter landscapes of ice-skaters frolicking on Amsterdam's canals, but the four-century period known as the Little Ice Age was, in fact, a time of erratic weather, devastating storms, unpredictable crops, famine, epidemic disease, and die-out in Europe's established population centers. As Grimalt points out, the climate shift in the fourteenth and fifteenth centuries was a minor blip compared to the abrupt changes in the global climate system that left their mark in the sediments of the world's oceans during the last three glacial cycles. The most recent, known as the Younger Dryas, occurred about 11,000 years ago, as Earth was emerging from the last ice age, and flipped the global climate system back into glacial mode for another thousand years. And yet statistical analysis of the pacing of the Little Ice Age suggests a common external trigger for it and the earlier, more severe, abrupt changes.

This was a more precariously balanced climate system than anyone had suspected. As paleoceanographers began exposing and deciphering its record in the 1990s, the climate scientists who had been trying to model the physical and chemical processes that determine climate, making ever more complex calculations on ever larger computers to predict the effects of human additions of greenhouse gases to the atmosphere, started paying closer attention and using the long records of Pleistocene climate change to tune and calibrate the models. What tripped the switch and flipped the climate from one state to another so abruptly? What determined the extent and severity of the changes? Was it possible that man's intemperate appetite for black gold—that burning hundreds of millions of years worth of Nature's alchemy to CO_2 in a couple of centuries—might trigger one of these sensitive features of the system and flip the climate suddenly into an unexpectedly extreme condition?

The climate scientists had found that when they increased the amount of atmospheric CO_2 in their models, the initial warming in the Northern Hemisphere resulted in a complete reorganization or slowing of the deep sea circulation, which, in the contemporary ocean, functions like a giant conveyor belt, moving large amounts of heat from the tropics to high latitudes. This conveyor takes more than a thousand years to make its round and depends on the sinking of particularly dense water—cold and saline—at certain strategic points near the poles, its movement along the bottom of the deep ocean basins, and eventual upwelling and return flow in warmer, wind-driven surface currents.

Deep ocean circulation had long been considered one of the more stable features of climate, gradually readjusting itself over millions of years to fit the changing geography of the continents and topography of the seafloor. Theoretically, however, it could function in several distinctly different modes even within the constraints of the current geography and topography. If the water in those strategic locations near the poles changed in density, then the conveyor might slow or strengthen; if, for example, the water warmed or became less salty, the conveyor might shut down altogether, leading, ironically, to dramatic cooling in the North Atlantic and Europe. What had the climate scientists worried was that the new paleotemperature evidence from ice and sediment cores indicated that the great ocean conveyor had, in actuality, changed modes or pace, abruptly and repeatedly, in the past few hundred thousand years.

The precise cause and nature of these changes in deep sea circulation were unclear, but they appeared to be tied to the most dramatic of the abrupt climate

Ocean circulation: the great conveyor belt

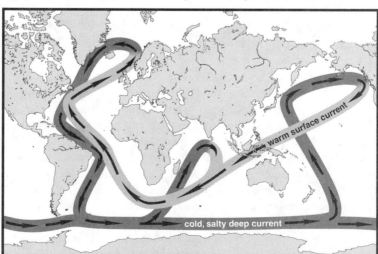

flickers and to the glacial–interglacial cycles, part of a complex web of biological and chemical changes that were linked to the earth's orbital cycles and, apparently, to the internal cycles of its moody sun, amplifying or modulating the effects of both. One of the most tangled and least understood strands of that web was one that geologists had generally failed to recognize until the 1970s: the productivity of microscopic algae living in the sunlit surface waters of the world's oceans. In most regions of the open ocean, there is little mixing between the warm lens of surface water and the cold deep water, such that essential phytoplankton nutrients such as nitrate and phosphate become depleted at the surface, where all the photosynthetic action is, and enriched in the deep water where bacteria break down the sinking detritus and remineralize the organic matter. Nutrients are replenished to the surface waters by dust-bearing winds or river runoff from the continents, or by upwelling of the nutrient-rich deep water. The latter occurs along certain coastlines, where the winds create offshore surface currents that pull the deep water to the surface, and along the equator, where divergent surface currents have a similar effect. Changes in wind direction or intensity, changes in seasonal weather patterns, or changing sea level and the resultant shifts in continental runoff during the waxing and waning of the ice sheets—all would have affected availability of nutrients in the surface waters. Changes in the availability of nutrients meant changes in marine productivity which, according to the so-called "biological pump" hypothesis, might have amplified all these effects by generating fluctuations in atmospheric CO_2 with associated warming or cooling as algae "pump" CO_2 out of the surface waters and convert it to organic matter, some small fraction of which persists in deep waters or sinks to the seafloor and is buried.

Paleoclimatologists and oceanographers have yet to decipher all the patterns of cause, effect, and feedback that comprise the earth's climate history, but they have made headway in documenting some of its many components, either from analyses of the Greenland and Antarctic ice cores, where air bubbles trapped in the ice provide a record of the concentrations of CO_2 and methane in the atmosphere, or from the longer mineral and fossil record preserved in a growing inventory of deep sea sediment cores. In the 1990s, they were

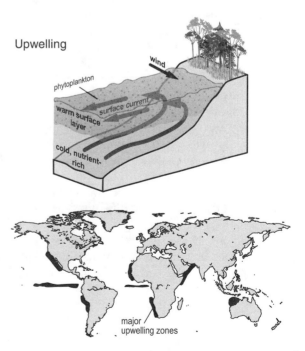

Upwelling

phytoplankton

wind

warm surface layer

surface current

cold, nutrient-rich

major upwelling zones

joined in their endeavors by a new generation of organic geochemists: buoyed by the success of the alkenone paleothermometer and the growing realization that life was not a passive bystander, but a shaping force of the earth's atmosphere, oceans, and rocks, the so-called "stamp collectors" were gaining a foothold in earth science and oceanographic institutions and were no longer relegated to the fringes of chemistry departments or constrained by the narrow interests of the petroleum industry. Jan de Leeuw moved his laboratory from Delft to the Netherlands Institute of Sea Research on the North Sea island of Texel, and "the Dutch group" became known as the NIOZ group. Jürgen joined the Institute of Chemistry and Biology of the Marine Environment at the University of Oldenburg; John Hayes, who was busy developing what would prove to be the most important new analytic technique for biomarker research since the GC-MS, created an official biogeochemistry group under the auspices of the geological sciences department at Indiana University; Geoff moved on to emeritus status at Bristol and joined the Woods Hole faculty as an adjunct scientist; and James Maxwell stayed put as Bristol's guiding light amid a steady flow of students and postdocs from oceanography and earth science departments around the world. Decades of molecular exploration were beginning to pay off. Hundreds of molecular fossils had been identified in sediment extracts, some semblance of a lexicon for interpreting them had been developed, and one could begin to use them, along with paleontological and mineral-based information, to reveal new facets of earth history, including how marine productivity and ecosystems have changed over the past few million years of oscillating climates.

One of the keys to detecting and understanding changes in marine productivity lies, ironically, in being able to identify organic matter that came from the continents, and here, of course, Geoff's old buddies the leaf wax *n*-alkanes play a major role. Darwin had noted in *The Voyage of the Beagle* that large amounts of wind-borne dust could be found far out to sea in the Atlantic Ocean, an observation that had always intrigued Geoff. By the early 1970s, when he started asking around about getting dust samples to analyze, oceanographers and geologists had already devoted quite a bit of effort to studying its movements and mineral properties, but they still couldn't always pinpoint its origin. Wind-borne dust from desert areas was composed mostly of mineral matter, but it could also come from exposed soils and dried lake beds, which contained some component of highly oxidized organic matter, or from vegetated areas where the wind had picked up pollen and plant spores. The clay particles and trace metals in the dust provided some information about the source of the mineral matter and the patterns of the winds; likewise, the different types of plant pollen and spores might have something to say about the vegetation the winds had passed over. Geoff's first question was, as usual: What sort of organic compounds are in there? Shouldn't there be some molecular remnants of land plants, and then could they tell *which* land plants? Quite fortuitously, around this same time Al Burlingame sent him a student who had been working on the DSDP cores in Berkeley. Bernd Simoneit suspected that much of the organic matter he found in the marine sediments, even in the middle of the Atlantic, had actually come from the continents, and was curious about its provenance. He also happened to be married to an airline stewardess who could get him free tickets to all sorts of exotic places, so he was able to fly all over the world to set up dust-collecting stations and visit the odd handful of scientists who had assembled collections for one reason or another.

Not surprisingly, the long-chain *n*-alkanes, *n*-alcohols, and fatty acids from leaf waxes were among the most prominent organic constituents of this material, more prominent even than they were in river-borne sediments. Bob Gagosian and colleagues at Woods Hole studied collections of dust from the South Pacific, with a similar result, finding these leaf wax compounds in dust collected from an atoll more than 5,000 kilometers from the nearest land source. Apparently, the wax particles could be blown off the leaves by strong winds, or rubbed off by the abrasive mineral particles in the dust. But the organic matter that came from the soils and dried lake beds had already undergone extensive decomposition and oxidation—more so than that in sediments from rivers and runoff, which often contain whole pieces of decaying plant matter. When found in marine sediments, then, the refractory leaf wax hydrocarbons—in particular, the C_{29} and C_{31} *n*-alkanes and the C_{28} *n*-alkanol—can give a measure of how much organic matter has been transported from the continents to the ocean, either blown out to sea on the wind or deposited along continental margins by rivers. Simoneit's analyses of the leaf wax hydrocarbons in dust, and the subsequent years of sediment analyses, also revealed recognizable geographical patterns in the chain-length distributions of these long-chain compounds that seemed to correlate with the sources of the dust. And, finally, in recent years it has emerged that the chain-length distributions of leaf wax hydrocarbons do systematically reflect plant type, though at a much more general level than Geoff had searched for in his initial studies of Canary Island succulents: plants adapted to relatively warm, dry environments make more of the longer chain compounds, and those adapted to cool, humid environments make more of the short-chain ones.

All the interesting and potentially informative triterpenoids that have been isolated from plants seem to be largely absent, or present in trace amounts in dust, presumably having succumbed to oxic degradation in the soils. But what about in the sediments deposited by rivers and runoff along the continental margins? This was something Jürgen wondered about for years. Of the many pentacyclic triterpenoid structures that had been isolated from higher plants, the only carbon skeleton that showed up regularly in ancient sediments and oils was oleanane, the reduced analogue of β-amyrin. Oleanane, in fact, was inexplicably abundant and widespread. Natural products chemists had isolated β-amyrin from the leaves of some flowering plants—but they had also isolated lupeol, friedelin, taraxerol, and a host of other pentacyclic triterpenoids. Why so much oleanane, then, and so few of the other carbon skeletons that one might expect? Jürgen started puzzling about this during his first years in Dietrich Welte's lab in Jülich, but not until the late 1980s, when he started analyzing Ocean Drilling Program sediments from Baffin Bay, did he and his colleagues stumble on an answer. Here, finally, buried hundreds of meters below the surface in the thick layers of sediments deposited by rivers and runoff from Greenland and Canada, they found some of the missing land plant pentacyclic triterpenoids—along with a plethora of triterpenes and related aromatic compounds that showed every indication of having been produced by dehydration and reduction of the biological alcohols and ketones, much as had been observed for sterols. Jürgen was working with two postdocs at the time, Lo ten Haven from NIOZ and Torren Peakman from Bristol, and they were struggling to identify all the triterpenes in the gas chromatograms in time for the ODP deadline on postcruise reports when they ran the hydrocarbon fraction of one of their extracts through a molecular sieve to remove the *n*-alkanes, one of which was coeluting with the triterpenes. To everyone's surprise, the sieve not only removed the *n*-alkanes, but changed the composition of the triterpenes: the compound they'd been trying to quantify, taraxerene, disappeared, and the amount of oleanene increased. A survey of the literature revealed laboratory experiments that showed how taraxerene and a number of other triterpenes were converted to oleanene in the presence of an acid catalyst—in this case, it seemed, the acidic zeolite of the molecular sieve. In a mechanism similar to the one that had been proposed for the rearrangement of sterenes to diasterenes, addition of a proton to the double bond generates a reactive charged intermediate wherein the double bond of the triterpene is able to migrate around the ring structure, with concurrent rearrangement of the attached methyl groups.

taraxer-14-ene   reaction intermediate   olean-12-ene

By the late 1980s, the Dutch had started doing theoretical molecular mechanics calculations to determine which of the sterenes produced by rearrangement of steroids were the most stable and energetically probable, and application of their findings to the triterpenes in the Baffin Bay sediments indicated that acid-catalyzed rearrangement of most of them would lead to one product: olean-12-ene. Careful quantitative analysis of the various triterpenes in the layers of sediment from the Baffin Bay core then revealed the answer to the oleanane riddle: in the relatively acidic clay-mineral–rich sediments generated by erosion of landmasses, triterpenes formed during the first stages of diagenesis converged on olean-12-ene, which further experiments indicated would slowly be reduced to the same mixture of oleanane isomers observed in more mature organic matter, only the most stable of which would survive in crude oils. It was a wonderfully eloquent bit of chemistry to discover in Nature's messy sedimentary laboratory, and it reinforced what the Dutch had been seeing with their theoretical studies: in early diagenesis, when kerogen didn't confound the issue, the reactions in the sediments could be predicted from first principles of chemical reactivity and thermodynamics.

Solving the oleanane riddle: proposed scheme for the
diagenesis of pentacyclic triterpenoids from plants

If, as appears to be the case, triterpenoids are largely absent from wind-borne dust, but present in river runoff, oleanene found in open ocean sediments can be used as an indicator that river-borne sediments from the continental shelf have been transported out to sea by bottom currents or submarine landslides. One even wonders if oleanene in coastal sediments might be used together with the *n*-alkanes to gauge changes in the relative input of organic matter from rivers and winds, or to detect a shift in the ecosystems on land, from triterpenoid-producing conifers, say, to grasses or deciduous forest. Another compound that is largely absent from dust and may be used to distinguish river-borne from both algal and wind-borne organic matter in marine sediments is lignin, the large polymer that provides structure and rigidity to land plants, most notably in the wood of trees. Unlike the steadfast *n*-alkanes, however, these compounds present serious problems with quantification and interpretation, as lignin breaks down into a dizzying array of difficult-to-analyze phenols, and the pentacyclic triterpenoids may well vary in their rate of survival during the sojourn from plant to river.

In an attempt to estimate past changes in marine primary productivity directly, the Bristol team returned to the roots of the biomarker concept and went straight to the heart of the matter: the green, light-absorbing chlorin heart of chlorophyll that Alfred Treibs had so cleverly linked to the red porphyrins in petroleum in the 1930s. Unlike phytol from chlorophyll's tail, chlorins break down rapidly in the presence of light and oxygen, and on land they are destroyed in the dying leaves or soil, long before they can be transported to marine sediments. So the chlorins in marine sediments come only from the chlorophyll of marine organisms—algae or photosynthetic bacteria—unlike phytane, which could have come from land plants or even, for that matter, from the lipids of nonphotosynthetic microbes. Chlorins are also rapidly broken down in the marine environment, but some fraction persistently reaches the sediments and can be preserved for millions of years before being transformed to the refractory porphyrins found in ancient rocks and mature oils. The problem with trying to use chlorins as a measure of marine productivity in deep sea sediment cores from the Pleistocene—besides that they break down if samples are exposed to light or oxygen—is that they aren't volatile enough to run on a GC-MS system. But in recent years Maxwell and his students have developed a method that uses high-performance liquid chromatography (HPLC)—a modern, automated version of the old column chromatography, where the stationary phase is a solid powder packed into a narrow steel column, the solvents are pumped through at high pressure, and both the rate and the mixture of solvents can be carefully controlled and gradually changed. Using such a system, chlorins are easy to separate from the mixture in a sediment extract and can be detected and quantified by an ultraviolet or fluorescence detector connected to the column outlet. The absolute concentrations of chlorins in marine sediments, then, can be used as an index of increases and decreases in marine primary productivity over the past few million years.

Specific information about which photosynthetic organisms are doing the producing and how the ecosystem changes over time comes from some of the many algal biomarkers that have been identified, often combined with paleontological inventories of the fossil shells and cysts. The amounts of alkenones can provide an index of coccolithophore activity, and dinosterol provides a relatively unambiguous record of dinoflagellate activity. Diatoms, which comprise almost half of the world's total marine productivity in the contemporary ocean, are somewhat more difficult to pin down. Like coccolithophores and dinoflagellates, they form hard encasements that can persist as fossils in the sediments. But the delicate diatom frustules are made of silica, which is even more prone to disintegration than the calcium carbonate coccoliths or the organic dinoflagellate cysts, and their fossil record is even more patchy and undependable. Nor can they, to date, boast a "perfect" biomarker. Numerous sterols have been found in various groups of diatoms, but they all overlap to some extent with sterols made by plants and other algal classes: brassicasterol has also been found in coccolithophores,

24-methylenecholesterol—with its double bond between the 24- and the 28-carbon—has been found in some dinoflagellates, fucosterol in brown algae, β-sitosterol in higher plants. In Pleistocene marine sediments, where the sterols have yet to succumb to diagenesis, brassicasterol is often used as an indicator of diatom activity, but it appears to be made by only a few of the many diatom genera and, like all of these compounds, is best measured as part of an array of biomarkers. In older and more mature sediments where the sterols have been reduced to their sterane skeletons, only dinosterane with its distinctive 4-methyl group preserves its source-specific structure, but one can get a rough idea of the relative contributions of higher plants, marine algae, and zooplankton from the side-chain substitution patterns and relative amounts of 29-carbon, 28-carbon, and 27-carbon steranes, respectively. The carotenoid pigment fucoxanthin is particularly abundant in diatoms, and under some conditions its cyclic components are preserved in the relatively stable and distinctive loliolides, which can thus provide clues to periods when diatoms were particularly productive.

Another ambiguous but potentially useful biomarker for diatoms was identified by Jaap Sinninghe Damsté and the NIOZ group when they went on a natural products offensive and started growing their own cultures of algae. Among the hundreds of species of diatoms they cultured, they finally identified a likely source for the alkane diols that had been frustrating de Leeuw for almost two decades—turning up in recent and ancient sediments and showing every promise of being great biomarkers, if one could only figure out *what*, exactly, they were markers for. Volkman had officially liberated the series of compounds from orphan status in 1992 when he identified some of the homologues in several species of yellow-green algae. But this was obviously not the whole story, because these yellow-green algae were generally restricted to brackish and freshwater environments and not likely to explain the widespread occurrence of alkane diols in sediments from open ocean environments. The NIOZ group came a little closer with the alkane diols in its cultures of marine diatoms from the genus *Proboscia*, which is common enough to explain their ubiquity in ocean sediments—except that the compounds in *Proboscia* are C_{28} and C_{30} homologues with their mid-chain hydroxyl groups at C-14, whereas the most abundant alkane diols in marine sediments are C_{30} and C_{32} compounds with hydroxyl groups at C-15. There have already been some attempts to use alkane diols as biomarkers, but until we have a more complete understanding of the distributions of different homologues in diatoms and other organisms, such work remains speculative at best.

Another group of long-orphaned compounds that has finally been traced to diatoms and seems to offer a more reliable indicator of their past activities is that comprised of the T-shaped highly branched isoprenoids, the HBIs. Steve Rowland, who was a student of Maxwell's in the early 1980s, says he got interested in these compounds when he was at Bristol simply because they turned up in everyone's sediment extracts, but no one knew where they came from. He and his first student at the University of Plymouth identified the structures of the most commonly occurring HBIs in the mid-1980s, but they had no idea of their sources until 1994, when Volkman finally isolated them from two species

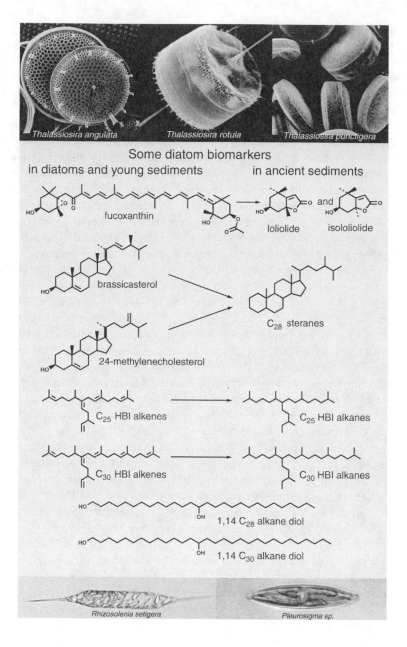

Some diatom biomarkers

in diatoms and young sediments   in ancient sediments

fucoxanthin

loliolide   isololiolide

brassicasterol

C_{28} steranes

24-methylenecholesterol

C_{25} HBI alkenes C_{25} HBI alkanes

C_{30} HBI alkenes C_{30} HBI alkanes

1,14 C_{28} alkane diol

1,14 C_{30} alkane diol

Thalassiosira angulata   *Thalassiosira rotula*   *Thalassiosira punctigera*

Rhizosolenia setigera   *Pleurosigma sp.*

of diatoms—20 years after they were first detected in sediments. Rowland and his team in Plymouth have now isolated the full contingent of C_{25} and C_{30} HBIs from a variety of diatoms and are finally able to account for the distribution and abundance of these compounds in sediments. The HBIs are apparently limited to four of the hundreds of genera of diatoms analyzed, but these four happen to include one of the most prominent diatom groups in the modern ocean and in the marine fossil record of at least the past 50 million years. The HBIs found in the algae are generally unsaturated, with two to four double bonds at different positions, and the distributions vary among species and depend, to some extent, on growth conditions. In the sediments, the double bonds are susceptible to rearrangement and cyclization during early diagenesis, but the distinctive HBI skeleton, with its sturdy T-shaped isoprenoid construction, is particularly difficult to break down, and the alkane fossil molecules, as well as a host of compounds formed by the incorporation of sulfur at the double bonds, can persist for tens of millions of years, offering a good general indication of changes in the productivity of this large, important group of algae. And, though most of the more specific information that might be reflected by distributions of the compounds is lost with the rearrangement of their double bonds, the Plymouth group has discovered one particular HBI that may provide clues to an elusive element of climate change in the relatively recent past.

The physiological role of the HBIs in diatoms remains a mystery, but having noted that the amount of unsaturation tended to decrease with decreasing growth temperature, the Plymouth group hypothesized that diatoms living within sea ice in polar regions might produce an HBI with just one double bond. These researchers proceeded to identify just such a monounsaturated C_{25} HBI in Arctic sea ice and associated sediments, and a convincing set of data now indicates that it is produced exclusively and in great abundance by sea ice diatoms, persists in the sediments for at least 10,000 years—and might conceivably be developed into a proxy measure for the amount and whereabouts of sea ice, one of the key variables in past, and future, climate change.

Despite uncertainties about the sources and distributions of many of the algal biomarkers that have been identified to date, Stu Wakeham's ongoing studies of lipids in sediment trap samples from different regions and from various depths in the water column illustrate that these compounds do in fact reflect phytoplankton distributions and nutrient supplies. But using them to understand the history of life, shifting ecosystems, and climate requires that we gather explicit information about the chemical environments in the water and sediments when they were produced and preserved. Not even the best of biomarkers, not even the *n*-alkanes or alkenones, were buried in the sediments in the same quantity as their precursors were produced by organisms. In the open ocean, only 1–2% of the organic matter produced makes it to the seafloor and survives the first onslaught of bacteria. Sediments in such areas are made up of carbonate, silicate, and clay minerals and typically contain less than 0.1% organic matter by weight. In certain protected basins or along some of the continental shelves where conditions converge to discourage decomposition and encourage preservation, as much as 15% of production might be preserved in the sediments. For the refractory alkenones, these percentages are likely to be higher, and for the chlorins lower. Whatever the case, if the percentages were constant over time—if for a given area they were the same now as, say, during the last glacial period 20,000 years ago—then one could indeed calculate the average annual production at any time in the past based on the chlorins or, moving back in time, their mature counterparts, the porphyrins, or one could calculate production by coccolithophores based on the alkenones, and so on. But, of course, the ocean environment *has* changed, and with it, the percentage of biological production that was preserved in the sediments.

As the early studies with DSDP cores and sediment traps showed, the degree of organic matter preservation depends not only on the structures of the various molecules, but also on the availability of oxygen in the deeper layers of the water column and sediments. The bacteria that consume organic matter use oxygen, so the sediments become anoxic as they accumulate. When there is a lot of biological productivity in surface waters and the supply of organic matter to the sediments is high, there is more bacterial activity, and oxygen is depleted rapidly in the uppermost layer of the sediment. In open ocean areas where there is free circulation of deep water and productivity at the surface is consistently low, the sediments don't become completely anoxic until about 10 centimeters depth. In some areas where there is a perennial lack of nutrients for phytoplankton at the surface, production is so low that the sediments never become anoxic. But in areas where primary production is exceptionally high and the water is highly stratified, the surface sediments, and even the water itself, can become anoxic. In the contemporary ocean, where deep currents bring in fresh oxygen to replace that consumed in respiration by organisms, completely anoxic water is rare, found mainly in areas where water circulation is limited—parts of the Arabian and Indian Seas, where the water just above the sediments is anoxic, or the Cariaco Basin off the coast of Venezuela and Europe's inland Black Sea, where the water is anoxic from the seafloor to within a few hundred meters of the surface. In some shallow areas along the continental margins, the so-called oxygen minimum—the layer of water beneath the photic zone where bacteria and zooplankton tend to congregate—is effectively isolated from any source of oxygen and can also become anoxic. Whenever the supply of oxygen in the water column or surface sediments is limited, a greater proportion of organic matter escapes decomposition and accumulates in the sediments. Likewise, preservation is enhanced when organic matter is buried quickly or "smothered" by sediments—when intense biological productivity in the surface water or dust and river runoff from land create a high sedimentation rate, or in places on the continental margins where submarine landslides or slumps have suddenly buried down-slope sediments.

Ever since the first DSDP cores were pulled up from the ocean depths and paleoceanographers noted that the amount of organic matter deposited and preserved in the ocean sediments had changed over time, they have been arguing about how to interpret those changes. Had the amount of biological productivity in the surface water changed? Or was there a change in ocean circulation, wind patterns, or rainfall that affected the supply of oxygen and preservation of organic matter in the sediments? Though the argument rages on in some circles, it has long been apparent that production and preservation both play a role and are only partially independent. The real question is: How can we distinguish their relative influence in a given sediment sequence? Clearly, we need independent proxies for both primary productivity *and* sediment oxicity if we are to understand the history of either.

Organic matter accumulation and preservation

Organic geochemists observed in the early 1970s that the relative amount of pristane and phytane in sediments seemed to be related to the amount of oxygen in the depositional environment, with more pristane than phytane in oxic environments and less pristane than phytane in anoxic ones. This fit with what the Bristol group was learning about the diagenesis of phytol, wherein a number of reaction pathways were likely to produce both pristane and phytane in the sediments—but all of the ones that produced pristane required oxygen, whereas those that produced phytane did not. In a first attempt at devising a proxy measure for past periods of water and sediment anoxia, they proposed that the ratio of pristane to phytane could be used as a rough indicator, with a ratio of less than 1.0 indicating an anoxic environment and anything greater than 3.0 signaling an environment where oxygen was readily available in the water and surface sediments. A number of factors besides oxygen availability may come into play, including the presence of other sources of pristane and phytane, but pristane–phytane ratios can indeed prove useful when taken together with other clues about depositional environments—including a number of similarly imperfect, but useful, biomarker proxies for anoxia.

In the presence of oxygen, the algal pigment fucoxanthin that is so abundant in diatoms is completely degraded and disappears without a trace, generating the more persistent loliolides only in thoroughly anoxic sediments. The presence of loliolides in some organic-matter–rich sediments from the Pleistocene thus indicates not only that diatom productivity was high, but also that the surface sediments or water had become anoxic.

The NIOZ group has suggested that gammacerane—among the first triterpenoids identified in the Green River shale and often abundant in sediments from highly saline environments—can be used as an indicator of highly stratified, partially anoxic water. Until recently, the only known sources of tetrahymanol, the presumed biological precursor to gammacerane, were the marine ciliates that tend to congregate at the sediment surface and other places where food is plentiful, eating anything small enough that falls into their paths. Ciliates happen to be easy to grow and manipulate in the lab, the unicellular equivalent of the biologist's fruit fly, so quite a lot is known about their biochemistry—including the small detail that they produce tetrahymanol as a sort of surrogate sterol when, and only when, they can't get sterols from their diet. The NIOZ group has reasoned that in highly stratified waters, the ciliates would feed mostly on the sterol-lacking bacteria that concentrate at the interface between the oxygen-containing surface water and the anoxic deep water. The story may, however, be more straightforward, as a number of other sources for tetrahymanol have recently been discovered, among them purple photosynthetic bacteria that can only live in anoxic environments.

Oxygen-shy bacteria do, in fact, provide a few explicit biomarkers for water column anoxia. Among the compounds that the Strasbourg group found during its explorations of the Paris Basin shales in the 1970s was isorenieratene, an unusual carotenoid pigment with aromatic rings at either end. Natural products chemists had isolated it from green sulfur bacteria—photosynthetic organisms that use sulfide instead of oxygen and employ special pigments that are sensitive to long wavelengths—which can thrive in the penumbra beneath the sunlit photic zone, up to 200 meters below the sea surface in clear waters. They require a thoroughly anoxic and sulfidic environment, and live in places like the Black Sea—where isorenieratene has been found in sediment trap material from the water column as well as in the sediments—and, apparently, the shallow sea that once covered the region around Paris. Oil-producing shales commonly form in such environments, where organic matter is both plentiful and well-preserved, and in the late 1980s, Roger Summons, working at Australia's now-defunct Baas Becking Geobiological Laboratory, identified isorenieratane and other reduced versions of the pigments in ancient crude oils. Green sulfur bacteria also contain a unique set of chlorophyll pigments, with characteristic patterns of side chains attached to the central chlorin ring system—patterns that Maxwell's student Kliti Grice found preserved in the ancient rocks she was studying in the 1990s. She determined the structures of the nitrogen-bearing ring fragments of chlorins, produced by its oxidation in the laboratory and, apparently, in the sediments—structures called maleimides that included one with a distinct isobutyl sidechain found *only* in the bacteriochlorophyll *e* made by green sulfur bacteria. The green sulfur bacteria thus announce their passing quite unambiguously, which is very convenient, given their specific needs: the presence of isorenieratene, or the closely related chlorobactene—or their reduced fossil versions—or

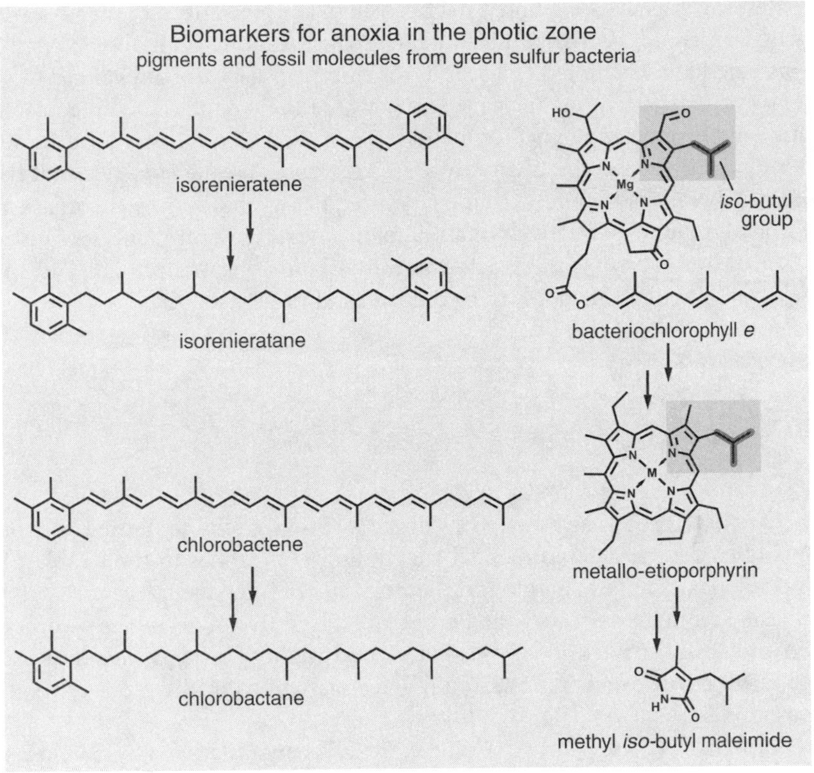

Biomarkers for anoxia in the photic zone
pigments and fossil molecules from green sulfur bacteria

isorenieratene

isorenieratane

chlorobactene

chlorobactane

bacteriochlorophyll *e*

iso-butyl group

metallo-etioporphyrin

methyl *iso*-butyl maleimide

of the unique methyl isobutyl maleimide that Grice identified clearly indicates that the water was anoxic up to a couple hundred meters from the surface at the time of sediment deposition.

Determining how much oxygen was present in the sediments beneath the oxic waters typical of most marine settings is somewhat more problematic—and yet crucial to understanding past changes in primary productivity, organic matter preservation, deep sea circulation, and climate on a global scale. The most promising techniques to date make use of pairs of biomarkers that differ markedly in stability under oxic conditions, and either come from the same sources or have had a constant rate of input to the sediments over the period studied. Joan Grimalt and his Barcelona cohort used the ratio of C_{18} unsaturated to C_{18} saturated fatty acids to compare the oxicity of Pleistocene sediments deposited at different times and places, taking advantage of the knowledge that unsaturated fatty acids decompose faster than their saturated counterparts in the presence of oxygen. Long-chain alcohols break down more rapidly than the corresponding *n*-alkanes and have a common source in the leaf waxes, so the ratio of C_{26} or C_{28} *n*-alkanol to C_{29} *n*-alkane has provided an even more reliable measure of changing oxicity and decomposition rates in areas with significant inputs from the continents.

Researchers have long noted relatively high concentrations of the 40-carbon acyclic isoprenoid lycopane in sediments, rocks, and oils from anoxic environments, and have attempted to use it as an empirical measure of changes in sediment oxicity, though it's unclear exactly where the compound comes from. A highly unsaturated analogue, lycopene, is a common component of the carotenoid pigments in algae and cyanobacteria, and there is some evidence that lycopane or its precursor comes from oxygen-requiring photosynthetic organisms. Possibly, lycopane is relatively abundant in anoxic environments because its unsaturated precursor breaks down rapidly when oxygen is present and there simply isn't enough time for it to be reduced and preserved.

lycopane

The NIOZ group, together with Jürgen and Sonja Schulte in the Oldenburg lab, studied the surface sediments beneath upwelling zones in the Arabian Sea and off the coast of Peru and found that the ratio of lycopane to the resistant C_{31} n-alkane accurately reflected the degree of oxygen depletion in the sediments, suggesting that the lycopane/C_{31} n-alkane ratio might be used as an oxicity index for ancient environments if one could gauge variations in the input of the two compounds.

Though there is clearly no single, reliable quantitative way to measure the history of marine productivity, oxicity in the sediments, and the export of carbon from the atmosphere to the deep water and sediments, multiproxy studies that include an array of biomarkers and mineral or paleontological indicators are beginning to provide qualitative records of many aspects of climate and associated ecosystem change over the past few million years. There is a certain opportunistic element to these studies, as they require sediments that can be dated reasonably accurately and that contain a relatively large amount of organic matter. A number of studies have focused on the thick, organic-matter–rich sediments deposited in the anoxic Cariaco Basin over the course of the last glacial to interglacial shift. Analysis of the algal biomarkers indicates that the last and most severe of the abrupt climate changes to interrupt warming about 11,000 years ago, the Younger Dryas, had a marked effect on primary productivity even in this tropical region. Concentrations of dinosterol, brassicasterol, fucosterol, β-sitosterol, cholesterol, and C_{37} alkenones all spiked at this time, indicating a transient increase in productivity that was in keeping with the changes in the tropical wind system and increased upwelling during the thousand-year cold snap. But the event also seems to have marked an enduring shift in the structure of the phytoplankton community, because the relative proportions of these compounds shifted dramatically, and irreversibly, with diatoms dominating during the Younger Dryas and coccolithophores and dinoflagellates taking charge for the rest of the Holocene.

Temperature proxy records in sediments and ice indicate that the last glacial period, from 50,000 to 20,000 years ago, was punctuated by numerous abrupt climate changes. At a subtropical site in the western Atlantic, Grimalt and his coworkers found that the accumulation rate of both *n*-alkanols and *n*-alkanes exhibited a pattern of abrupt change that was closely correlated with sea surface temperature changes, with the leaf wax compounds increasing during the cold spells—putative evidence that the northwesterly winds had abruptly increased in intensity, bringing more dust from North America. The relative amounts of alkanol and alkane didn't change, indicating that the rate of decomposition and amount of oxygen in the water and sediments at this open ocean Atlantic site had remained constant. But in sediments from the western Mediterranean, the ratio of alkanol to alkane fluctuated in beat with the climate changes, decreasing during the cold snaps—evidence, according to Grimalt and crew, that these events involved influxes of cold, oxygen-rich Atlantic deep water into the Mediterranean.

Meixun Zhao, Geoff, and their colleagues at Dartmouth College and Bristol studied sediment sequences spanning the past 160,000 years and two glacial cycles at a site in one of the upwelling zones off the Atlantic coast of northern Africa, near Cap Blanc. Large fluctuations in the accumulation of chlorin marked dramatic changes in overall productivity, and changes in the relative amounts of dinosterol, alkenones, alkane diols, and silica showed that the structure of the phytoplankton community had varied significantly. But the group was hard-put to discern the relationships between these changes and the changing climate regimes. There was no correlation with sea surface temperatures, so changes in the intensity of upwelling, which brings nutrients and cold water to the surface, were not, apparently, responsible. Productivity did vary with the amounts of leaf wax *n*-alkanes and *n*-alkanols, presumably because the phytoplankton were stimulated by nutrients from minerals in the dust and river runoff that carried them. Geoff says he'd often noted that increases in alkenone content in Atlantic sediments coincided with increases in leaf wax *n*-alkanes, presumably for the same reason. But at a coastal site like the one at Cap Blanc, the changing inputs of leaf wax compounds could be reflecting any of a number of climatic factors: simple changes in the intensity of the dust-carrying winds; shifting weather patterns that produced more rain in certain regions on the continent and more runoff into the ocean; or the opposite, increasing aridity or a change in wind direction so that winds came from a more arid region and picked up more dust.

The biological pump hypothesis maintains that relatively high rates of marine productivity in key regions of the world's oceans might have been responsible for lowering the concentration of atmospheric CO_2 during glacial periods, as recorded in the ice cores. An alternative hypothesis suggests that the structure of the phytoplankton community may have shifted in favor of coccolithophores during the warm periods, and formation of their calcium carbonate coccoliths would have liberated CO_2 to the water and atmosphere. Evidence for or against either hypothesis has been elusive, and one might hope that these detailed histories of marine primary productivity in various parts of the globe would provide some enlightenment. But neither the changes in productivity nor the variations in the structure of the phytoplankton community at the Cap Blanc site showed correlations with global climate patterns. Similar studies of sediments from the Indian Ocean and Arabian Sea indicate that in these areas the amount of productivity and the structure of the phytoplankton community both remained relatively constant over the past two ice age cycles. Analysis of the chlorins in a core from the South China Sea, northwest of Taiwan, revealed clear oscillations in the overall productivity of phytoplankton that were apparently related to oscillations in the intensity of the winter monsoon season in the region, with productivity at its highest toward the end of the last glacial period, at the onset of warming. Clearly, such studies are still too few and too incoherent to be integrated into geochemists' models of the global carbon cycle—and, for various reasons, the organic geochemists may be looking in the wrong places.

One region where changes in the intensity of the biological pump might have had a concerted effect on the concentration of atmospheric CO_2 during glacial periods is in the Southern Ocean encircling Antarctica, where, in the current interglacial period, the major nutrients, nitrogen and phosphate, are readily available and primary productivity appears to be limited by a lack of micronutrients such as iron—which might well have been supplied by additional inputs of wind-borne dust from the South American pampas, Africa, or Australia during glacial periods. This is a difficult region for oceanographic cruises, however, and appropriate sediment cores have been relatively hard to come by. One of the few multiproxy biomarker studies was done by a couple of Grimalt's former students, working with the group that Roger Summons had helped build in Australia. They analyzed sediment cores from the Tasman Sea, south of Australia, determining concentrations of long-chain *n*-alkanes, brassicasterol, dinosterol, and alkenones in sediments deposited over the last 350,000 years. The concentration of *n*-alkanes at this site clearly fluctuated in concert with the glacial–interglacial cycles and correlated with the content of dust trapped in the Antarctic ice cores, increasing during glacial and decreasing during interglacial periods. Dinosterol and brassicasterol show a positive correlation with the Milankovitch cycles and the *n*-alkanes, but the alkenones show none. The brassicasterol fluctuations are particularly pronounced, and the authors postulate that silica in the dust may have given the diatoms a particular advantage during glacial periods. They propose that the high productivity of diatoms and dinoflagellates, specifically, during the glacial periods—and not of coccolithophores, whose calcium carbonate shells would have liberated CO_2 and countered the effects of enhanced organic matter production—provides nominal evidence that the biological pump was operating in the hypothesized manner in this key region, and may indeed have played a significant role in regulating the fluctuations in atmospheric CO_2. But, of course, such studies are still just a drop in the bucket of evidence needed to reconstruct and understand the patterns of cause, effect, and feedback that comprise the earth's climate history on a global scale.

Most multiproxy biomarker studies have focused on sediments that are particularly rich in organic matter, for the obvious reasons that the organic compounds are better preserved and more easily detected. But such sediments often pose a mystery in and of themselves. Sediment cores from the Mediterranean Sea exhibit bands of black, organic-matter–rich sediments periodically interspersed in pale brown and yellow sediments with less than 0.1% organic matter. From a few centimeters to a few decimeters thick, spanning up to 10,000 years and containing up to 30% organic matter, these black sediments, which geochemists call sapropels, began appearing about five million years ago, during the early Pliocene when the first ice sheets were forming in the northern hemisphere. There must have been drastic, periodic changes in the Mediterranean environment to produce them—but what?

A lack of fossils from bottom-dwelling foraminifera hinted that oxygen may have been lacking in the surface sediments and deep water of the basin, and recent biomarker analyses offer more evidence: loliolides are abundant, and in at least two of the Pliocene sapropels, isorenieratene is present, implying that the anoxia extended into the lower photic zone, as in the contemporary Black Sea. Had circulation in the Mediterranean somehow changed? Was it shut off from the Atlantic? Or did the input of organic matter suddenly increase so dramatically that all the oxygen in the deep water was used up by decaying organic matter? An increase in the ratio of long-chain *n*-alkanols to *n*-alkanes within the sapropels indicates that preservation of organic matter increased significantly during the shift from an oxic to an anoxic environment, as might be expected. Dramatic increases in the accumulation rates of organic matter, and of phytoplankton biomarkers such as loliolide, brassicasterol, alkenones, alkane diols, and dinosterol, are also observed

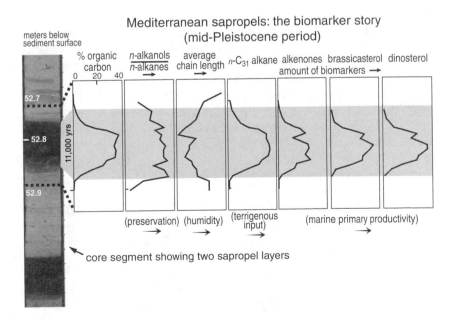

core segment showing two sapropel layers

and must, to some extent, be the result of enhanced preservation. But there are also indications that primary productivity increased, and the greater diversity of phytoplankton biomarkers within the sapropels reflects a more diverse community, as is generally observed when primary productivity is high. An increase in the absolute amounts of leaf wax *n*-alkanes signals an increase in the input of material from land, with its associated nutrients, which would have fueled phytoplankton growth. And in at least one of the sapropels, the chain-length distribution of the leaf wax *n*-alkanes shifts to shorter chain lengths, implying that they came from vegetation growing in a more humid region.

The emerging story, based on these and other chemical and geologic data, is that a periodic intensification of the monsoon system and more precipitation in Europe and Africa—in sync with the earth's 23,000-year cycle of precession on its spin axis—brought more river runoff and nutrients into the Mediterranean, spurring phytoplankton growth, an excess of organic matter production, decomposition, and anoxia, and resulting in the periodic deposition of the sapropels. That all makes a fine story for the Mediterranean, which is a restricted basin where a few nutrients and a little extra organic matter production go a long way...but some of the first DSDP cores in the 1970s had revealed that, some hundred million years earlier, similar layers were deposited in the middle of the open ocean.

Like the Mediterranean sapropels, the Cretaceous black shales appear as dark bands of organic-matter–rich sediment in the midst of long stretches of pale gray or white mudstones and limestone. Laminations in the sediments and the sudden disappearance of fossils of bottom-dwelling foraminifera indicate that oxygen in the bottom water was in short supply. But these black shales were deposited over longer stretches of time, with no apparent pattern in the timing, between 140 and 80 million years ago. Most remarkably, they appear contemporaneously in cores drilled thousands of miles apart in areas of the world's oceans where it is hard to conceive how so much organic matter could possibly have accumulated. Five such sediment layers have now been identified in the middle and late Cretaceous, and one way back in the early Jurassic period. They present an even greater enigma than the Mediterranean sapropels, manifesting episodic perturbations in the chemistry and biology of the world's ocean—which oceanographers initially dubbed "oceanic anoxic events"—that lasted from a few hundred thousand to more than two million years.

The biomarkers in these black shales tell a story of anoxia that may have been almost as severe as that in the present-day Cariaco Basin or Black Sea—but on an oceanwide scale. High concentrations of lycopane and steroids relative to long-chain *n*-alkanes and a low ratio of pristane to phytane corroborate other evidence that oxygen was in low supply in the sediments and deep water of the Atlantic during these anoxic events. But the NIOZ group and others have also found significant concentrations of isorenieratane and other molecular fossils of isorenieratene and chlorobactene in black shales from at least two of the events, and the Bristol group has found the telltale isobutyl maleimides and related porphyrins—all indications that green sulfur bacteria were thriving.

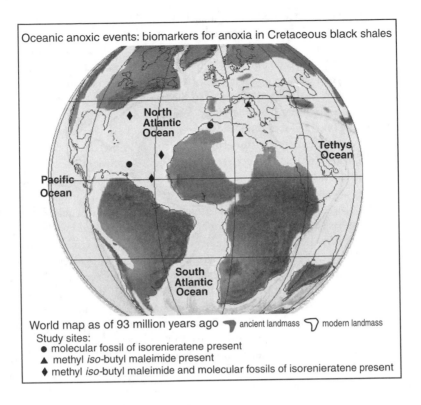

Oceanic anoxic events: biomarkers for anoxia in Cretaceous black shales

North Atlantic Ocean

Tethys Ocean

Pacific Ocean

South Atlantic Ocean

World map as of 93 million years ago 🢒 ancient landmass 🢖 modern landmass
Study sites:
● molecular fossil of isorenieratene present
▲ methyl *iso*-butyl maleimide present
♦ methyl *iso*-butyl maleimide and molecular fossils of isorenieratene present

Either the deep water was entirely anoxic from the sediments to within 200 meters of the surface, or there was a mid-depth zone of decomposition, an oxygen-minimum layer that was entirely anoxic, as in parts of the contemporary Arabian Sea. Of course, the oceans and continents looked very different during the Cretaceous than they do today: the Atlantic Ocean was still a relatively young, narrow waterway, recently formed by seafloor spreading, and the ancient Tethys Ocean on the other side of Africa was shrinking. It's easy to imagine these oceans with obstacles to the circulation needed to replenish oxygen in their deep water, or so shallow that the bottom of the photic zone coincided with the sediment surface and made for a particularly intense zone of microbial activity and oxygen consumption. Relatively low concentrations of leaf wax *n*-alkanes and the absence of oleanene or oleanane indicate that little of the organic matter came from the continents, while high concentrations of HBIs and of 27- and 28-carbon steroids attest to a particularly productive phytoplankton community. In the Atlantic and Tethys oceans, at least, it would seem that a spurt in marine productivity generated enough additional organic matter and oxygen-consuming decomposition to send a warm, nearly stagnant and stratified ocean over the edge into a state of anoxia that, in modern times, exists only in a handful of isolated basins. But black shales are also found in the Pacific Ocean, which by all accounts was a vast, largely unencumbered expanse of water during Cretaceous times. In recent studies of black shales deposited in the middle of the Pacific about 120 million years ago, Simon Brassell's group, now at Indiana University, found low ratios of pristane to phytane and relatively high lycopane concentrations, implying that low-oxygen conditions, if not photic zone anoxia, may have existed during this event even in the open ocean. Had the Pacific Ocean also stagnated? Or could marine productivity have increased so drastically, for such extended periods, around the entire globe? And, most problematic of all, why and how?

There always seems to be something missing in these multiproxy studies, yet another parameter that will constrain the others—a mineral that no one has the wherewithal to analyze, or the particular biomarker that couldn't be analyzed because it is too difficult to separate on the GC column used or because the student doing the project ran out of time; or perhaps some crucial bit of sediment core got lost during the coring operations, or worms in the sediment had mixed up the layers during or shortly after deposition and the time periods were not well resolved.... Perhaps if these organic geochemists had gotten a few more data points beyond the juicy black shales and sapropels they're so fond of, or had pushed their analytical techniques to the max and provided information about the biomarker composition of the 0.1% organic matter in the mudstone that was deposited during the 5,000 years before and after, or if they had moved away from their favorite coastal upwelling sites and analyzed the Pleistocene sediments from the places that paleoceanographers and climatologists consider key to understanding the global cycles of nutrients and primary productivity... The problem, of course, is that the organic matter is so poorly preserved in these well-oxygenated, open ocean areas that what one finds is not at all representative of the overall biological input and may not be particularly informative—but the cumulative data from such sites may be. Eloquent simplicity is not necessarily a forte when it comes to using biomarkers in paleoenvironmental studies. Rather, "the more the merrier" would seem to be the better guiding principle, the full exploitation of the expanding molecular lexicon and increasing instrumental sensitivity—along with appropriate statistical treatments and comparative ratios to turn the jumble of data produced into meaningful parameters that can be integrated into the larger context of paleoceanographic and paleoclimate studies. But first, we must consider another dimension of molecular information, one that started adding ballast to the lexicon in the 1980s and would eventually promote the interpretation of biomarker data to a higher art.

6

More Molecules, More Mud,
and the Isotopic Dimension
Ancient Environments Revealed

The limits of my language, mean the limits of my world.
—Ludwig Wittgenstein, 1889–1951, Austrian philosopher
  From *Tractatus Logico-Philosophicus* (1921)

*The attitude of man towards the earth is still, on the whole, that of a
parasite. For a parasite, nevertheless, the life of the host is of prime
importance.*
—Lourens G. M. Baas Becking, 1895–1963, biologist and inventor
  of the term "geobiology"
  Quoted in Anton Quispel, *International Microbiology* 1 (1998)

Throughout the 1980s, while the molecule collectors were busy exploring the
ocean sediments, tracking their finds into the past, and learning to read the
messages hidden in the carbon skeletons, one analytical chemist *cum* geochemist at Indiana University was finding that important elements of the lexicon lay
not only in the molecular structures, stereochemistry, and distributions of the
carbon skeletons, but in the carbon atoms themselves.

John Hayes had done his graduate work at MIT in the mid-1960s under Klaus
Biemann, one of the doyens of mass spectrometry who, like Carl Djerassi, was
interested in natural products with biomedical applications. When Hayes told
Biemann he wanted to do his doctoral thesis on the organic constituents in meteorites, Biemann was uninterested, to say the least. Forty years later, Hayes can still
quote the eminent scientist's response to his proposal, replete with thick Austrian
accent: "Don't talk to me about zat junk." Biemann walked away from the discussion without another word, and Hayes was so mortified by his own foolishness
that he couldn't bring himself to tell his wife about the incident. When he went
into the lab the next day, he was convinced that his graduate career was over—but
Biemann had done some homework and had a change of heart. "It seems we can

get lots of money for zat junk!" he exclaimed as soon as he saw Hayes. NASA was, at the time, offering generous funding for such projects.

For all his skepticism, Biemann was eventually seduced by the extraterrestrial "junk" and even ended up designing the mass spectrometer for the Viking Mars mission. Hayes remembers him commenting, a couple of years into the meteorite project, that it was actually "much more interesting than the thousandth alkaloid in the thousandth tree," though Hayes himself says his doctoral thesis was unexceptional, completed before the Murchison meteorite fell and things really got interesting. While he was working on it, however, struggling to divine how and where the organic compounds he detected in his samples had formed, he had an idea that would, years later, form the catalyst for some of his most remarkable accomplishments: it occurred to him that the carbon atoms themselves should bear witness to a molecule's provenance.

As Hayes tells the story, the idea began to evolve during the countless hours he spent staring at the forests of little black lines, the so-called "peaks," in countless mass spectra. There were the peaks of interest, corresponding to the molecular ion and fragment ions generated when the molecule was hit by an electron and broken up. But lined up on the heavy side of each of these main peaks was a minority population of little ones, corresponding to the same ion but containing one or more atoms of the rare heavier isotope of carbon. In the early days of organic mass spectroscopy, the spectra submitted for publication were often cleaned up to remove the black fur of peaks produced by traces of impurities in the sample, occasionally even doctored to the point that the isotope peaks disappeared and only the "useful" peaks remained. Anyone who worked with mass spectroscopy on a regular basis, however, knew that no naturally occurring substance, even an absolutely pure one, could generate such a clean spectrum. For his part, Hayes spent so many long hours communing with "real," undoctored mass spectra that he decided the messy isotope peaks must be "good for something" in their own right.

Naturally occurring heavy versions of most of the elements that make up organic compounds—carbon, hydrogen, oxygen, nitrogen, and sulfur—had been identified in the 1920s and 1930s. Generated by the nuclear reactions within stars and inherent to the elemental makeup of the planet, they comprise a small percentage of these elements' total inventories on Earth: 0.02% of oxygen is ^{18}O, and the heavier ^{13}C isotope comprises 1.1% of the earth's inventory of carbon. These scarce heavy isotopes have an extra neutron or two in their nuclei but the same number of protons as ^{16}O or ^{12}C, and they are involved in all the same reactions as their mainstream counterparts. But the slight difference in mass makes for slight differences in the ease with which a given physical or chemical process occurs in molecules containing different isotopes, resulting in the isotopes being preferentially distributed or fractionated between different substances. With the development of mass spectrometers capable of accurately quantifying small differences in the abundance of these atoms, it became possible to observe the effects of this fractionation over the course of earth history.

The isotopes of a given element are not evenly distributed between different materials, but, rather, ice and snow are depleted in ^{18}O relative to seawater (the reason the ice ages had such a pronounced effect on the $\delta^{18}O$ of foraminifera), plants and organic-matter–rich rocks are depleted in ^{13}C relative to carbonate rocks, and so forth. Some of these differences can be predicted directly from the fundamental physical and thermodynamic properties of the atoms: a chemical bond to a heavier isotope has a lower vibrational energy and is ever-so-slightly stronger than a bond to its lighter counterpart; molecules containing a heavy isotope usually react more slowly in chemical reactions than their lighter versions do; and in a reversible reaction at equilibrium, the heavier isotope is concentrated in the compound where it is bound most strongly. These theoretical considerations account for the observations that water containing ^{18}O evaporates less readily than the usual $H_2{}^{16}O$ does, and the bicarbonate in seawater is enriched in ^{13}C compared to the dissolved carbon dioxide.

The ratio of ^{13}C to ^{12}C in biologically produced organic molecules depends on the ratio of ^{13}C to ^{12}C in the material that provided their carbon atoms and on any isotope fractionation that occurs during their assembly and should, in principle, differ from one compound to the next and from one environment to the other. Indeed, Phil Abelson and Thomas Hoering did experiments with amino acids at the Carnegie Institute in 1961 and showed that ^{13}C was distributed unevenly even *within* the same compound. John Hayes was impressed when he read their paper as a student: presumably, the different steps in the biosynthesis of organic compounds involved varying degrees of fractionation, and this was registered by the distribution of ^{13}C. Some positions in the molecule must be more enriched in ^{13}C than others, and the little isotope peaks in the molecule's mass spectrum should, in principle, show this! They *were* good for something! If you could measure the ratio between the isotope peak and the main peak for each of a molecule's fragment ions, then you could figure out the precise distributions of isotopes in any organic molecule you could analyze by GC-MS. And if one could determine the precise distribution of isotopes in the organic compounds in an extract from a meteorite or Precambrian rock, then one would have a very specific record of where and how those organic compounds had formed—the question of the day.

It was a naive proposition, a student's fantasy—not much was known about isotope fractionation during specific biosynthetic reactions, and one could not, even with Klaus Biemann's sophisticated array of mass spectrometers, detect the minute differences in the sizes of the fragment ion isotope peaks. Nevertheless, Geoff says that when Hayes was in Bristol as a postdoc, his idea that one could, in principle, determine the precise isotopic makeup of individual compounds was the subject of many long conversations—not in connection with the lunar samples they were analyzing at the time but, rather, with the homely old leaf wax hydrocarbons. True to form, Geoff was enthusiastic and encouraging: if one could determine the isotopic makeup of the *n*-alkanes extracted from sediments, it would add a whole new layer of information as to their origins, maybe even the conditions under which they were biosynthesized!

Warren Meinschein, who had finally answered the call of academic science and left Esso, was likewise interested, and when Hayes joined him at Indiana University in 1970, they made a first attempt to develop the idea. Hayes approached the engineers at the Finnigan-MAT instrument company in Germany about designing a new instrument, and they began what would turn out to be a career-long collaboration—despite the nominal failure of their first enterprise. Hayes has a vivid memory of spending Thanksgiving Day 1972 in the lab eagerly trying out the new prototype from Germany... only to discover that his grand experiment, if not the instrument itself, was a total failure. The instrument combined the precision of the isotope ratio mass spectrometers used by geologists to analyze carbonates with the capabilities of the mass spectrometers used by organic chemists. It was supposed to allow determination of ^{13}C to ^{12}C ratios for each of a molecule's fragment ions—but there were unforeseen problems with hydrogen atoms moving around when the molecules fragmented, and the isotope ratios were variable and unreliable.

Discouraged but not dissuaded, Hayes resorted to dismantling the molecules chemically, carbon atom by carbon atom, and using the new mass spectrometer to measure the isotopic composition of each piece. The method was useless for geochemical samples, but it did eventually reveal the distinct isotopic composition of a fatty acid molecule. And it did, in turn, inspire a fascinating medley of investigations into the various biosynthetic processes responsible for distributing the isotopes of carbon both within and among organic molecules. Pat Parker and his team of microbiologists in Texas, plant physiologists and biophysicists in Australia and California, and geochemically inspired biochemists in Japan, not to mention Hayes's own group in Indiana, all contributed bits of insight.

The carbon isotope composition of a material is generally expressed as "$\delta^{13}C$," defined as the amount that the ^{13}C to ^{12}C ratio measured in a substance deviates from that of the CO_2 generated by dissolving an international calcium carbonate standard—originally from a rock formation in South Carolina—expressed in parts per thousand (per mil or $^0/_{00}$). It is, however, not the $\delta^{13}C$ values per se that are of interest, but *comparisons* of isotopic compositions of different materials, the differences between $\delta^{13}C$ values in the various components of a system: between the different kinds of molecules in a cell, or between the organisms in an ecosystem and their inorganic carbon sources, or between the various carbon-containing components of a geologic deposit—between carbonate rocks and bulk organic matter, or between specific fossil molecules in the organic matter.

As Hayes showed with his fatty acid dismantling experiments, biological isotope fractionation can occur during each step of a biosynthetic process. This results in the uneven distribution of isotopes within molecules, and it creates significant differences in the *overall* isotopic compositions of different classes of molecules. In algae, for example, the sugars generated directly by photosynthesis are generally less depleted in ^{13}C than the lipids are, with differences ranging from 2 to 10 per mil, depending on the species and the specific compounds; isoprenoid lipids tend to be slightly less depleted than *n*-alkyl lipids; and so forth. The most pronounced carbon isotope fractionation in biology, however, one that is registered in the $\delta^{13}C$ values of an organism's entire biomass, generally occurs during the initial transfer of carbon from geosphere to biosphere. Plants, algae, and microorganisms all exhibit a preference for the lighter, ^{12}C-containing version of their carbon sources—carbon dioxide or bicarbonate for photosynthetic organisms, or methane for certain methanotrophic microorganisms. Thus a plant will be significantly depleted in ^{13}C relative to the CO_2 in the air—more than 20 per mil in the case of some land plants—and an alga will be depleted relative to the dissolved CO_2 or bicarbonate in the water where it grows, and so on. Moving up the food chain to animals and other heterotrophic organisms that use biologically prefabricated organic matter as their carbon source, the fractionation goes in the opposite direction: as organic matter is respired and burned to CO_2, carbon–carbon bonds are being broken and ^{12}C is released more readily than ^{13}C. Thus the compounds in heterotrophic organisms will generally be slightly *enriched* in ^{13}C relative to their food sources, with each step up the food chain resulting in about 1–1.5 per mil enrichment compared to the one before it.

By the mid-1980s, many of the details of biological isotope fractionation that Abelson and Hoering, John Hayes, and others had only been able to theorize about in the 1960s were documented, and it was apparent that the isotopic composition of fossil molecules might provide precisely the sort of information Hayes had imagined as a starry-eyed graduate student, even without resort to the intramolecular details. Limited at first to fossil molecules that could be isolated and purified—the most plentiful compounds in the most organic-matter–rich rocks—Hayes teamed up with Pierre Albrecht in a study of the Strasbourg group's favorite rock, the Messel shale. The group in Strasbourg extracted and purified the compounds, determined their molecular structures, and sent them to Hayes in Indiana. Hayes and his students burned each compound and injected the CO_2 gas produced into a conventional isotope ratio mass spectrometer to determine its ^{13}C content—much as oceanographers did when they determined the $\delta^{13}C$ of bulk organic matter. Hayes says he had some exceptional students around this time—Kate Freeman in particular, he claims, was blessed with "golden fingers"—and it wasn't long before the researchers contrived an interface that allowed them to connect a gas chromatograph directly to an isotope ratio mass spectrometer. As each compound emerged from the GC column, it was converted to a puff of CO_2 and fed directly into the mass spectrometer, where its ^{13}C content could be determined on the fly, so to speak. They called the technique gas chromatography–isotope ratio monitoring–mass spectrometry, which quickly reduced to the acronym GC-irm-MS. With it, they could determine the isotopic composition of any compound in a mixture that could be cleanly separated on a GC column.

The Indiana-Strasbourg Messel shale studies of the late 1980s illuminated the full potential of compound-specific isotopic information to elucidate the ecology of an ancient ecosystem, in this case a large lake during the Eocene epoch. The Strasbourg group had identified the structures of more than a dozen porphyrins, some of which clearly derived from algal chlorophylls or one of the bacteriochlorophylls, and some of which were of ambiguous origin. The $\delta^{13}C$ values of the bacteriochlorophylls were consistently 2 per mil more negative than those of the algal chlorophylls, and the $\delta^{13}C$ values for most of the ambiguous porphyrins bore the distinct signature of one or the other, allowing their tentative assignment to bacterial or algal sources. The $\delta^{13}C$ of dinosterol was slightly more depleted

Typical $\delta^{13}C$ values in organisms, the environment, and geologic deposits

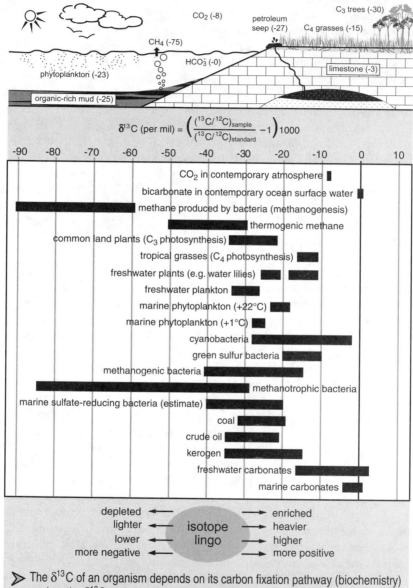

$$\delta^{13}C \text{ (per mil)} = \left(\frac{(^{13}C/^{12}C)_{sample}}{(^{13}C/^{12}C)_{standard}} - 1 \right) 1000$$

> The $\delta^{13}C$ of an organism depends on its carbon fixation pathway (biochemistry) and on the $\delta^{13}C$ and concentration of its carbon source (environment).

> The $\delta^{13}C$ of a specific compound depends on its biosynthetic pathway. Lipids can be up to 30 per mil more negative than biomass.

than that of the algal porphyrins, reflecting the different biosynthetic pathways that presumably linked them to a common source of CO_2 in the surface water. Isoarborinol, a triterpenol found exclusively in land plants, had a $\delta^{13}C$ very similar to that of contemporary land plants, and visible fossils in the shale indicated that organic matter from higher plants had made a significant contribution to the sediments, apparently washing into the lake from the surrounding watershed. None of the porphyrins, however, had a $\delta^{13}C$ within the range expected for land plants, which supported the hypothesis that land plant chlorins were unable to survive the exposure to oxygen during the journey from land to lake sediment.

One long-chain acyclic isoprenoid structure that was present in high concentrations in the Messel shale had recently been linked to the unusual membrane lipids of methanogenic microorganisms, which required completely anoxic conditions and were often found deep in the sediments. These methanogens obtained their energy by reducing CO_2 or acetic acid to methane, a process which involved one of the largest carbon isotope fractionations in nature: the methane produced in laboratory experiments with the organisms was 40–95 per mil more depleted in ^{13}C than the CO_2 or acetate that the organisms imbibed. Fractionation during the assimilation of carbon for molecule-building purposes was apparently more moderate, however, and the methanogens themselves were only slightly more ^{13}C depleted than photosynthetic algae grown from the same carbon source. Of course, in a place like the Messel Lake, methanogens and algae weren't likely to have used the same carbon source, because the algae lived near the lake surface, using CO_2 that was equilibrated with the air, whereas methanogens could only have lived in the anoxic environments deeper in the lake or sediments, using the ^{13}C-depleted CO_2 or acetate generated by decomposition of organic matter. In the shale, Hayes and crew found that the unusual long-chain isoprenoid was about 8 per mil more depleted in ^{13}C than pristane, whose $\delta^{13}C$ value matched that of the algal lipids. They inferred that the pristane had come from the decomposition of chlorophyll, whereas the long-chain isoprenoid and most of the phytane—which was similarly depleted in ^{13}C and seemed to be a likely diagenetic product of the newly identified methanogen lipids—derived from methanogens. Remarkably, two other long-chain acyclic isoprenoids and several of the hopanoids were so depleted in ^{13}C, with δ values ranging from –60 to –75 per mil, that they seemed to have been made by organisms that *used* the methane released by the methanogens. Precisely what those organisms were and the amazing ways they operated would not be apparent for more than a decade, but the distribution of porphyrins implied that green sulfur bacteria may have been present and the lake water anoxic, which meant that methane could have been produced throughout much of the water column, as well as in the sediments.

The most significant revelation of the Indiana-Strasbourg team's studies in the 1980s resulted from their comparison of the $\delta^{13}C$ values for the individual microbial and algal biomarkers—the hopanoids, acyclic isoprenoids, steroids, and porphyrins—with the $\delta^{13}C$ of the bulk organic matter, both kerogen and bitumen, in the shale. Despite the rich, diverse flora and fauna that apparently lived in and around the ancient Messel Lake and left such well-preserved fossils for posterity, the supreme rulers of the lake's ecology and regulators of the flow of carbon and dissolved gases within its margins were, in fact, the heretofore invisible microbes.

Another geochemist who was becoming acutely aware of the pivotal role that microbes played in the movements of carbon through the biosphere and geosphere, and who, like Hayes, had been thinking about putting the isotopic dimension of fossil molecules to use in understanding it, was Roger Summons in Australia. Summons had spent the early years of his career studying biochemical processes in plants, and his first forays into organic geochemistry were under the auspices of the Baas Becking Geobiological Laboratory—Australia's homage to one of the first microbiologists to recognize the importance of microbial life in regulating geochemical cycles and forming geologic deposits. In the 1970s and 1980s, the Baas Becking Laboratory was one of the few places in the world where geochemists and microbiologists worked side by side on the same fundamental research questions. Even before Hayes and crew started their compound-specific isotope measurements on the Messel shale porphyrins, Summons was painstakingly isolating what appeared to be fossil remnants of isorenieratene from crude oil samples, determining their isotopic makeup, and noting that they were distinctly enriched in ^{13}C. He happened to be aware of recent biochemical studies showing that green sulfur bacteria not only employed a sulfur-based version of photosynthesis and contained unusual carotenoid pigments, but also captured CO_2 via a unique sequence of reactions that resulted in relatively little carbon isotope fractionation and created a biomass that was enriched in ^{13}C relative to that of most other phytoplankton. The isorenieratane in the crude oils was 8 per mil enriched in ^{13}C compared to the other hydrocarbons in the oils, putative proof of the compound's provenance from the green sulfur bacteria carotenoid—and of the oils' provenance from source rocks formed in an anoxic depositional environment. It was the first so-called compound-specific isotope analysis, in which the distinct isotopic signature of a source organism's biochemical system was used to track the origin of a fossil molecule—something that Hayes's new GC-irm-MS would soon make possible for tiny traces of compounds that couldn't be readily isolated.

In 1988, as the Indiana group was beginning its first explorations with its new instrument, Hayes arranged to spend his Guggenheim fellowship in Summons's lab in Australia, initiating years of collaborative isotopic explorations and establishing a tradition of trans-Pacific winter migrations for several generations of Indiana students. Ironically, that was the same year that the Baas Becking Geobiology Laboratory met its demise at the hands of Australian bureaucrats—just as geochemists and microbiologists in other parts of the world were beginning to recognize the value of disciplinary intermarriage. Summons ended up working for the petroleum-oriented Bureau of Mineral Resources, but managed to continue the studies he'd nurtured at Baas Becking, working with a carefully chosen, interdisciplinary collection of collaborators around the world. Despite the misguided perceptions of Australian bureaucrats in the late 1980s, interest in the coevolution of biochemical and geochemical systems was growing, not just among a few starry-eyed geochemists and microbiologists but throughout the international oceanographic and geologic communities—with a particular emphasis on understanding how the global carbon cycle had changed throughout earth history.

An increasing amount of evidence from the Ocean Drilling Program cores indicated that the distribution of carbon between atmosphere, ocean, biosphere, and rocks had changed significantly—and in concert with the climate—over the course of the past 200 million years. Geologists had noted that the amount of isotope fractionation between the carbonate and the bulk organic matter in Cretaceous period sediments and rocks was consistently greater than in recent times, and that the organic matter in Cretaceous sediments was some 5 per mil more depleted in ^{13}C than that in more recent sediments. Was this due to some pervasive difference in ocean ecology, radically different distributions of species, or the imprint of some yet undiscovered bacterial process? Or was it, as the geologists speculated, due to universally enhanced isotope fractionation by plants and algae during the Cretaceous?

Hayes and company found that porphyrins isolated from Cretaceous marine sediments were even more depleted in ^{13}C than the bulk organic matter was—the opposite of the situation in more contemporary sediments and an indicator that the enhanced fractionation was broadly associated with phytoplankton, just as some geologists suspected. How and why were a little more difficult to discern. Most plants incorporate CO_2 via the cycle of biochemical reactions that Melvin Calvin elucidated in the 1950s. Though a handful of other pathways have been discovered in bacteria, and in tropical grasses and plants, the Calvin cycle was the only pathway known in marine algae, and it seemed unlikely that a different species distribution could have caused such a large change in isotope fractionation. There were, however, indications from both geochemical data and laboratory studies that the amount of fractionation during photosynthesis by algae and photosynthetic bacteria might depend on the concentration of CO_2 in the water where they grew—specifically, that they might be more discriminating in their choice of isotopes when the CO_2 concentration was high. One of the most plausible explanations for the Cretaceous period's warm climate invoked a greenhouse effect from high levels of atmospheric CO_2, and paleoceanographers had found some evidence of extensive volcanism, which might have released large amounts of CO_2—but such evidence was circumstantial, at best, because there was no way to measure the atmospheric pressure of CO_2 millions of years ago, no paleo-CO_2 proxy on par with the $\delta^{18}O$ climate proxy or the newly developed alkenone sea surface temperature proxy... or was there? If the $\delta^{13}C$ of the organic matter in the rocks was low because algae responded to an excess of dissolved CO_2 in the ocean surface waters, then perhaps that effect could be quantified or calibrated.

Hayes had toyed with this idea since the late 1970s, when he first started trying to analyze the isotopic makeup of organic molecules. His oceanographer friends were more than a little intrigued by the idea of a paleo-CO_2 proxy, but not until 1985—when a friend at Columbia University mentioned an unpublished Ph.D. thesis he was an external examiner for in New Zealand—did it begin to seem at all plausible. A student at Waikato University had done a study on natural mixed populations of algae that he'd collected from lakes in New Zealand and grown in the laboratory under carefully controlled conditions. Bruce McCabe had measured the $\delta^{13}C$ of the algal biomass and that of the dissolved CO_2 they used, and then determined how the difference between the two varied as he changed the concentration of dissolved CO_2 in the water. It was the first time anyone had documented a predictable quantitative relationship between the amount of fractionation by the algae and the amount of available CO_2. Whether or not one could turn that relationship into a calibration that would hold for organisms that lived millions of years ago, under variable climates and conditions, remained to be seen. That the $\delta^{13}C$ of organic molecules generated by photosynthesis depended on the CO_2 concentration in the water was clear. But there were also indications that isotope fractionation depended on a number of other environmental and physiological factors.

In Australia, the plant physiologist Graham Farquhar had just come up with a mathematical model for the isotope fractionation that accompanies uptake of CO_2 during photosynthesis:

$$\varepsilon_P = \varepsilon_t + ([CO_2]_{inside}/[CO_2])(\varepsilon_f - \varepsilon_t)$$

Here, the overall fractionation between the CO_2 the plant or alga used and the organic matter produced by photosynthesis, ε_P, was expressed as a function of fractionation during the transport of CO_2 across the cell membrane, ε_t; fractionation during the formation of the carbon–carbon bond as mediated by the carbon-fixing enzyme, ε_f; and the relative amounts of CO_2 inside the cell, $[CO_2]_{inside}$, and in its environment, $[CO_2]$.

The marine algae and photosynthetic bacteria that live in the surface waters of the ocean use different versions of the rubisco enzyme that fixes CO_2, so they

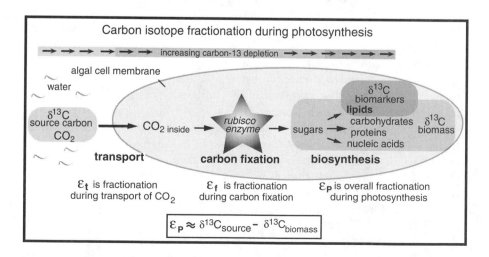

were likely to have different fractionation factors, ε_f. In algae where CO_2 moves into the cell by simple diffusion, the amount of fractionation ε_t and the concentration of CO_2 inside the cell, $[CO_2]_{inside}$, might be predictable. But some algae complicate matters by actively transporting CO_2 into their cells or making use of bicarbonate as well as dissolved CO_2. And the concentration of CO_2 inside the cell would also depend on growth rate, with fast-growing cells using it up faster and maintaining a lower concentration.

In his empirical calibration with the New Zealand lake algae, McCabe had simply found the best mathematical fit between the amount of CO_2 in the environment and the amount of photosynthetic fractionation. If one wanted to use such an equation to determine the amount of CO_2 in some other environment, one had to make the assumption—or hope—that the physiological factors that Farquhar had identified were either constant or had a relatively minor effect. The porphyrins that Hayes and Freeman had been analyzing in Cretaceous rocks were derived from chlorophylls made by a wide variety of photosynthetic organisms, but if one could determine $\delta^{13}C$ values for a more specific biomarker, from a single species or group of algae, then this might not be such a bad assumption and one might indeed be able to estimate past levels of CO_2. One would need the perfect biomarker, something particularly stable, from some extant group of algae with a long history and a wide distribution. And, of course, one would also need a way to determine the $\delta^{13}C$ of the dissolved CO_2 that the algae had imbibed.

In the summer of 1986, when Hayes presented his study of porphyrins in Cretaceous marine rocks at the Gordon Research Conference in New Hampshire, there was at least one Woods Hole graduate student in the audience who had been doing a lot of thinking about the "perfect biomarker." The U_{37}^K index had hit the press earlier that year, and alkenones had been the talk of the town in Woods Hole—but John Jasper had been using them in his thesis research, even before the temperature proxy was developed. In comparing the changing inputs of organic matter in the Gulf of Mexico during glacial and interglacial periods, he had needed a relatively persistent, quantifiable index of specifically marine input, and the alkenones had won the prize hands down. Earlier that summer, Jasper and the other students had invited Geoff to Woods Hole for a special two-week seminar course, much of it spent in long, contemplative beach walks and seaside brainstorming sessions about other ways to use the new biomarkers. So as Jasper sat listening to Hayes talk about the relationship between CO_2 and the amount of isotope fractionation by algae and plants, it was only natural that he should be thinking about the alkenones and his Gulf of Mexico core, which contained a very detailed record of glacial–interglacial climate change and was chock-full of alkenones.

If one could measure the $\delta^{13}C$ of the alkenones and the $\delta^{13}C$ of carbonate shells from planktonic forams, then one could determine a very explicit ε_p value from sediment extracts—specifically, for *Emiliania huxleyi* and a few closely related species of coccolithophores—avoiding most of the problems Hayes described. Jasper's Gulf of Mexico cores covered 100,000 years, from the beginning of the last ice age to the present, and there was a good record of the CO_2 content of the atmosphere for a portion of that period in the Greenland ice core. The ocean surface water was presumably in equilibrium with the air in most areas, and he would only need to know the temperatures, which he could obtain from $\delta^{18}O$ or alkenone U_{37}^K measurements, to determine the amount of CO_2 dissolved in the water. He could then calibrate the relationship between ε_P and CO_2 based on the alkenone $\delta^{13}C$ measurements, and the changing concentrations of dissolved CO_2 over the past glacial–interglacial cycle! Alkenones had been found in sediments dating back tens of millions of years, and though no one knew precisely what had made them then, one might assume that a limited number of closely related species of coccolithophore were also responsible. So one might, in theory, apply such a calibration to rocks dating all the way back to the Cretaceous...Jasper was so excited by these ideas that he went up to talk to Hayes right after the presentation. The problem, he admitted, was that he didn't have any way of measuring the $\delta^{13}C$ of alkenones in the complex mixtures he extracted from his sediments. But Hayes was enthusiastic—there was an instrument in his lab in Indiana that could do precisely that. They just needed to write a proposal so Jasper could work at Indiana University for a year or two when he finished his doctorate at Woods Hole.

The initial results from the Gulf of Mexico core were exciting. Jasper measured the $\delta^{13}C$ of the diunsaturated C_{37} alkenone and of the carbonate in foram shells from a single species known to live at approximately the same water depth as *Emiliania*. Alkenones are typically 3.8 per mil more depleted in ^{13}C than the total biomass of the algae, so this amount was added to the alkenone $\delta^{13}C$ to give values for the $\delta^{13}C$ of the Emily biomass. The photosynthetic fractionation, ε_P, was then determined from the difference between the carbonate $\delta^{13}C$ and the $\delta^{13}C$ of the algal biomass. Jasper made a crude numerical calibration of the relationship between ε_P and CO_2 concentration using a few of these ε_P measurements and dissolved CO_2 concentrations calculated from the measurements of CO_2 in the Greenland ice cores. He then used this calibration to calculate dissolved and atmospheric CO_2 concentrations based on the ε_P values from the rest of his core samples. While he was working on the project, analyses of the CO_2 content of air trapped in the long Antarctic ice core had been published, enabling a comparison between his entire 100,000-year alkenone proxy record of CO_2 from the sediments and the record from ice core measurements.

Like the first appearance of the alkenone paleothermometer a few years earlier, the alkenone paleobarometer caused a stir when the first results appeared in *Nature* in 1990. The sawtooth pattern of the ice ages and the correlation between the glacial–interglacial shift and the amount of CO_2 in the atmosphere were as apparent in the alkenone $\delta^{13}C$ data as they were in the direct measurements of CO_2 in the air bubbles trapped in the ice core. Some differences were also apparent, but their significance was yet unclear. The calibration itself was still crude, and the correlation between dates in the layers of ice and sediments was imperfect. Also, comparison of atmospheric CO_2 measurements, as measured in the air from the ice cores, with dissolved CO_2 measurements in the surface water, as recorded by the alkenones, assumed that the CO_2 in atmosphere and surface

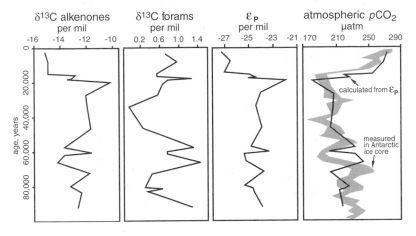

Developing a proxy for atmospheric CO_2, 1990
Estimates of CO_2 concentrations during the last ice age cycle based on the $\delta^{13}C$ of alkenones and foraminifera in a Gulf of Mexico core

water were in equilibrium—but occasional upwelling of CO_2-rich deep water at the Gulf of Mexico site might have disturbed that equilibrium. In fact, one of the beauties of the method was that analysis of cores spanning the same time period in different geographic locations might distinguish regional anomalies in the dissolved CO_2 concentration of the surface ocean and determine where it had been *out* of equilibrium, either burping CO_2 into the atmosphere or sucking it down into the ocean—one of the keys to understanding the relationship between ocean circulation, productivity, and atmospheric CO_2 levels.

Throughout the 1990s, attempts to calibrate CO_2 and ε_P values were made from alkenones in surface sediments and suspended organic matter, and from laboratory cultures of *Emiliania huxleyi* and the other most common alkenone producer, *Gephyrocapsa oceanica*—not only by Hayes's Indiana group, but also by teams of ex-students and postdocs who had scattered around the world, by plant physiologists in Hawaii, and by Stuart Wakeham, who had moved with his sediment traps to the Skidaway Institute of Oceanography in Georgia. These calibrations only partially justified initial hopes that the effects of the other environmental and physiological variables that determined isotope fractionation would be small enough to disregard. In some situations, the effect of variations in algal growth rates on photosynthetic fractionation was as pronounced as the effect of changing CO_2 concentration. A reformulation of Farquhar's equation relates growth rate to the concentration gradient across the cell membrane:

$$\varepsilon_p = \varepsilon_f + ([CO_2] - [CO_2]_{inside})(\varepsilon_t - \varepsilon_f)(1/[CO_2])$$

In empirical calibrations from laboratory cultures or natural populations of algae, this has been simplified to

$$\varepsilon_p = \varepsilon_f + b/[CO_2]$$

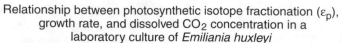

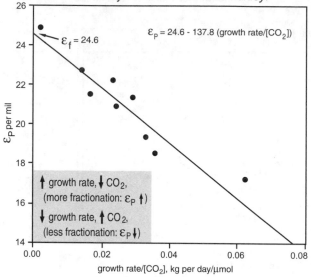

Relationship between photosynthetic isotope fractionation (ε_p), growth rate, and dissolved CO_2 concentration in a laboratory culture of *Emiliania huxleyi*

where all of the physiological effects are bundled into one numerically determined value, *b*. In laboratory experiments with cultures of *Emiliania huxleyi* and several species of diatom, *b* did indeed vary with growth rate, and it was actually the *ratio* of growth rate to CO_2 concentration that yielded the best correlation with ε_p values: isotopic fractionation increases when CO_2 in the environment increases and/or when growth rate decreases.

In the broad surveys of algae and suspended organic matter collected from surface waters around the world, growth rate was difficult to measure, but the *b* factor was found to vary with phosphate concentrations in the surface water—presumably because phosphate is one of the main growth-limiting nutrients—and alkenone-based ε_p values were best correlated with the ratio of phosphate concentration to CO_2. This raises the possibility that ε_p values might provide clues to changes in the amount of marine primary productivity over geologic time, which is almost as titillating—and as frustrating—as the possibility of obtaining information about past CO_2 concentrations. Titillating, because the history of marine productivity and its relationship to atmospheric CO_2 concentrations, ocean circulation, and climate remains elusive. Frustrating, because one needs to know CO_2 levels before one can determine growth rates, and vice versa.

Kate Freeman and her students at Pennsylvania State University—most notably Mark Pagani, now with his own biogeochemistry laboratory at Yale—have been bravely attacking this double-headed beast from all angles. They have acquired ε_p records from alkenone and foram $\delta^{13}C$ measurements in a wide range of ancient marine sediments. Using data from the many calibrations of CO_2 versus ε_P, and employing a variety of techniques and oceanographic information to constrain ancient surface water phosphate concentrations and the related b factor, they have estimated dissolved and atmospheric CO_2 concentrations all the way back to the early Eocene epoch.

For the Miocene epoch, 25 to 5 million years ago, Pagani and his colleagues analyzed sediment cores from open ocean sites where nutrients are consistently low and there are no cycles of upwelling or input from adjacent landmasses, so it was fairly safe to assume that average growth rates had not changed much, and they used actual measurements of surface water phosphate concentrations to approximate values for the b factor. In older sediments, things become more problematic. Alkenones are relatively rare in sediments from the Eocene and Oligocene epochs, and Pagani had to patch together data from a variety of different ocean environments, with varying nutrient and current regimes. Here the *similarity* of the observed trends in alkenone ε_P values at the different sites provides some evidence that they were primarily determined by changes in atmospheric CO_2 concentration and not by changes in nutrient cycling and availability, which would surely have been different at each site. Of course, the coccolithophores were changing and evolving during these long time periods, and the alkenone-producing organisms that predated *Emiliania huxleyi* or *Gephyrocapsa oceanica* may have changed in ways that affected other components of the b factor, such as the alga's cell size or its preference for obtaining carbon from dissolved bicarbonate rather than CO_2. Though there is no conclusive way of eliminating this possibility, Pagani and colleagues argue effectively that the observed trends in ε_P are unlikely to be the result of such evolution.

According to their provocative though still dubious reconstruction of atmospheric CO_2 levels, climate and CO_2 were not always as closely linked as they were during the past 100,000 years of ice age climates. A general, irreversible decline in atmospheric CO_2 concentration during the Eocene and a dramatic drop at the beginning of the Oligocene period appear to be coupled to a steadily cooling climate, and may well have been responsible for it. But during the Miocene, atmospheric CO_2 levels remained relatively low and unchanged, while the climate proxies attest to significant, irreversible changes in the global climate. This is in keeping with the hypothesis that major global climatic trends during this period were determined by tectonic plate movements and opening of the Drake Passage between South America and Antarctica, which initiated the circumpolar current that isolates Antarctica and led to formation of the East Antarctic ice sheet.

It is the beauty of the Ocean Drilling Program that Pagani and his colleagues have been able to extract such an amazing amount of information from what is, in essence, an ambiguous measurement. A more rigorous reconstruction of the history of atmospheric CO_2 levels must await development of reliable independent proxies for growth rate or surface-water nutrient concentrations and, perhaps, more insight into the alkenone-producing organisms that predated Emily, or the calibration and use of ε_P values derived from $\delta^{13}C$ values for other algal biomarkers, such as dinosterol and dinosterane or the HBIs. Richard Pancost,

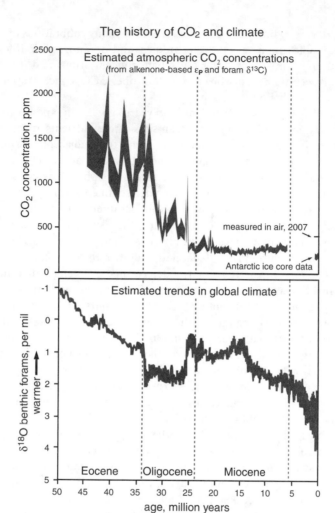

The history of CO₂ and climate

another of Freeman's students, worked with her and Wakeham to explore such possibilities, measuring the $\delta^{13}C$ of various diatom sterols and dinosterol in suspended organic matter from surface waters off the coast of Peru, where upwelling creates a gradient of dissolved CO_2 and nutrient concentrations as one moves away from the coast. But their attempts to calibrate ε_p values determined for diatoms and dinoflagellates, with dissolved CO_2 concentrations and growth rates approximated from nutrient concentrations, met with limited success. Diatom cell size and carbon transport mechanisms vary greatly among species and genera, and none of the diatom sterols were specific enough to provide a good proxy. And for reasons that are still unclear, the dinosterol $\delta^{13}C$ values in the Peru samples were nearly constant, little influenced by either changing nutrient or CO_2 concentrations. But even as biomarker paleo-CO_2 and growth rate proxies await redemption or discovery, the ability to determine the isotopic composition of individual biomarkers has opened doors to understanding relationships between climate and ecology on many other fronts.

As soon as Geoff saw that Hayes and his students had come up with a mass spectrometer that allowed one to determine the ^{13}C content of specific compounds in a sediment extract, he found a British instrument company that had developed a prototype of a similar instrument and started trying it out on his leaf wax *n*-alkanes, using the $\delta^{13}C$ values and chain-length distributions to distinguish among species of trees and track their inputs in lake sediments. The most useful application of such data to date, however, involves a much broader distinction among plant types than that among species or genera, or even families, a distinction that reflects fundamental biochemical differences in their photosynthetic pathways.

Like algae, most trees, shrubs, and northern grasses operate with what's known as C_3 photosynthesis, employing the Calvin cycle's rubisco enzyme for the initial assimilation of CO_2 into a 3-carbon sugar. But tropical grasses and some salt-marsh grasses use C_4 photosynthesis, where the initial capture of CO_2 is facilitated by an enzyme with a greater affinity for CO_2 and a less discriminating taste in carbon isotopes than rubisco's. In these plants, the CO_2 is incorporated into a four-carbon acid and ferried into a special set of cells within the leaf, where it accumulates, isolated from the atmosphere. Here the Calvin cycle takes over and the finicky rubisco, now supplied with a high concentration of CO_2 and protected from oxygen in the atmosphere, operates quite efficiently and with less isotope discrimination. These C_4 plants are generally well adapted to drought and high temperatures, and on the contemporary earth they dominate the huge areas of arid grassland and savanna in subtropical Africa, South America, Australia, and South Asia. Their complicated system of photosynthesis apparently requires more energy to run than C_3 photosynthesis, putting them at a disadvantage in cool, moist climates.

Geoff's group's initial compound-specific isotope investigations of *n*-alkanes in the leaf waxes of C_4 plants revealed that their $\delta^{13}C$ values were consistently 10–15 per mil less depleted in ^{13}C than the corresponding compounds in C_3 plants, reflecting the less finicky nature of the CO_2-fixing enzyme in C_4 plants. In what would appear to be a lifelong extension of the project he started in Tenerife in 1960, and true to the legacy of the Cambridge biochemist whose cigarette tins full of leaf waxes launched his career as a natural historian of molecules, Geoff has been working with Jürgen's student Florian Rommerskirchen on a new collection of leaf waxes from African plants. With the isotope information added, the chain-length distributions are more distinct than they ever were in the Tenerife study. Surveys of the *n*-alkanes and *n*-alkanols in leaf waxes from hundreds of species of plants and grasses have confirmed that these compounds in C_4 plants are consistently less ^{13}C depleted, and their homologue distributions more skewed toward longer chain lengths, than they are in C_3 plants. The *n*-alkanes in C_4 plants have an average $\delta^{13}C$ of about –22 per mil, as opposed to –34 per mil for C_3 plants, and an average chain length of 31 carbon atoms, compared to 29 for the C_3 *n*-alkanes. The isotopic difference between the two plant types clearly

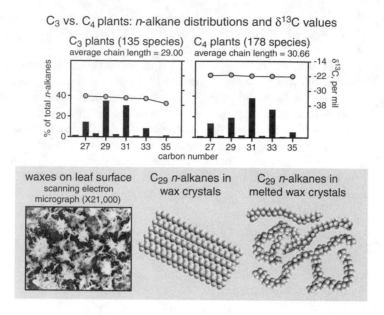

C_3 vs. C_4 plants: *n*-alkane distributions and $\delta^{13}C$ values

C_3 plants (135 species)
average chain length = 29.00

C_4 plants (178 species)
average chain length = 30.66

waxes on leaf surface
scanning electron
micrograph (X21,000)

C_{29} *n*-alkanes in
wax crystals

C_{29} *n*-alkanes in
melted wax crystals

stems from their different methods of assimilating carbon. The chain-length difference may be directly related to C_4 plants' competitive success in warm, arid climes. When skies are clear and the sun shines directly on a plant's leaves, leaf surfaces can become so hot that the waxes melt and lose their protective qualities, including the ability to guard against drought. But the compounds with longer carbon chains have slightly higher melting points, and their relative abundance in C_4 plants may be just enough to keep the molecules aligned in the wax crystals under these conditions.

Unlike algae, which assimilate bicarbonate or CO_2 directly from the water in their environment, the cells that fix carbon in land plants are protected within leaves, and CO_2 collects in the leaves' intercellular spaces before it is assimilated. The amount of CO_2 available for assimilation is regulated to some extent by the opening and closing of pores in the leaf, so the direct effect of changes in atmospheric CO_2 concentration on carbon isotope fractionation is less evident than it is in algae. Nevertheless, at very low atmospheric concentrations, C_3 plants are hard-put to maintain the relatively high concentrations of CO_2 that the rubisco enzyme needs to function efficiently, and C_4 plants have a distinct advantage. Analyses of sediment cores from several mountain lakes in Kenya revealed that *n*-alkanes in sediments deposited during the last glacial period had relatively positive $\delta^{13}C$ values and—taking into account the $\delta^{13}C$ of atmospheric CO_2 from the period, as measured in air trapped in the Greenland ice core—low fractionation factors, like those typical of C_4 grasses. The onset of the interglacial period in this region was marked by a shift to the more pronounced fractionation and negative $\delta^{13}C$ values typical of C_3 plants. According to similar analyses of sediments from a lake in Guatemala, C_4 plants also had the upper hand in Central

America during the last glacial period. Given the C_4 predilection for warm climes, this might seem counterintuitive. But conditions in these regions were not only cooler during glacial times, but considerably more arid, and the concentration of CO_2 in the atmosphere was dramatically lower than during the interglacial period—enough, it seems, to have given the C_4 plants a competitive edge. Sediments from a lake in northern Mexico, however, tell a different story: in this region, C_3 plants prevailed throughout the glacial period just as they do today, perhaps because extensive rainfall counterbalanced any advantage the C_4 plants gained at low concentrations of CO_2.

Paradoxically, some of the best information about regional changes in continental climate regimes and vegetation is coming not from lakes, but from marine sediments. Geoff has worked with Jürgen's group and an interdisciplinary mix of northern German colleagues for years, analyzing Atlantic sediment cores from a north–south transect along the west coast of Africa in an attempt to map the changing ecosystems on the southern half of the continent during the last ice age cycle. They combined data from their analyses of isotopic compositions and distributions of leaf wax hydrocarbons in the sediments with a sharp-eyed Bremen palynologist's counts of plant pollen grains and spores—blown out to sea with the dust, and taxonomically distinguished by their different forms— and made use of geographers' catalogues of vegetation types and atmospheric scientists' calculations of wind trajectories to estimate the sources of dust at their study sites.

The ecosystems so determined for both the last glacial and the current interglacial periods reflect a general north–south trend from humid to arid conditions, with the average chain lengths of the n-alkanes and n-alkanols increasing and their $\delta^{13}C$ values becoming less negative as one moves south from the equator and the ecosystems shift from C_3 plant–dominated rain forest and woodlands to C_4 plant–dominated grasslands and deserts. But at the height of the last glacial period, both the drier climes and the C_4 grasslands clearly extended farther north into the equatorial region than they have during the Holocene interglacial.

Going back beyond the Holocene to a much earlier period of ice age cycles, the NIOZ scientists have studied the waxing and waning of C_4 plant dominance in southern Africa between 1.2 million and 450,000 years ago. Focusing on a single long core from the Angola Basin, they chose the C_{31} n-alkane as an indicator, estimating the percentage of C_4 plants from its $\delta^{13}C$ values and the intensity of the trade winds and dust generation from its rate of accumulation in the sediments. Comparing these data with sea surface temperatures determined from alkenone U^K_{37} values, and ice expansion and retreat from the foram $\delta^{18}O$ values, they came to the conclusion that fluctuations in C_4 plant prevalence for the region had been tied to the changing intensity of the monsoon system and degree of aridity, rather than to low CO_2 concentrations during the ice ages.

Paleobotanists who attempt to reconstruct the evolution of ecosystems in a region from the fossil plant record are often unable to quantify the relative importance of different groups of plants—they may well see some trees in the

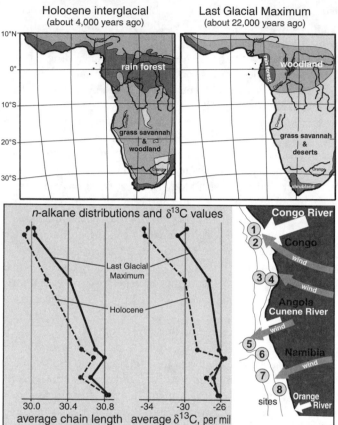

Mapping with biomarkers and plant pollen
Comparing interglacial and glacial African ecosystems

forest, but the forest itself eludes them. The fossil record of C_4 plants is particularly skimpy, because most of them lack fossil-forming woody parts. Biomarker and pollen studies of lake and coastal marine sediments now offer some recourse, providing a view of changing ecosystem type over an entire watershed or wind source region. One thing that has become clear in these studies is that for a given level of atmospheric CO_2, there is a climate optimum where C_4 plants can compete efficiently with C_3 plants and dominate an ecosystem—up to a certain point. Farquhar's laboratory experiments with CO_2 uptake indicate that even in an exceptionally dry, hot climate, if atmospheric CO_2 were to rise above 500 ppm, it would be impossible for C_4 grasses to compete with C_3 plants.

The oldest fossil evidence of C_4 plants is from the mid-Miocene, and there is some evidence from the fossil teeth of herbivores that C_4 grasses thrived during the late Miocene period. According to the ice-core records and alkenone ε_P values, atmospheric CO_2 concentrations have not exceeded 500 ppm during the past 25 million years. But Pagani's somewhat more tentative estimates from earlier time periods suggest that it rose above this level during the Oligocene and was well above 1,000 ppm during the Eocene. The even cruder estimates for the Cretaceous period—based on models of the carbon cycle, ancient soils, ε_p values of porphyrins, and an array of marine algal biomarkers—vary widely but are all well above the 500 ppm limit, ranging from 900 ppm all the way up to 3,300 ppm. It would thus seem unlikely that C_4 plants could have thrived much before the Miocene, and some theories of plant evolution hold that the C_4 photosynthetic pathway in tropical grasses was, in part, an adaptation to decreasing atmospheric CO_2 levels at this time. It may, however, have evolved repeatedly in different types of plants before it found its contemporary home in the grasses, and there is some evidence that it was already in existence during the Cretaceous.

Multiproxy investigations of Cretaceous black shales indicate that CO_2 levels may have dropped dramatically during the oceanic anoxic events, in which case C_4 plants might have seized the opportunity for their moment in the sun, just as they apparently did during the ice ages. Given the global distribution of these shales and their high content of organic matter, a huge amount of organic carbon was buried in the sediments during these events—carbon that would otherwise have been recycled to CO_2 in the surface water. Values of $\delta^{13}C$ for both the marine organic matter and the carbonates formed from bicarbonate typically shifted to more positive values over the course of black shale deposition, as more and more organic matter was buried and ^{12}C was partitioned into the sediments. Such a massive burial of organic matter over half a million years would, presumably, have resulted in CO_2 being sucked out of the atmosphere and might have been responsible for occasional cooling trends in the midst of an otherwise warm Cretaceous period. Isotope fractionation factors determined for porphyrins, sulfur-bound phytane, and C_{27} and C_{28} steranes in black shales deposited during one of the most pronounced oceanic anoxic events, about 93.5 million years ago, all suggest that there was a marked decrease in atmospheric CO_2 at this time—though precisely how much is a matter of some contention, with estimates varying across studies, from 20% to 80%. The NIOZ group found additional evidence in the $\delta^{13}C$ values of leaf wax n-alkanes, which shifted over the course of shale deposition from values in the –30 to –38 range typical of C_3 plants toward the more positive values typical of C_4 plants. They estimated that the change from C_3 to C_4 plant ecosystems occurred within 60,000 years of the onset of organic-matter–rich sediment deposition and proposed that this was the time required for burial of enough carbon to reduce the atmospheric CO_2 concentration below the 500 ppm mark and allow C_4 plants to get the upper hand.

The extraordinary dark bands of sediment discovered by the Deep Sea Drilling Project in the 1970s are finally providing something more than molecules for the collectors. But the realization that a bit of productive exuberance in the surface waters might have pushed the deep waters of a slowly circulating ocean system over the edge into anoxia for more than 100,000 years at a stretch—or that a pronounced CO_2-based greenhouse effect that had persisted for most of the long Cretaceous was interrupted by this event, and, incidentally, that C_4 photosynthesis may have evolved in land plants more than 90 million

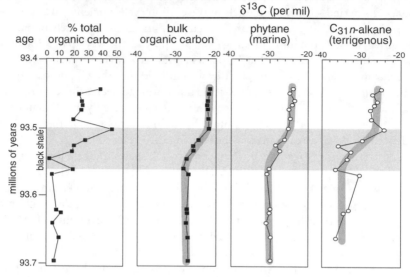

Did Cretaceous oceanic anoxic events allow C_4 land plants to become more prevalent? Evidence from a DSDP core off the coast of northwest Africa

years ago—is only the beginning. The real surprises are coming not from the biomarkers of land plants and algae, but from those of smaller, more unassuming and sometimes downright anonymous organisms, which, as we shall see, have finally attracted the attention of microbiologists.

As the investigations of the Messel shale demonstrated, addition of the isotopic dimension to the structural information in the molecular lexicon makes it easier to link fossil molecules to source organisms, and to differentiate among possible sources for the same compound. But perhaps most important for understanding the contributions of microorganisms to an ecosystem, the isotopic makeup of a biomarker can link it directly to a biochemical *process*. The prevalence of organisms that obtain their sustenance by different means—C_3 or C_4 photosynthesis, respiration, fermentation, or one of the various energy- and carbon-harnessing processes that microorganisms have construed—can now be assessed and environmental conditions inferred, even when those organisms leave no other trace of their existence. Even, indeed, when they comprise organisms that have completely eluded detection—unknown, hitherto unimagined extinct and, miraculously undiscovered, extant species. Commercially built GC-irm-MS has joined the ranks of organic geochemists' most coveted tools, and compound-specific isotope analysis has become a crucial component in their investigations. With such tools in hand, they have joined ranks with microbiologists and molecular biologists, and the rich, varied, microbial life of natural environments has thrown open its doors to exploration, yielding astonishing discoveries at every turn.

7

Microbiologists (Finally) Climb on Board

By this, the Earth itself, which lyes so neer us, under our feet, shews
quite a new thing to us, and in every little particle of its matter, we
now behold almost as great a variety of creatures as we were able
before to reckon up on the whole Universe itself.
—Robert Hooke, 1635–1703, British natural philosopher, referring to his view of life
    through one of the first microscopes
    From *Micrographia: Or Some Physiological Descriptions of Minute Bodies Made by*
    Magnifying Glasses (1667)

With the most rational philosophers an increase in their knowledge
is always attended by an increased conviction of their ignorance.
—Georg Christoph Lichtenberg, 1742–1799, German natural philosopher
    From *Aphorisms* (1990 translation)

If I could do it all over again and relive my vision in the twenty-first
century, I would be a microbial ecologist. Ten billion bacteria live
in a gram of ordinary soil... almost none of which are known in
science. Into that world I'd go with the aid of modern microscopy
and molecular analysis.
—Edward Osborne Wilson, 1929– , American biologist
    From *Naturalist* (1994)

In the half century since paleontologists began finding putative microfossils in Precambrian sedimentary rocks, it has become apparent that not only is most of life's history absent from the visible fossil record, but huge sectors of extant life remain to be discovered. Until the early 1980s, the only way to identify and study species of microbes—many if not most of which are morphologically indistinct from each other—was by growing them in the laboratory, isolating the separate colonies of organisms that developed as they reproduced, and then noting differences and similarities in what they consumed and produced. But the capacity to read the information in microbial genes that was developed in the 1980s and 1990s opened an entirely new world for study—much as the invention of the microscope had in the eighteenth century—and laid bare the unnerving fact that the vast majority of microbes on the planet had been boycotting the microbiologists' carefully prepared cultures.

Microbiologists had spent almost a century painstakingly cultivating, isolating, and classifying microorganisms, and yet they had failed to identify the most abundant microbes in natural waters, sediments, and soils. Indeed, it now appears that the hundreds of thousands of microbes named and maintained in the world's bacteria zoos, or "culture collections," invaluable as they are, comprise but a tiny and somewhat random sampling of the microbial world. These microbial cultures provide the only means by which biologists can directly manipulate and study the biochemistry and physiology of this huge sector of life in the laboratory—but they are distinguished more by their ability to prosper under laboratory conditions than by their importance in natural ecosystems. In the past few years, application of new techniques from molecular biology has resulted in the discovery of thousands of strange new types of microorganisms, and there is the implication of countless more: in the twenty-first century we find ourselves unexpectedly gathered at the threshold of a new world, looking not to Mars or Jupiter or to some distant galaxy, but gazing awestruck at the mud beneath our feet and the water in our seas.

The two paradigm-challenging discoveries that would bring us to that threshold were both made in 1977, but had nothing to do with each other. One was a spectacular discovery made by oceanographers exploring the seafloor of the eastern Pacific Ocean in a tiny, three-person research submarine. It immediately galvanized the wider scientific community and, for the first time, brought microbial ecosystems that had nothing to do with disease or food into the limelight. The other discovery was made by a lone molecular biologist poring over tables of gene analyses in a small laboratory in Illinois, and several years would pass before microbiologists and biologists recognized its profundity.

On the seafloor just east of the Galapagos Islands, at one of the mid-oceanic ridges that separate tectonic plates, oceanographers discovered incredible regions of hydrothermal activity, places where geysers of hot water spouted from every crack and the seafloor was festooned with a veritable fantasia of strange mineral formations. This was more or less what oceanographers, geologists, and geochemists expected, if, as all evidence indicated, the ridge was a center of seafloor spreading, with hot magma rising from deep within the earth to form new oceanic crust. Any seawater that percolated through fissures in the rocks would be heated, producing hydrothermal activity. The superheated water would dissolve chemicals in the rocks, and as the water spouted back into the icy deep sea, these would precipitate and form minerals. The scientists in their submersible could actually see the shimmering hot water from the vents turning milky white or pale blue as it cooled and the minerals precipitated. Further exploration revealed places where the geysers were literally black with the fine particles of metal sulfides that precipitated as the hydrothermal fluids cooled. What no one had expected, however, what shocked oceanographers, geologists, and most of all, biologists, was the cornucopia of life they found swarming around the hydrothermal vents.

Biologists had long written off the deep sea as a cold, lightless desert, occupied only by the few scavengers that could survive on the meager supply of organic matter from the surface, or by a lone fish that could navigate large distances in search of sustenance. But these hydrothermal oases were surrounded by lush communities of giant mussels, dinner-plate–sized clams and crabs, anemones, pale pink fish, and dense stands of bizarre red-capped tubeworms, all thriving in absolute darkness beneath 2,500 meters of water. Further exploration in the research submersibles revealed that these hydrothermal vents, with their lush ecosystems populated with previously unknown species of invertebrates, were scattered all along the mid-ocean ridges. But what was sustaining them? Where did the energy for all that life come from, if not sunlight and photosynthesis in the surface waters? There couldn't possibly be enough bits of organic matter raining down from the surface to feed all those mussels and clams and worms. What could they eat? For that matter, *how* did they eat? The strange tubeworms were just hollow tubes with no mouths, stomachs, or anuses!

Samples of the hot water coming out of the vents were swarming with bacteria; indeed, every hard surface in the vicinity of the vents was caked with thick mats of white or pink or yellow microbes. These, it seemed, were the primary producers for the entire fantastical ecosystem, a resourceful collection of minute autotrophic bacteria that had the wherewithal to harness chemical energy by oxidizing the sulfide in the hydrothermal waters. Many of the mobile vent animals seemed to graze on the microbial mats, like so many deer in a meadow. But what about the immobile, gutless tubeworms and mussels? Geochemists compared the carbon isotope content of tissue from the mussels and tubeworms around the vents with that of mussels from other areas, and found that they were wildly different: the vent organisms clearly did not feed on detritus from the surface, which would have borne the isotopic signature of photosynthetic organisms. In 1980, Colleen Cavanaugh, a young microbiologist at Harvard, proposed that the tubeworms, which obviously couldn't graze, must have sulfide-oxidizing bacteria living inside their tubes. Cavanaugh was still a graduate student at the time, and it took some doing for her to obtain samples from the strange creatures— but when she did, she found a whole factory of sulfide bacteria living in symbiotic bliss within their tubes. It turns out that the feathery red plume at the end of the tube is full of hemoglobin and serves as a sort of gill, absorbing dissolved compounds from the hydrothermal vent fluid and providing the bacteria with oxygen and with a ready supply of the sulfide they need for energy. The bacteria can then fix carbon and manufacture sugars for their host worm's consumption. Similar symbiotic relationships were found in many other vent animals, such as mussels, clams, and shrimps, which can actively mine the metal sulfides from the rocks to fuel their bacterial factories. More than 300 new species of invertebrates have been discovered at deep sea hydrothermal vents since 1977, but it's the incredibly productive chemoautotrophic bacteria at the base of their food chain that has really transformed thinking about life, shifting the focus of biology from plants and animals to microbial life and its interaction with the mineral world.

A hydrothermal vent landscape (1980, East Pacific Rise)

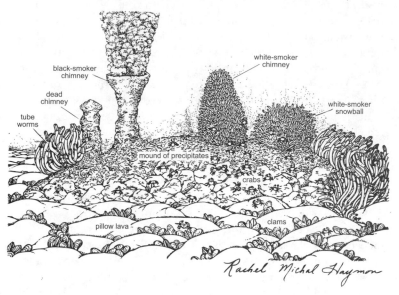

Biologists were still puzzling over the enigma of so much life in the deep sea, and buzzing with excitement about the first new species of invertebrates, when Carl Woese published a paper announcing that the supposedly fundamental dichotomy of life—undifferentiated unicellular organisms called prokaryotes or, at the time, bacteria, versus compartmentalized cells with membrane-enclosed genetic material, the eukaryotes—was not, in fact, a dichotomy. Woese had been trying to determine the evolutionary relationships among microorganisms and develop a universal system of taxonomic classification that reflected those relationships, something that had eluded biologists for more than a century. Rather than basing classification on scant morphological variations or metabolic differences that might or might not be discerned in bacterial cultures, he turned to the ultimate form of chemical taxonomy: differences between organisms' most essential molecules, their genes. By painstakingly comparing sequences of nucleotides in ribosomal RNA molecules from hundreds of species of bacteria, Woese identified the particular nucleotide sequences that were common to all, thus defining them as bacteria...and then he came across a handful of species that were missing those sequences. They were all simple undifferentiated cells that looked and acted like bacteria, but according to their genes they were as different from bacteria as they were from lions or pine trees. What then? He assigned them a group of their own and soon found more species that fit right in, including all the known species of methane-producing organisms, as well as an odd assortment of extremists that thrived in bizarre environments where it seemed impossible that anything at all could live: halophiles living in extremely saline brines, thermophiles thriving in the near-boiling water of hot springs, and thermoaciophiles, who like their hot water spiked with acid. Woese named his new group archaebacteria, and would

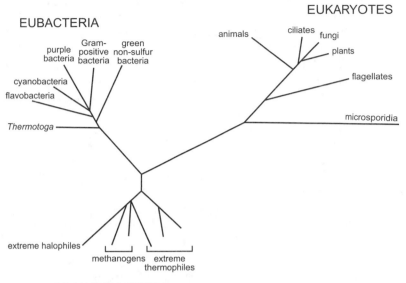

Woese's universal phylogenetic tree, 1987

eventually propose a new taxonomy based on a universal phylogenetic tree that better reflected evolutionary relationships between *three* fundamental domains of life: Eubacteria or simply Bacteria, Archaebacteria or Archaea, and Eukarya.

In 1977, when Woese proposed the division of the prokaryotes—broadly defined as organisms whose cells lack internal membrane-bound compartments—into two distinct taxonomic groups based on their rRNA sequences, many microbiologists in the United States ignored or pooh-poohed the distinction as insignificant. But microbiologists in Germany had already noted that the cell walls of methanogens were different from those of other bacterial cells, and gradually other evidence of fundamental differences began to accumulate...including one that caught the immediate attention of the organic geochemists in Strasbourg and Bristol. In 1979, just three years after Michel Rohmer tracked the hopanoids to bacterial lipids, biochemists in Canada and Italy confirmed that the methanogenic, halophilic, and thermophilic microorganisms in Woese's archaea all had cell membranes with an entirely different structure and lipid makeup than either the eukaryotes or the bacteria. In place of the usual phospholipids, where a glycerol molecule is linked to fatty acids by ester bonds, the archaeal membranes consisted of acyclic isoprenoid chains attached to glycerol by ether linkages. Ether bonds are generally more stable than ester bonds, so at first glance this appeared to be an adaptation to the extreme environments where archaea thrived. But it soon became apparent that the strange lipids were indicative of a more profound and universal difference between the Archaea and the other two domains of life.

One of the most widespread compounds was a diphytanyl glycerol diether with two 20-carbon phytanyl chains, which would come to be known as archaeol. Another common compound was a *di-bi*phytanyl *di*glycerol *tetra*ether wherein two 40-carbon biphytanyl chains were bound at both ends by ether links to glycerol molecules, so that instead of dangling free, the carbon chains were constrained in a long, narrow rectangular sort of structure. The biphytanyl chains themselves were distinct in that they consisted of two phytane molecules connected by a *head-to-head* link, instead of the usual head-to-tail isoprenoid link. Stranger still, the biphytanyl chains in the tetraethers of some thermophilic organisms contained one or more five-carbon rings, where the methyl side chain of an isoprenoid group had looped around to form a cyclic structure within the chain. How such compounds were assembled to form a cell membrane was not entirely clear, but there was some evidence that the tetraethers traversed the entire membrane to form a relatively dense, solid monolayer, rather than the fluid phospholipid bilayer of eukaryote and bacterial membranes. Whatever the case, it was the sturdy, idiosyncratic isoprenoid chains that caught the attention of organic

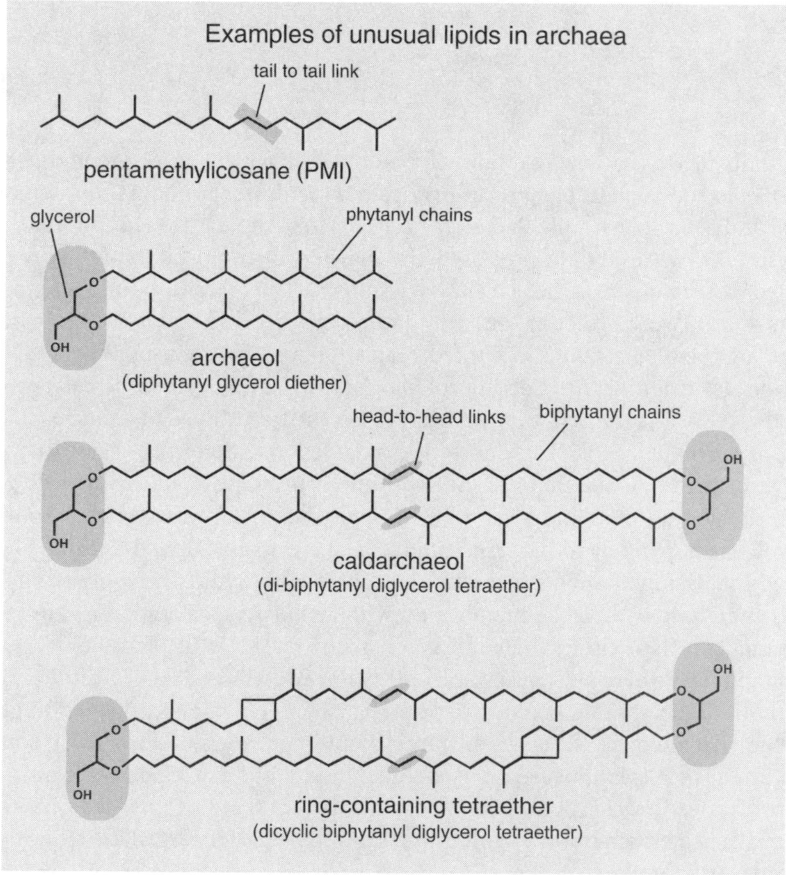

Examples of unusual lipids in archaea

tail to tail link

pentamethylicosane (PMI)

glycerol   phytanyl chains

archaeol
(diphytanyl glycerol diether)

head-to-head links   biphytanyl chains

caldarchaeol
(di-biphytanyl diglycerol tetraether)

ring-containing tetraether
(dicyclic biphytanyl diglycerol tetraether)

geochemists in Bristol and Strasbourg. Acyclic isoprenoids in general seemed to be an archaean specialty. Among the nonpolar lipids found in sediments and crude oils, regular acyclic isoprenoid hydrocarbon chains with 15 to 35 carbon atoms, as well as a couple that had *tail-to-tail* links in the middle like carotenoid pigments, were found in such quantity that it seemed they must have some significant role to play in cell membranes—but to date there is no direct evidence that hydrocarbons lacking a polar group can, as such, form part of a membrane.

When the archaeal membrane lipids started making news in 1979, there were, as it happened, a number of molecules languishing in Strasbourg's so-called molecule orphanage that bore a striking resemblance to the weird biphytanyl components of some of the archaeal lipids. They had been discovered in the Messel shale and relegated to anonymity by a visiting German postdoc back in 1972, when Pierre Albrecht and the Strasbourg crew were trying to get a better idea of what, exactly, was in the shale's kerogen. Walter Michaelis was trying out various chemical treatments to break the bonds in the kerogen and then running the compounds he generated on the GC-MS and figuring out their chemical structures. The reagent boron tribromide, which was strong enough to break all the oxygen bonds, released a preponderance of acyclic isoprenoids, including 40-carbon head-to-head linked biphytanes—some of which had five-carbon ring structures woven in along their chains. Michaelis says that neither he nor anyone else in Strasbourg could find any clues as to the source of these strange compounds, and when his fellowship ended, he left Strasbourg disappointed, assigning them to orphanhood.

In 1979, however, not long after the Italian biochemists who were studying archaeal membranes reported the structures of the biphytanyl diglycerol tetraethers, Moldowan and Seifert at Chevron found a series of ring-containing biphytanes in their petroleum samples. The biochemists had found the ring-containing compounds only in thermophilic organisms, which, the Chevron group postulated, could have lived in the maturing sediments when they reached oil formation temperatures. But what about the compounds Michaelis and Albrecht had found in the immature kerogen of the Messel shale? It had never reached oil formation temperatures, and its plethora of well-preserved animal and plant fossils were ample evidence that the ancient lake itself had not been particularly hot or overly acidic.

Michaelis arranged for a year's leave from his position at Hamburg University and went back to work with the Strasbourg researchers. This time they worked on sediment extracts and immature oils, rather than kerogen, choosing samples of various ages that had been deposited in low-oxygen or anoxic environments resembling that of the ancient Messel Lake, places like the Cariaco Basin and the Black Sea. They isolated the ethers by thin-layer chromatography, used their chemical treatments to cleave the oxygen bonds—and obtained an entire series of acyclic isoprenoids with a clear structural heritage from the archaeal lipids the microbiologists had isolated. The ether bonds were surprisingly resilient, as the biphytanyl diglycerol tetraethers persisted intact even in immature oils and Cretaceous age sediments.

Identifying methanogen lipids in these sediments was exciting, but not particularly surprising, because the samples came from the sorts of anoxic environments where methanogens thrived. But tetraethers with ring-containing biphytanyl components were also prominent—and, again, none of the environments sampled showed any signs of having ever been hot enough to support thermophilic organisms. In a speculative review of microbial biochemistry published in 1982, Guy Ourisson, still marveling that a bunch of old rocks and mud had led the way to discovery of hopanoids in bacteria and had such an impact on biochemical and evolutionary theories, predicted that these ring-containing biphytanyl tetraethers in the sediments would eventually lead to discovery of new species of methanogenic archaea.

In the Bristol lab, news that some of the acyclic isoprenoid alkanes in ancient sediments might be fossil molecules of anaerobic, methane-producing microorganisms, rather than remnants of chlorophyll or carotenoid pigments, was enough to trigger a reassessment of the scientists' collection of DSDP sediment samples. If they treated the polar fractions of their extracts to cleave the ethers, would they find the head-to-head linked biphytanyl chains or, perhaps, the distinctive isoprenoid alkanes with the tail-to-tail links? Biologists had known since the nineteenth century that there were microorganisms that produced methane and lived in waterlogged environments where there was lots of organic debris and no oxygen. In many marshy areas, the gas could actually be detected bubbling out into the atmosphere. But only recently, in the late 1970s, when marine chemists began analyzing the dissolved gases in water trapped in the pores of DSDP sediments, had it become apparent that such organisms also lived deep within marine sediments along continental margins, and even deeper beneath the sea floor in the open ocean. Sure enough, when the Bristol group focused on DSDP samples from core sections where methane had been detected, the nonpolar hydrocarbon fractions of the extracts consistently boasted large amounts of two distinctive acyclic isoprenoid alkanes: the 30-carbon squalane, and the 25-carbon pentamethylicosane known as PMI, both with a tail-to-tail link in the middle, and both among lipids of methanogens identified by the microbiologists. Here, finally, was a distinctive series of fossil molecules that seemed to reflect a particular group of microorganisms that could only live in certain environments—anoxic ones. But though PMI, and the diphytanyl glycerol diether known as archaeol, and the simple ringless biphytanyl diglycerol tetraether known as caldarchaeol, with its distinctive head-to-head isoprenoid linkages, were all added to the molecular lexicon in the early 1980s, it would be another 15 years before anyone employed that lexicon in a hypothesis-driven attempt to understand the production—and consumption—of methane in such environments.

Organic geochemists in the 1980s were almost as ignorant of sediment microbial life and its leavings as they were in the 1960s, when Pat Parker first noted that the *iso* and *anteiso* fatty acids in his Texas mud had been made by heterotrophic bacteria. They had gained some insight into diagenesis, much of it the result of microbial processes, and marine chemists working with the

DSDP and ODP had been making routine measurements of dissolved gases and minerals—microbial waste products and energy sources—in the top few hundred centimeters of marine cores. But direct knowledge of the actual organisms or processes responsible was limited. It was impossible to link the compounds found in the DSDP stamp-collecting endeavors with specific groups of sediment microorganisms, or to disentangle the molecular fossil record of life and the environment in ocean surface waters from that of the sediment microbes. It had taken almost a decade, not to mention a good deal of serendipity, just to trace the hopanoids to a bacterial source—and then most of the specific structural variations in the compounds that Rohmer isolated were in the polar side chain, lost during early diagenesis and of little use in distinguishing bacterial processes that had been prevalent in ancient environments. The few microbiologists who had denounced their discipline's subservience to medicine and set out to explore the microbial world in nature were still developing tools for their enterprise. Much of their work focused on microbial mats, partly because these were relatively simple ecosystems to study, and partly because they contained enthrallingly exotic organisms that could live in salty brines or near-boiling water and might, potentially, be analogues to ancient life-forms.

The motivations for Geoff's long-standing interest in microbial mats were not much different, though of course they had a molecular slant. Ever since working on Melvin Calvin's Precambrian rocks in the 1960s, he had wondered if they might yet find molecular fossil clues to early life in the ancient stromatolites, if only they knew better what to look for—and what better place to find that out than in living stromatolites or, for that matter, microbial mats in general? On a more pragmatic front, microbial mats constituted relatively simple environments where different types of microbes were, presumably, associated with visibly distinct layers that were dominated by a limited variety of organisms. This arrangement made it easier to link the mats' molecular content with specific groups of photosynthetic or heterotrophic microorganisms, which, in turn, they hoped would help in interpreting the emerging molecular fossil records in marine sediments. But, finally, if one pushes Geoff to explain the logic of his microbial mat studies, he will admit that microbial mats are rather intriguing, slimy-looking things, usually growing in bizarre places where nothing much else will grow, and, as the refrain of 50 years goes, "We just wondered what was in them!" His first real encounter with the slimy things was in 1968, during a teaching stint with Pat Parker in Texas. Over the next decade, he analyzed samples whenever he could find an interested microbiologist to work with, and in the early 1980s he and a group of like-minded geochemists—Jan de Leeuw, Jaap Boon, and Al Burlingame among them—teamed up with a trio of Israeli, German, and Danish microbiologists who were studying the oldest living microbial mat ever discovered, deposited over the past 2,500 years in a shallow tidal lagoon known as Solar Lake, on the Sinai Peninsula.

The Solar Lake mat consisted of more than 70 centimeters of microbial organic matter, in which the geochemists identified a variety of carotenoid pigments,

Microbial mats from around the world

Geoff harvesting a "slimy thing,"
Texas, 1972

Core from the Solar Lake mats

Mat builders: cyanobacteria as
seen through the light microscope

Cyanobacterial mat from
Yellowstone National Park.

short-chain *n*- and *iso*-alkanes and fatty acids, and the mid-chain branched alkanes that appeared to be a cyanobacterial specialty. While they attempted to map the different distributions of compounds in the layers of mat, the microbiologists tried to determine "who" was doing what in each layer—no easy task. This was a molecular-scale world, where a slight chemical gradient in the concentration of a nutrient or gas might be the only salient indication of activity, and the microbiologists had designed tiny electrodes that could be inserted directly into the living mat to measure oxygen concentrations, acidity, and sulfide. The mat layers, they found, were not only less homogeneous than they appeared to be, but also entirely interdependent—the leftovers from one layer were the next one's raw material, and one organism's waste or refuse was another's most coveted energy source. Ecology, it seemed, was key. The microbes lived in tight-knit, medieval-style communities whose members were as reliant on each other as weaver and tailor, or winemaker and tippler—isolate them in a laboratory culture, and they were likely to change operations or cease to function altogether.

Though most of the compounds the geochemists identified in the Solar Lake mat were either too unspecific or too rapidly transformed by diagenesis to be of much use as biomarkers in the geologic record, their patterns and distributions could, potentially, offer quite a few clues about the living organisms, something that caught the attention of the project's microbiologists—particularly that of the Dane, Bo Barker Jørgensen, who would collaborate with organic geochemists off and on for the next 20 years. The lipid analyses also impressed a microbiologist who had nothing to do with the Israeli mats at all but was, rather, working at Montana State University and studying microbial mats in Yellowstone National Park. Like the Solar Lake microbiologists, Dave Ward had long suspected that laboratory culture techniques were not telling the whole story when it came to microbial communities, even apparently simple ones like those in the mats he was studying. He says he'd been generally skeptical of laboratory cultures ever since he'd taken a class in aquatic chemistry: the range of temperatures, pH, nutrients, and other chemical constituents in one small sample of water from a local lake were so far removed from the carefully contrived conditions he had to use to get lake bacteria to grow in the laboratory that it was hard to believe his cultured organisms could have played much of a role in the microbial life of a real lake.

When Ward heard about the Solar Lake studies in 1982, he was studying the microbial mats that grow in and around Yellowstone's hot springs, where the extreme heat and bizarre hydrothermal chemistry presumably limited the range of organisms that could survive. And yet Ward suspected that there was more to the mats than met the eye—that they were far more complex than they appeared under the microscope, and that the few organisms that adapted well to life in a laboratory culture were misleading him. He was thinking about trying to extract the fatty acids from thin slices of mat, to see if the distributions would allow him to differentiate and quantify the organisms living in different layers, when he went to a conference at Woods Hole and heard Geoff's report on the Solar Lake mats. "It was awesome," Ward says, summing up his original impression, which was that there was a lot more to the molecular composition of the mats—and the information it might provide—than he had realized. He immediately queried Geoff about taking a year's sabbatical leave in Bristol to learn the methods.

Geoff, of course, was keen on any project that involved something as beautiful and exotic as Yellowstone's hot springs, not to mention close collaboration with a young and upcoming microbiologist who knew the hot spring mats as well as they could, to date, be known. He and Ward scheduled a sabbatical stay for Ward in Bristol, and a visit to Montana for Geoff—but the summer before the sabbatical, Ward received a call from a well-known molecular biologist who also wanted to visit Yellowstone and, as it turned out, was developing a technique for in situ study of microbes that looked even more promising than lipid biomarkers.

Norman Pace was one of Woese's most stalwart supporters and a principal contributor to his growing, rRNA-based phylogenetic scheme. Like Ward and the Solar Lake group, Pace and his colleagues were trying to develop ways to characterize populations of microorganisms within their natural environments, and it had occurred to them that they might be able to apply their rRNA techniques directly to the mixed populations in water and sediments. They wanted to collect water samples from the Yellowstone hot springs, filter out the microbes and extract their nucleic acids, isolate and sequence the rRNA, and compare the sequences to those in the phylogenetic database of microbial species. If the sequences belonged to known species of microbes, then they could determine the relative abundances of those microbes in the hot spring microbial community. If they discovered new sequences, they would gain some knowledge of the diversity of organisms in the hot spring environment; and if they could fit those new sequences into the existing phylogenetic tree of evolutionary relationships, they might gain a few clues to the physiology and metabolism of the new organisms. That was the idea, anyway. It was precisely what Ward was looking for: he arranged to spend most of his sabbatical at Pace's lab in Denver, and feeling a bit chagrined, apologized to Geoff and scaled back his Bristol stay to a few months. That was enough, however, to launch a long series of collaborative projects with organic geochemists—first with Bristol, and then later with NIOZ.

The Yellowstone hot springs were typical of the strange environments that archaea were known to be fond of, and several species had already been isolated by microbiologists. When the Montana-Bristol alliance started work in 1984, both Geoff, in his molecule-collecting fervor, and Ward, who had spent years studying methanogenic archaea, were eager to look for distinctive isoprenoid ether lipids. They started with an alkaline hot spring mat where methanogens were clearly active. Working with students and postdocs from both universities, they identified archaeol and caldarchaeol, both among the lipids in the one species of methanogenic archaea that had been isolated from the mat and, it was beginning to seem, common to most if not all archaea. But they also found the strange ring-containing biphytanyl compounds, which none of the cultures of organisms grown from the mat contained. The same compounds turned up again in an acidic hot spring mat, but again, none of the known mat organisms contained them. It seemed that even these simple, well-studied, contemporary microbial mats harbored organisms that Dave Ward and the various other microbiologists who'd spent years scrutinizing them were completely unaware of.

Much of the interest in the Yellowstone microbial mats focused on their intriguing assortment of primary producers, photosynthetic organisms that were likely to bear a strong resemblance to those that colonized a young, unsettled earth. Years of culture work and microscopic observation had led microbiologists to conclude that the top few millimeters of each mat was dominated by one or two species of photosynthetic organism, depending on the temperature and chemical makeup of the particular hot spring: eukaryotic algae in relatively cool acidic springs, thermophilic cyanobacteria in alkaline hot springs, and thermophilic anoxygenic photosynthetic bacteria—which utilize reduced sulfur compounds and don't generate oxygen—in sulfidic springs. The lipid analyses indicated that the only mats to contain significant amounts of sterols were those formed by eukaryotic algae, in keeping with observations that, of the three domains of life, only members of the Eukarya made sterols. The productive layers of mats formed by anoxygenic photosynthetic bacteria were characterized by distinct series of C_{29}–C_{33} alkenes and of C_{28}–C_{38} wax esters. The mats made by cyanobacteria sported the same set of mid-chain branched 17-carbon alkanes that had been

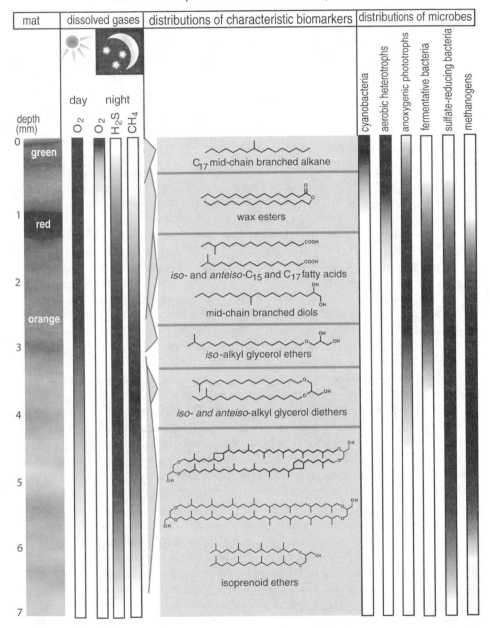

A microbial mat community from an alkaline hot spring: schematic representation of a vertical profile

found in the Solar Lake mat. The alkenes and wax esters weren't of much use to geochemists trying to read the geologic record, because the distributions were quickly lost to diagenesis and would, in any event, be obscured by inputs from eukaryotes in more complex environments. But the mid-chain branched alkanes were distinctive and persistent, and had even been identified among the hydrocarbons in Precambrian rocks and stromatolites—putative evidence for the presence of cyanobacteria on the early earth.

For microbiologists, one of the most important contributions of the Montana-Bristol alliance, and later of the Montana-NIOZ alliance, was in understanding the structure of microbial communities and how they depended on the flow of energy. Just as the Solar Lake studies had shown, the mat layers were not as distinct as they appeared, and organisms could actually move up and down in the mat, following the diurnal to nocturnal shift in chemical gradients created by the photosynthetic organisms. But the distributions and relative amounts of compounds did reflect the distribution of organisms from various links in the food chain—photosynthetic organisms enjoying the sunshine at the top of the mat, aerobic heterotrophic community members performing the first round of decomposition beneath them, and then anaerobic fermentative microbes breaking up the energy-rich sugars and carbohydrates, while the sulfate-reducing bacteria and methanogenic archaea mopped up the leftovers at the bottom. The lipid distributions also provided clues to the microbes' respective metabolic roles and the flow of energy in the mat. In an alkaline hot spring mat where primary production was shared by cyanobacteria and anoxygenic photosynthetic bacteria, mid-chain branched alkanes, wax esters, and alkenes, as well as straight-chain C_{16} and C_{18} fatty acids, were in greatest abundance. The C_{15} and C_{17} *iso* and *anteiso* acids typical of aerobic heterotrophs and fermentative bacteria were next in abundance. And the alkyl glycerol ethers associated with sulfate-reducing bacteria, which depend on small two- and three-carbon organic compounds from the heterotrophs and use sulfate instead of oxygen, were present in trace amounts, as were the isoprenoid glycerol ethers typical of methanogens, which compete with sulfate reducers for small molecules from the heterotrophs or use the CO_2 they generate.

The lipid work on the Yellowstone mats served as the basis for what is now a useful and actively growing lexicon of microbial lipid biomarkers—in large part because of the addition of the isotopic dimension, on the one hand, and the gene-based techniques that Pace and his colleagues had been developing, on the other. Geoff was at the Montana lab when Ward obtained his first rRNA sequence from one of the surface layers of the mats they had been studying. Ward excitedly read off the new sequence of nucleotides, and Geoff compared it to that from the one species of anoxygenic phototroph that had been isolated from the mat—a distant relative, as it turned out, of the one Ward had just discovered. By 1990, Ward's survey of the rRNA gene sequences in the mats' surface organisms made it clear that what had appeared to be virtual microbial monocultures were, in fact, composed of many different, hitherto unknown species. In one of the most thoroughly studied mats, the species of cyanobacteria and anoxygenic phototrophs that had appeared to share exclusive rights to the mat's primary production were not even the most prevalent primary producers. As Ward had long suspected, the organisms he and his colleagues had isolated in cultures were misleading them about the diversity and ecology of organisms in even these most rudimentary of microbial communities.

Meanwhile, some headway had been made in identifying microorganisms responsible for the breakdown of organic matter in marine sediments, though understanding was still largely based on what could be deduced from the chemistry of the sediment pore water, namely, changes in concentrations of the dissolved salts and gases that could be used and produced by microbes. These studies had identified a general hierarchy of microbial processes similar to that in the microbial mats, except that primary production and the first steps of decomposition occurred in the surface waters of the ocean and were separated by thousands, rather than thousandths, of meters from the sediment processes. Moving downward into the sediments, the hierarchy of processes was characterized by the changing availability of oxidants available to burn the various microbial "fuels"—electron acceptors and electron donors, respectively. When all the oxygen has disappeared, the next best oxidant, nitrate, was used, and then manganese, ferric iron, sulfate, and, finally, the fermentative bacteria could use components of the organic matter to oxidize other components, and the methanogens could even use CO_2 as an oxidant. It was the combination of available oxidants and fuels that called the shots for microbial activity, and as the best oxidants disappeared, the amount of energy available decreased and the fuels that could be used became more limited. Again, it was soon apparent that one

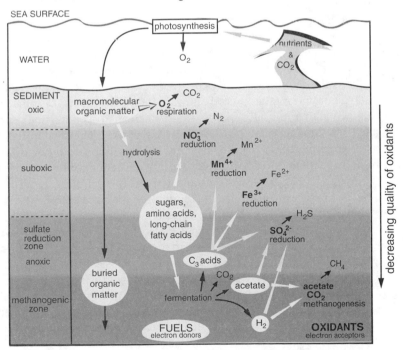

Microbial breakdown of organic matter in marine sediments

organism's waste was another's fuel, and in some cases, what was an oxidant for one process could be a fuel for another.

Even before microbiologists had a good idea of precisely what organisms were doing and how, the importance of sediment microbial ecosystems in regulating the chemistry and biology of the ocean was apparent. One of the big discoveries of the DSDP and ODP in the 1970s and 1980s was that huge amounts of methane were being produced in the anoxic sediments just below the zone of sulfate reduction, particularly in areas with a high input of organic matter such as along the continental margins. Geologists had long known that there were large stores of the ultimate, one-carbon hydrocarbon trapped deep in the rocks, but they'd assumed it came mostly from petroleum source rocks that had been buried and heated so extensively that all the carbon–carbon bonds had cracked in some portion of the bitumen. Methane that was clearly associated with petroleum deposits and clearly thermogenic had $\delta^{13}C$ values around −40 per mil. Most of the methane in marine sediments, however, turned out to be decidedly more depleted in ^{13}C, with $\delta^{13}C$ values between −90 and −60 per mil. The studies of isotopic fractionation by methanogens that John Hayes and others did in the 1980s showed that methane generated by methanogens was about 60 per mil more depleted in ^{13}C than the CO_2 or small acids they consumed—both of which are generated by decomposition of organic matter in the sediments and already somewhat depleted in ^{13}C relative to the CO_2 in the surface waters. So methane with a $\delta^{13}C$ value that was more negative than −60 was pretty likely to have come from microorganisms, and not from the thermal cracking of oil.

The chemists, geologists, and oceanographers of the DSDP and ODP explorations were more interested in what happened to all the methane being generated in the sediments than they were in the details of the organisms that produced it. Their measurements indicated that the dissolved methane accumulated until the sediment pore water became saturated, and then free gas began to accumulate, as well. As the sediment was buried and the pores contracted, the methane was squeezed out and migrated slowly upward in the sediments. In some areas along continental margins, large deposits of methane had accumulated beneath the surface from organic-matter–rich sediments that were deposited millions of years ago; sometimes dissolved methane and bubbles of gas could be detected rising from the seafloor, forced to the surface by compaction, or by seawater circulating through cracks created by tectonic activity. If conditions were right in these areas, methane could also be found immobilized as a strange icelike solid in the top few hundred meters of sediment. This methane hydrate, where molecules of methane were trapped in a cage of water molecules, had long been known to chemists as a curious chemical phase—ice that could catch fire!—which formed under very specific conditions of gas concentration, temperature, and pressure. But the explorations of the 1970s and 1980s revealed that the requisite combinations of high methane production, low temperatures, and high, but not *too* high, pressure attain in the sediments of many of the world's oceans: along continental margins, near places where tectonic plates converge, in polar seas, and in some shallow basins, sediments are loaded with flammable ice, sometimes embedded

Methane hydrates (2002)

up to 500 meters deep, sometimes exposed on the sediment surface, and occasionally piled up in great mounds.

The sheer magnitude of this cache of potentially accessible natural gas was a surprise to geologists and petroleum companies alike. In a 1995 review, Keith Kvenvolden, who had left NASA for a job with the U.S. Geological Survey in the mid-1970s and spent years exploring for and studying methane hydrates, estimated that they contained twice the amount of carbon and energy potential of all known reserves of fossil fuels. But the thing that had geochemists and microbiologists puzzled for more than two decades was that most of the methane they detected rising through the sediment pores neither froze into methane hydrate nor seeped into the sea, but rather disappeared altogether. Something, it seemed, was eating a good portion of the methane produced by the methanogens.

Microbiologists had isolated a number of species of methane-consuming bacteria from soils and freshwater lakes, places where methane generated by methanogens in anoxic sediments seeps up into the oxic sediments or water. But none of these methane munchers, the so-called methanotrophs, could function without oxygen, and the methane in the marine sediments disappeared well *within* the anoxic sediments, far from any trace of oxygen. Furthermore, microbiologists contended that an organism that consumes methane in the absence of oxygen was absolutely inconceivable, that the process simply couldn't generate enough energy to be biochemically viable—that it was, in short, impossible. But impossible or not, the geochemists said, such an organism must exist, because methane was disappearing en masse, and there was simply no place else for it to go. In the mid-1980s, their studies showed in no uncertain terms that this disappearing act occurred when the methane reached the band of suboxic and anoxic sediments where sulfate-reducing bacteria live. When Danish microbiologists finally honed in on the methane munchers in organic-matter–rich sediments from the Kattegat Strait, between Denmark and Sweden, they suggested that these organisms were somehow dependent on the sulfate reducers—but the organisms themselves eluded detection, and exactly how they accomplished their impossible task remained a mystery.

In the meantime, Hayes and his colleagues were finding anomalous depletions in the ^{13}C content of the organic matter in ancient rocks—well beyond differences of a few per mil associated with changes in CO_2 concentrations or phytoplankton productivity—which suggested that there were places and times in the earth's history when the highly depleted methane generated by methanogens was being consumed by methanotrophic organisms of some sort and channeled back into the food chain. The first compound-specific isotope analyses in the Messel shale had revealed hopanoids and unidentified isoprenoids with δ^{13}C values as low as –60 per mil, which had to include a sizable contribution from methanotrophic organisms of some sort. A similar study of the Green River shale that Roger Summons and one of Hayes's winter migrant workers did in Australia turned up hopanoids that were even more severely depleted in ^{13}C—including one that was not only isotopically but structurally distinct, with a δ^{13}C of –85 per mil and a methyl group attached at the number 3 position of the A ring. Rohmer had found this structural quirk in the hopanoids of only two of the hundred strains of bacteria he'd analyzed—and both of them were methanotrophs. Presumably, such organisms had lived in the oxic surface waters of the ancient Green River and Messel lakes, consuming methane that bubbled up from the sediments or deep water, much as they did in the modern lakes where they'd been discovered.

Hayes figured that if, as appeared to be the case, there were methane-munching organisms that lived in anoxic sediments, then they would also leave behind lipids that were severely depleted in ^{13}C and, with any luck, had chemical structures that provided clues to the identity of mystery microbes—though what sort of structures and what sort of clues those might be, he had no idea. He wrote up what he says was a shamelessly vague proposal for a collaboration and sent it to Bo Barker Jørgensen, one of the authors of the Kattegat Strait study. But Jørgensen didn't need much convincing. He obtained another Kattegat Strait core, and one of Hayes's students did an exhaustive survey of the lipids and their isotopic signatures in sediments from each zone of microbial action that the Danes defined. In his 1992 thesis, Liangqiao Bian notes that only one compound showed unequivocal signs of having been made by any sort of methane-consuming organism. It was present only in the narrow strip of anoxic sediment where Jørgensen detected methane oxidation, and it was even more radically depleted in ^{13}C than the methane in the sediments—but it wasn't a hopanoid like any of those found in aerobic methanotrophs. It was, rather, a branched alkane that was inseparable from phytane on the GC column and might have eluded detection altogether, if not for the isotope measurements and the carefully defined zones of microbial action in the

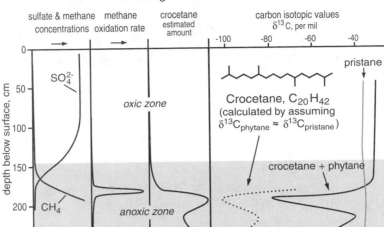

Methane oxidation and carbon-13 depleted crocetane in Kattegat Strait sediments

core. In Bian's chromatograms, a subtle change in the shape of the phytane peak for sediments from the zone of methane oxidation and an accordant shift in the $\delta^{13}C$ from −30 to −80 per mil hinted at the presence of a hidden compound.

Once they noted its presence, it was possible to obtain a mass spectrum and determine the structure as that of crocetane, a 20-carbon acyclic isoprenoid like phytane, but with a tail to-tail link in the middle, or like PMI but minus a five-carbon isoprene unit, depending on how one looked at it. Crocetane had never been isolated from any organism, but it could well have been present, even abundant, in sediments and oils and gone undetected, hidden by the omnipresent phytane. When the overlapping peaks were accounted for, the crocetane had a $\delta^{13}C$ that ranged from −60 to −100 per mil: it clearly hailed primarily from the organisms responsible for oxidizing methane in the anoxic sediments and was, in fact, the first real physical evidence of their existence. But *what* were they? Bian wrote a fine master's thesis in 1992 but immediately abandoned geochemistry for the pharmaceutical industry, and Hayes left the problem of anaerobic oxidation of methane to the microbiologists to figure out—or so he thought.

Meanwhile, Roger Summons had been inspired by the results of the Green River study, among other things, to form another cross-Pacific alliance, this time with Linda Jahnke, a microbiologist who was growing cultures of methanotrophic bacteria and cyanobacteria at NASA's Ames Research Center in California. Summons had been noting for some time that extended hopanes with an extra methyl group on the A-ring were found only in certain types of ancient rocks and oils, and the distinct methanotrophic isotope signature of the 3-methyl hopanes in the Green River shale fueled his mounting suspicion that the hopanoids in sediments and ancient rocks might harbor more information than diplopterol   3β-methylbacteriohopanetetrol

they were given credit for. Lipid analyses of Jahnke's bacterial cultures revealed that many cyanobacteria make extended hopanoids with a methyl group on the number 2 carbon of the A-ring, and the methanotrophs contain large quantities of extended hopanoids with the distinctive 3-methyl ring structure, as well as exceptionally large amounts of the common 30-carbon hopanoids diplopterol and diploptene.

Neither of these embellishments to the hopanoid ring structure had been found in other types of bacteria—and both clearly survived diagenesis and were preserved for posterity in ancient rocks and oils. Studies of isotope fractionation by the methanotrophic bacteria indicated that the hopanoids were as much as 30 per mil more depleted in ^{13}C than the methane the bacteria consumed—if the bacteria lived in marine sediments, their lipids might be expected to have δ^{13}C values of –90 to –120 per mil and be easy enough to spot. Further investigations of isotope fractionation in the cultures, however, would show that all methanotrophic bacteria were not created equal: the two known families employed two different carbon fixation pathways, and the amount of fractionation thus differed significantly between them. Like carbon fractionation by CO_2-consuming algae, fractionation by methanotrophs also depends on methane concentration. And the δ^{13}C values of the methane itself can be quite variable, with values of –30 per mil and even higher reported for thermogenic methane in a few areas. Extensive uptake by methanotrophs can also leave the residual methane that seeps into surface sediments or water bodies substantially enriched in ^{13}C. Under certain conditions, then, at low concentrations of relatively heavy methane, the δ^{13}C values of hopanoids from certain species of methanotroph could conceivably be as high as –35 per mil, making them hard to distinguish from the hopanoids of other bacteria. As John Hayes is fond of saying, "There are no magic numbers." Still, there was a reasonably good chance that an active population of methanotrophs in the sediments or water of a marine basin would have left a distinct hopanoid footprint in the sediments, much like that observed in the Green River shale—3-methyl hopanoids that were over 50 per mil more depleted than the biomarkers from photosynthetic algae—and that was something both Hayes and Summons were keen on tracking.

In the early 1990s, as Summons and Jahnke were beginning their studies of methanotrophic bacteria—and as Freeman, Jasper, and Pagani were correlating alkenone ε_p values with CO_2 concentrations, Kvenvolden was tallying the distributions of methane hydrates, and Hayes was puzzling over isotopic anomalies in the kerogen of Precambrian rocks—it was fast becoming apparent that the amount of methane in the atmosphere varied dramatically over the course of earth history and, alongside CO_2, played a principal, if not starring, role in determining climate. Analyses of the gases in the Greenland and Antarctic ice cores were showing that the amount of methane in the atmosphere had varied in concert with Milankovitch cycles and with abrupt climate changes over at least the past 220,000 years. And in 1995, a group of geologists in Michigan published a study in which they hypothesized that one of the most abrupt climate changes evident in the geologic record was due to the sudden release of a massive amount of methane from marine sediments.

In Michigan, Gerald Dickens and his colleagues were trying to ascertain what had happened to the earth's climate 55 million years ago, during the transition from the Paleocene to Eocene epoch—just prior to the long sustained cooling that led from a greenhouse into an icehouse world. $\delta^{18}O$ measurements indicated that temperatures rose dramatically, and globally, within a few thousand years, while the $\delta^{13}C$ values of foram carbonate dropped to significantly more negative values, implying that a large amount of ^{13}C-depleted carbon had been released into the oceans. Dickens proposed that a slight heat wave had turned into a cataclysmic global climate change when water temperatures in continental shelf areas rose above the limit for methane hydrate stability and massive amounts of free methane gas were suddenly released into the sediments, ocean, and atmosphere by disintegrating hydrates. The hypothesis was compelling, but there was no direct evidence for methane in the atmosphere or ocean at the time, and the authors could foresee no way of obtaining it.

John Hayes was preparing to move his laboratory from Indiana to Woods Hole when the Dickens paper was published, but its methane hydrate hypothesis was still nagging at him nearly two years later, when he was well settled into his new lab, replete with a new coterie of students and postdocs. Wouldn't such a massive release of methane into the world's oceans have resulted in a population explosion of methanotrophic organisms? And wouldn't they have left their mark in the sediments in the form of ^{13}C-depleted hopanoids? Someone should really take a look at the biomarkers in the Paleocene-Eocene sediments and see if there were any signs of methanotrophs. Jürgen's student Kai Hinrichs, who went to work on a postdoctoral fellowship with Hayes in 1997, says he couldn't figure out why Hayes kept giving him papers about methane to read when he was supposed to be working on the alkenone CO_2 proxy, expanding on John Jasper's work. Hinrichs had written a proposal to do alkenone ε_p measurements in Pleistocene sediments from different geographic locations and determine if and how oceanic sources and sinks for CO_2 had changed over time. Hayes had been enthusiastic enough about it when they'd talked the year before, but once Hinrichs got to Woods Hole, the project seemed to languish. For his part, Hayes was concerned that the emerging problems of distinguishing CO_2 concentration from growth rate in alkenone ε_p values would hamper the success of his postdoc's project... but he was also eager to follow up on Dickens's paper. Whatever the case, Hinrichs says, getting him interested in methane was one of John Hayes's most valuable contributions to his career—though it wasn't immediately apparent.

They requested as many Paleocene-Eocene sediment sequences as they could find in the DSDP and ODP core repositories, but, unfortunately, most of them had come from open ocean sites with low productivity and scant organic matter. "They were really boring," Hinrichs says. "No peaks, nothing to look at." A few months into the project, however, Hayes went to a meeting in San Francisco and was offered samples that were a little more interesting—not of Paleocene-Eocene sediments, but of contemporary sediments where methane hydrates were decomposing in real time and methane was actually bubbling into the water.

Oceanographers had been finding these methane seeps along the continental margins since the 1980s, and explorations with the early research submersibles had revealed that they supported oases of life similar to those at the deep sea hydrothermal vents. In the fall of 1997, the Monterey Bay Aquarium Research Institute was making preparations to explore one such seep in the Eel River Basin using an innovative new submersible—as Hayes learned when he got to talking with the institute's director at the American Geophysical Union meeting. This was a site where methane hydrates were right at their threshold of stability, actively releasing methane into the water—a small-scale, real-time version of the global event that Dickens had proposed to explain abrupt climate and ocean chemistry changes at the onset of the Eocene period. When the institute director offered Hayes samples, he jumped at the chance. Hunting for biomarker evidence of such an event in 55-million-year-old sediments was a little like hunting for fingerprints when you weren't even sure a crime had been committed, let alone what sort of creature the criminal might have been. But here was a chance to examine the scene of a crime-in-progress, so to speak. They could at least see how evidence of a massive release of methane from methane hydrates 55 million years ago *ought* to look.

There was plenty of organic matter in the Eel River Basin sediments that Hinrichs analyzed, and lots of methane…but no ^{13}C-depleted hopanoids. The only compounds of interest, he says, showed up as two large peaks with δ^{13}C values of –100 and –110 per mil. They eluted from the column where the hopanols should have been, but they weren't hopanols. He had treated the extract with a reagent that reacts with alcohols to produce derivatives that are more amenable to gas chromatography, and according to the mass spectra of these derivatives the two compounds were glycerol ethers, now rendered nonpolar and volatile enough to get through the GC column. Hinrichs says he called Jürgen in Oldenburg for help in pinning down the structures, and Jürgen directed him to a paper reporting the mass spectrum for archaeol—but then the isotope machine at Woods Hole broke down and he was left with nothing to do except contemplate his two compounds and read about methane. That was when he started thinking seriously about the lack of oxygen in the Eel River Basin sediments and suspecting that his ^{13}C-depleted compounds might have come from anaerobic organisms that were consuming methane in the anoxic sediments, rather than aerobic methanotrophic bacteria at the sediment surface or in the water.

Meanwhile, back in Germany, a graduate student at the University of Kiel was analyzing sediments from a spectacular methane seep system just to the north of the Eel River Basin, a place off the coast of Oregon where one submarine ridge was literally paved in methane hydrate. Though Marcus Elvert had studied organic chemistry, he ended up doing his Ph.D. with Erwin Suess, a marine geologist who had been exploring the Hydrate Ridge region since the 1980s. Like many of the methane seep sites on the continental margins, it was a subduction zone where two tectonic plates converged and the oceanic plate slipped beneath the continental one to be reabsorbed by the mantle. Suess was among the first to observe the methane bubbling up from the seafloor and the fantastic communities of giant clams, tubeworms, crabs, and carnivorous fish that were thriving in the vicinity, on the otherwise barren seafloor. He was also one of the first to realize that the carbon in everything, from the tissues of the mussels to the surrounding carbonate minerals, was severely depleted in ^{13}C. Suess was interested in understanding the geology of the area. He wanted to know how such systems had evolved over time and where on the globe they had existed in the geologic past. And he wanted to know what happened to all the free methane gas that was bubbling out of the sediments—how it was channeled back into the biosphere, and how the ^{13}C-depleted carbonate had formed. Despite the ecosystem's apparent likeness to a hydrothermal vent ecosystem and the thick mats of sulfide-oxidizing bacteria covering the exposed methane hydrates, this was clearly a place where methane, not sulfide, called the shots. Suess charged the young organic chemist he enlisted in 1996 with the specific task of identifying biomarkers for methanotrophic bacteria, in the hope that they might shed some light on both the biological and geological questions the methane seeps had raised.

Just as the hydrothermal vent organisms relied on symbiotic sulfide-oxidizing bacteria, at least some of the animals in the methane seep ecosystem clearly relied on symbiotic methanotrophic bacteria. Colleen Cavanaugh, who had discovered the tubeworms' symbiotic sulfide-oxidizing bacteria, found that the gills of mussels from a methane-seep community contained rRNA genes related to those of methanotrophic bacteria, and Summons and Jahnke had found relatively large amounts of ^{13}C-depleted hopanoids and diploptene in the gills of the same organisms. But the ecosystem's dependence on methane extended far beyond these symbiotic relationships. The fluxes of methane and sulfate in the Hydrate Ridge sediments indicated that, in addition to the methane bubbling and seeping out of the sediments, a lot of methane was being consumed by some mystery organism deep in the sediments—which, like the Eel River Basin sediments, were anoxic almost to the sediment–water interface—and there was, at the same time, an inordinately active population of sulfate-reducing bacteria that produced an excess of sulfide...which, presumably, fed the extensive mats of sulfide-oxidizing bacteria that covered the sediment surface.

Suess was a marine geologist who did most of his research at sea and had little in the way of laboratory facilities, so he sent Elvert to use the instruments in Michael Whiticar's lab in Canada, which included a new isotope-ratio-monitoring mass spectrometer. Whiticar himself had been an early student of Suess's and

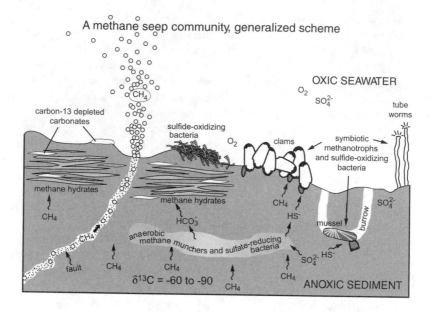

A methane seep community, generalized scheme

had been studying the production and consumption of methane since the 1980s. Unlike Hinrichs, Elvert was specifically looking for anaerobic methanotrophs before he even extracted his first Hydrate Ridge sample. Like Hinrichs, however, he started out looking for hopanoids. None of the anaerobic bacteria that Rohmer had examined contained hopanoids, and he and Ourisson had postulated that anaerobes were incapable of making them. There was, however, no specific requirement for oxygen in the biosynthesis of bacteriohopanetetrol, and Elvert thought the methane-munching culprits might be hopanoid-producing methanotrophic bacteria akin to the ones found in mussel gills, aerobic bacteria that had somehow adapted to anoxic conditions in the sediments. He wrote to Roger Summons to request samples from Jahnke's bacterial cultures—but not long after the cultures arrived, he analyzed his first Hydrate Ridge samples and realized that the hopanoids weren't going to help him. There simply weren't any. What there was, was a huge phytane peak—except the mass spectra showed clearly that it wasn't phytane. It was almost pure crocetane.

Elvert had to wait his turn to use the much-coveted isotope machine in Whiticar's lab, so weeks went by before he could determine the isotope content of his crocetane. But he knew he was onto something, because Whiticar had told him about a conference poster he'd seen of Bian's unpublished Kattegat Strait study, which reported ^{13}C-depleted crocetane in the methane-oxidizing zone of the sediments. Elvert changed strategies, turning his attention from hopanoids to crocetane. The Hydrate Ridge sediments were loaded with it, and when he finally got a turn at the isotope machine, he found it had a $\delta^{13}C$ of –107 per mil—a sure sign that it was made by methane-consuming organisms. But what were they?

This was the question Bian had been unable to answer in the Kattegat Strait study. Elvert, however, had a couple of extra clues. The Hydrate Ridge sediments contained two other isoprenoid alkanes that were equally depleted in ^{13}C: PMI, and an unsaturated counterpart.

Elvert says he read everything he could get his hands on, trying to make sense of his data. Except for Bian's work, crocetane was unheard of in organisms or sediments. But PMI was apparently synthesized by methanogenic archaea, and the Bristol researchers had found it in their assay of methane-rich marine sediments. Since then, microbiologists had managed to culture methanogens from marine sediments, and the NIOZ group had just reported finding various unsaturated versions of PMI in one of them. But what was going on in the methane seep sediments? Elvert's PMI was too radically depleted in ^{13}C to have been made by a methanogen. It had to come from something that was *eating* methane. But how could an organism be eating its own waste product? It was like a human suddenly inhaling CO_2 and exhaling oxygen. Finally, he came across a paper by a group of microbial ecologists who had posited just that and offered an explanation: they proposed that the consumption of methane required a consortium of microbes working together, rather than a single organism, and that this consortium consisted of methanogens working in reverse with sulfate reducers to help them along. In other words, the methanogens could oxidize methane back to CO_2 and hydrogen, if the sulfate-reducing bacteria immediately sucked up the hydrogen, effectively pulling the reaction in an otherwise energetically unfavorable direction.

Elvert was excited—the hypothesis explained what geochemists and marine chemists had been observing for years, the close association between methane oxidation and sulfate reduction that the Danes had demonstrated in the Kattegat Strait sediments. Now he had direct physical evidence to support it: lipids from methanogens that were eating methane. It seemed quite a coup, especially for a graduate student doing organic chemistry in a geology department! As he was preparing his paper for publication, Elvert says, he e-mailed Summons in Australia and asked what he thought of the idea. Summons didn't take much convincing. "Yes," he replied, immediately, "I think you're right." And then he added the crushing news that Elvert's discovery was not, as he'd thought, entirely original. The reason Summons was so readily convinced was that he'd just been to a conference where he'd spoken to two scientists who had also found ^{13}C-depleted methanogen lipids in methane seep sediments and had come to similar conclusions. One was Kai Hinrichs at Woods Hole. And the other was Volker Thiel, a postdoc working just down the road, in Walter Michaelis's Hamburg lab.

By the time their three papers appeared in 1999, the three scientists had started to swap information, much as the Bristol and Dutch groups had done with the alkenones, or Strasbourg and Bristol with the extended hopanoids decades earlier—though in this case, there was a definite sense that the three young researchers were racing for first publication of their discovery. The *Nature* and *Science* reviewers, however, weren't convinced by Thiel's evidence, or by Elvert's, and the two had to resubmit to other journals, whereas Hinrichs's particular assembly of data was finally so compelling that it skated past the *Nature* editors and appeared promptly—with the result that the race for publication was rendered irrelevant and the first paper completed was the last to appear. Their evidence, in any case, was more complementary than redundant. And, as it turned out, parallel work had been going on under Jaap Sinninghe Damsté's tutelage in the NIOZ lab, bolstering the case.

Thiel hadn't been studying an active methane seep at all, but rather a limestone outcrop that showed signs of having formed in a shallow basin where methane seeps were active during the Miocene epoch. Unlike Hinrichs's and Elvert's contemporary methane seep sediments, his samples contained significant amounts of a severely ^{13}C-depleted hopanoid. It appeared to be a diagenetic product of diploptene, with the double bond shifted two carbons over. Noting the preponderance of diplopterol and diploptene in the methanotrophic bacteria that Summons and Jahnke had been studying and the general lack of hopanoids in anaerobic bacteria, Thiel speculated that it came from aerobic methanotrophs that had lived in the oxic waters above the seep. But his samples also contained crocetane, PMI, and an unidentified isoprenoid ether lipid that were even more depleted in ^{13}C than the hopanoid, with δ^{13}C values of –105 to –115 per mil. Like Elvert and Hinrichs, Thiel concluded that these compounds had been synthesized by some hitherto unknown methane-consuming archaea. He also found a series of ^{13}C-depleted alcohols that must have come from organisms that had utilized a methane-derived carbon source—though not, apparently, from the methanotrophic archaea that made the isoprenoids. They had δ^{13}C values around –88 per mil, dramatically depleted compared to the rest of the organic matter, but decidedly more positive than the isoprenoid lipids from the putative methanotrophic archaea. Their carbon skeletons were the C_{14} to C_{18} *iso, anteiso,* and mid-chain branched ones typical of many heterotrophic bacterial lipids. Fatty acids with these structures had been found to be particularly plentiful in anaerobic sulfate-reducing bacteria, and noting the extensive evidence linking methane oxidation with sulfate reduction in anoxic sediments, Thiel speculated that these compounds derived from sulfate-reducing bacteria that coexisted with the methanotrophic archaea and somehow utilized carbon from their decaying biomass.

Hinrichs says that it wasn't until he figured out the structure of the second isoprenoid ether in his Eel River Basin samples that he realized he was seeing geochemical evidence of a biochemical phenomenon that neither he nor Hayes had considered. The compound Jürgen had helped him identify was archaeol, the simple glycerol diphytanyl ether that was common to most archaea. This second compound, though similar, was not so common and took him longer to figure out—partly because he didn't realize that the standard derivatizing reagent he'd used, which was supposed to react with the alcohol groups and make the molecule more volatile and amenable to GC analysis, had reacted with only one of the molecule's two alcohol groups. The mass spectrum was difficult to decipher, and Hinrichs actually misinterpreted its fragmentation pattern in his 1999 paper—but he deduced the correct compound, and that was enough to put him on the right geochemical and biochemical track. It was hydroxyarchaeol, with an OH group attached to the second phytanyl chain, a compound that had been found only in methanogenic archaea and was, along with various isomers, a major component only in the order *Methanosarcinales*. A bit of library research revealed that *Methanosarcinales* included the most versatile of the known methanogens, organisms that could consume almost any small molecule available from CO_2 to acetate, methanol, and…methane? If there was a methanogen that could reverse its chemistry to consume its own waste product, it would surely be related to the *Methanosarcinales*. Hinrichs, of course, was reading all the same literature as Elvert and Thiel…but he worried that his lipid evidence for such an organism was circumstantial at best. After all, gene-based explorations of microbes in natural habitats had been revealing a plethora of hitherto unknown archaea, thriving in places where no one would have expected archaea to thrive. None of these organisms had ever been isolated in cultures and had their lipids analyzed. How, then, could one argue that archaeol, PMI, crocetane, or even hydroxyarchaeol came from methanogens and not from some unknown, completely unrelated group of archaea? Efforts to isolate and grow the anaerobic methanotrophs in culture had failed, and the genetic techniques were of little help because one didn't know what processes were associated with the genes identified—until the microbiologists and geochemists started combining genetic techniques with lipid measurements and compound-specific isotope measurements.

Hinrichs recalls how Hayes "perked up his ears" when he told him what he'd found in the Eel River Basin sediments and his hypothesis about what it meant—and what he wanted to do next to find out if it was correct. Hayes had expected his postdoc to find ^{13}C-depleted hopanoids in the seep sediments, not hunt down the elusive organisms responsible for the anaerobic oxidation of methane! Hinrichs had been talking to a microbiologist at Woods Hole who'd agreed to extract and analyze the rRNA in sediments from the Eel River Basin and try to identify the organisms responsible for the ^{13}C-depleted hydroxyarchaeol—if Hinrichs could get more samples.

He was in luck: when he called the Monterey Bay Aquarium Research Institute, the director told him that he didn't need more samples because Ed DeLong, a microbiologist there at the institute, had already extracted and sequenced the RNA from sediments at the same site. When Hinrichs called DeLong and explained his hypothesis about hydroxyarchaeol-producing *Methanosarcinales* eating methane in the seep sediments, DeLong hadn't yet analyzed his data. But two days later, he called back, exclaiming into the telephone: "Wow, you're right!"

The ribosomal RNA sequences indicated the presence of two distinct groups of archaea. None of them matched any of the sequences in the public database that had grown out of Woese's phylogenetic scheme, but they all fell between or within three orders of methanogens, including *Methanosarcinales*, and one cluster of sequences formed a distinct group of its own. By this time, the very concept of "species" of microbes had fallen by the wayside, and the Linnaean system of taxonomy had been replaced by phylogenetics: rather than assigning a grandiose Latin scientific name to the new group of putative anaerobic methanotrophs,

DeLong and his colleagues celebrated their discovery with the inglorious acronym "ANME 1." The association between ^{13}C-depleted lipids and phylogenetic group was still circumstantial, but it was enough to convince Hinrichs and his coworkers to submit their paper—and, apparently, to convince the persnickety *Nature* editors to accept it.

More definitive evidence for a link between the methanotrophic archaea and sulfate-reducing bacteria would come from the NIOZ lab, where Kate Freeman's student Richard Pancost was working as a postdoc with Sinninghe Damsté. They were analyzing samples from active and extinct "mud volcanoes" in the Mediterranean Sea, places where tectonic pressure causes deeply buried, methane-laden sediments to erupt and flow out over the seafloor. Like Thiel, they found *iso* and *anteiso* compounds that were dramatically more depleted than sterols from the eukaryotes or the hopanol typical of heterotrophic bacteria, but less depleted than the acyclic isoprenoid lipids. This time, however, instead of alcohols, the compounds were *iso* and *anteiso* fatty acids—the same structures that had been isolated from a number of sulfate-reducing bacteria. Like Elvert, Pancost also found ^{13}C-depleted crocetane and PMI, and their unsaturated versions, and like Hinrichs he found archaeol and hydroxyarchaeol. These compounds were all severely more depleted in ^{13}C than any of the bacterial or eukaryotic lipids in the sediments...but the amount of depletion varied substantially and they appeared to come not from one, but from several species of archaea.

The 1999 papers had sparked a flurry of interest among microbiologists, much of it centered around the Max Planck Institute for Marine Microbiology (MPI) in northern Germany, now headed by Jørgensen. Though ANME continued to elude attempts to isolate it in culture, a group of researchers led by Antje Boetius at the MPI joined an expedition to retrieve fresh samples from the sediments of Hydrate Ridge—and attempted to photograph and count the ANME. They made fragments of RNA molecules that were complementary to the rRNA sequences identified in the Eel River Basin sediments, adding a fluorescent functional group so that they would appear red in the proper light. In what amounted to a needle-in-the-haystack search for the sulfate-reducing bacteria that appeared to be associated with ANME, they also made a set of molecules based on the rRNA of an assortment of sulfate-reducing bacteria, from all sorts of environments. When Boetius mixed these two sets of genetic probes into a Hydrate Ridge sample, they bonded with the complementary rRNA in their target organisms, rendering the archaeal cells red and the sulfate reducers green. The probes for DeLong's ANME 1 group marked only a scattering of cells. But a probe designed to target another of the Eel River sequences—the one most closely related to *Methanosarcinales*—produced clumps of red cells that made up more than 90% of the biomass in the Hydrate Ridge sediments. The sulfate reducers were more difficult to find. Boetius says that she had tried 11 of the 12 probes for sulfate-reducing bacteria with no success, when, shortly before midnight and a bit blurry eyed from the long day at the microscope, she looked at her last slide: there were the globs of glowing red spheres, now surrounded by a shell of bright green ones—the long-hypothesized consortium of methanogen-like archaea and sulfate-reducing

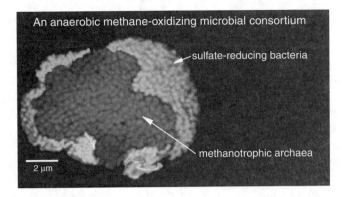

An anaerobic methane-oxidizing microbial consortium

sulfate-reducing bacteria

methanotrophic archaea

2 μm

bacteria. The successful bacterial probe was based on an rRNA sequence that had been detected in polar marine sediments, but the bacteria themselves had never been grown in culture and were just as mysterious as the archaea.

In its Hydrate Ridge samples, the MPI team measured sulfate reduction and methane oxidation rates, and counted cells, and there was every indication that the ANME consortia were responsible for the oxidation of methane. Likewise, the team found all of the lipids that had been identified with the methane-oxidizing archaea—crocetane, PMI, archaeol, and hydroxyarchaeol—with δ^{13}C values as low as –135 per mil, as well as *iso-* and *anteiso-*C_{15} acids with values of –63 and –75 per mil. But could one prove that the ^{13}C-depleted lipids actually came from the organisms in the consortia, that the archaea were, in fact, consuming methane? Hinrichs initiated another collaboration, this time with Christopher House at Pennsylvania State University. House had been working with a technique that allowed him to make carbon isotope measurements on the tip of a microscopic pin, even directly in a single cell. Victoria Orphan, a student of DeLong's, made fluorescent rRNA probes targeting both of the ANME organisms they had identified, and once the organisms were labeled and could be viewed on a microscope slide, House was able to determine the δ^{13}C of individual ANME cells: here, finally, in 2001, was *unequivocal* proof that the two groups of methanogen-like archaeal genes did indeed belong to methane-consuming organisms.

Could there be others? Did different organisms thrive in different areas, as the lipid evidence seemed to imply? How did they function, what compounds did the members of the consortium exchange, and how? And the question that had captivated John Hayes and Roger Summons: Could these lipids or their fossil counterparts provide clues to the role methane played in the intertwined history of life, the carbon cycle, and climate over geologic time?

Exactly how the methanotrophic archaea function and how they interact with their bacterial neighbors remains, to date, unclear. Recent work indicates that some methane-oxidizing archaea may contain modified methanogen genes *and* enzymes, which supports the hypothesis that they reverse the chemistry of methanogenesis. One study of methane seeps in the Gulf of Mexico suggests that some of these organisms can actually switch modes between methanogenesis and the oxidation of methane, depending on environmental conditions. But the microbiologists at the MPI have now disproved the initial hypothesis that the sulfate-reducing bacteria utilize hydrogen produced by the methanotrophs as an energy source. How electrons, energy, and carbon are shuttled between the archaea and their partner bacteria remains unclear. The ANME are clearly using methane as their carbon source, but what do the sulfate reducers use? Why are their lipids so much more depleted in ^{13}C than other heterotrophic organisms in their environment, and yet so much less depleted than the archaeal lipids? In sediments where the anaerobic oxidation of methane does not occur, sulfate reducers use small molecules left over from the decomposition of organic matter by other types of heterotrophic bacteria, and one hypothesis maintained that the sulfate reducers in the consortium derived their carbon from a mixture of this material and the ^{13}C-depleted biomass of the methane-oxidizing archaea. But the MPI scientists found that when they incubated the Hydrate Ridge sediments and added typical organic substrates, the sulfate-reducing bacteria made use of none of them. Another hypothesis is that these particular sulfate-reducing bacteria are autotrophic and derive their ^{13}C-depleted carbon from the CO_2 produced by the methanotrophs.

The genes of a number of different bacteria have now been associated with the anaerobic oxidation of methane, and at least three distinct groups of archaea have been identified. Orphan's extended studies with the fluorescently labeled rRNA probes showed that the tight, hierarchical relationship displayed by the Hydrate Ridge consortia in Boetius's first attempt is not the only way to go: some consortia are speckled balls where the two types of microorganisms are all mixed up, sometimes groups of methane-oxidizing archaea and sulfate-reducing bacteria live in separate neighboring clusters…and some ANME archaea have even been found living in lonely splendor, keeping their bacterial partners at arm's distance, so to speak.

The variety of distinct organisms involved in the anaerobic oxidation of methane is reflected by the variety of archaeal and bacterial lipids that have now been associated with the process in different environments. Clearly, a diverse assortment of organisms has, in fact, developed more than one strategy to accomplish the "impossible" task of oxidizing methane in different anoxic environments.

Microbiologists now seem to take such diversity and resourcefulness in stride. "It would be pretty unusual if there were just one bug, or one set of bugs involved," Mandy Joye told me when I met her at the Hanse Institute in 2003. "If these guys can do it, someone else probably can." A specialist in microbes that live in extreme environments, Joye was on sabbatical from the University of Georgia, working with Boetius's group at the MPI. Even a geochemist knows that "bug" is slang for microbe, but it's not so easy, these days, to know exactly what a microbiologist means by "one" bug. Microbiologists now speak of "strains" and groups of microbes, and it seems that the only way to pin them down to a precise designation of an organism is to listen to the long litany of its evolutionary

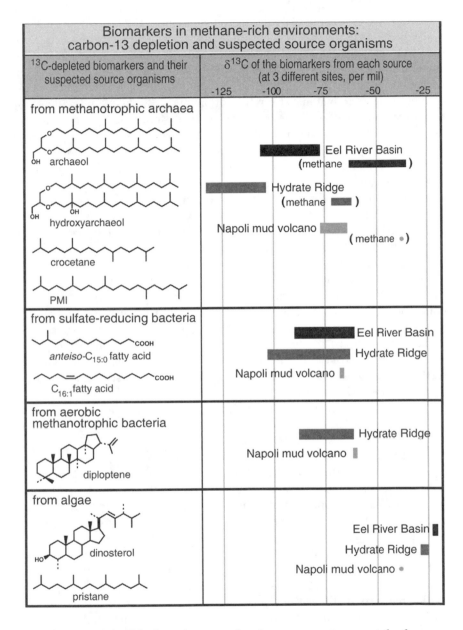

relatedness—sort of like listening to one's aging great aunt recount the four generations of divorce, remarriage, illegitimate children, incest, and intercontinental immigration that somehow connect to a second cousin, once removed, who lives in the outer reaches of Mongolia.

Whatever their precise species, strain, or "gene cluster" compositions and metabolic strategies, it is now apparent that anaerobic methane-munching consortia are capable of oxidizing methane at phenomenally high rates and, in some environments, producing a sizable amount of biomass. The MPI group's incubation experiments showed that both sulfide and bicarbonate are by-products of the consortia activities, which accounts for the extensive mats of sulfide-producing bacteria and precipitates of ^{13}C-depleted carbonates that are so often associated with methane seeps. In the early 1990s, Russian microbiologists did microscopic studies of the microbial mats from the bottom of the Black Sea and posited that they were made up of methane-oxidizing archaea. More recent seafloor explorations of the Black Sea have revealed incredible reefs up to 4 meters high, and analyses with genetic probes show definitively that they are composed of consortia biomass and ^{13}C-depleted calcium carbonate. This is a far cry from the marginal, energetically challenged, biochemically "impossible" organisms once imagined responsible for the disappearance of methane in anoxic marine sediments. According to recent estimates, they consume more than 80% of the methane that might otherwise make its way into the ocean and atmosphere. Aerobic methanotrophic bacteria play a relatively minor role, simply because methane is produced and stored in the deep anoxic sediments and must run the gauntlet of methane-munching archaea before it reaches the oxic surface sediments or water—there just isn't a lot left over for the aerobic bacteria. Taken together, these populations of methanotrophs act as an efficient control valve on the amount of methane released from the ocean to the atmosphere: despite the huge amount of methane produced and stored in marine sediments, the ocean currently contributes only 3–5% of the global methane flux into the atmosphere. Indeed, most methane now comes from ever-expanding human activities such as rice cultivation, livestock raising, and fossil fuel production, and from the dwindling natural wetlands. But what opens and closes the microbial control valve in the ocean? How would it respond—how did it and how will it respond—to additional inputs of methane from the sediments?

When gas measurements in the ice cores first revealed fluctuations in the methane content of the atmosphere during the last two ice age cycles, climatologists hypothesized that they were linked to a climate feedback effect: warming caused increases in tropical rainfall and an extension of wetlands where methanogens were active, and the methane they produced escaped into the atmosphere and caused accelerated warming. But some geologists found this unconvincing, because the fluctuations in methane were closely correlated with episodes of rapid and excessive global warming that didn't give enough time for wetlands expansion. They suggested that at least some of the fluctuations were caused by the sudden release of methane from "melting" hydrates in the sediments. There was, however, little compelling evidence to support either hypothesis until the spring of 2000, when researchers at the University of California published a study of sediment cores from the Santa Barbara Basin, off the coast of central California.

Geologist James Kennett and his colleagues presented evidence that there had been several sudden, episodic releases of methane into the basin during the past 60,000 years, and that these coincided with brief warm intervals during the same period. They attributed the methane to the sudden breakdown of methane hydrates in the sediments around the shallow edges of the basin and hypothesized that similar events of methane hydrate collapse in continental shelf sediments and shallow basins around the world were responsible for the bursts of methane evident in the ice core gases over the past 200,000 years—less cataclysmic, repetitive versions of the event Gerald Dickens had hypothesized to explain the sudden warming and change in carbonate chemistry at the turn of the Paleocene period 50 million years earlier. Kennett's hypothesis was controversial, and the evidence—pronounced ^{13}C depletion in the carbonate of planktonic forams in the sediments formed just prior to the warm periods—was ambiguous and limited. As Kvenvolden points out, it was not entirely clear if and how such large stores of methane hydrate could be regenerated rapidly enough to provide for the frequent releases postulated. The implications of the hypothesis were huge. Would the current, human-induced period of global warming trigger a similar collapse of methane hydrates, with a sudden release of methane into the atmosphere and catastrophically accelerated warming?

This time, Hinrichs didn't need John Hayes to egg him on. To what extent methanotrophs living in the water and sediments of the Santa Barbara Basin would have responded to the hypothesized bursts of methane, and whether their molecular remains would be apparent in the sediments, was still anybody's guess. But he now had a good idea of what those remains might look like. And, unlike the Paleocene-Eocene sediments he'd tried to analyze, the Santa Barbara Basin sediments were loaded with organic matter. He had done extensive analyses of that organic matter when he was a student with Jürgen, and it was anything but boring.

He chose a site in the middle of the basin, where there was no methane seepage and no methane hydrate at present, and a section of core that spanned a period in the middle of the last ice age, when, according to Kennett's hypothesis, there had been several releases of methane from the breakdown of methane hydrates in the sediments around the periphery of the basin. In most of the section, molecular fossils of both aerobic and anaerobic methanotrophs were absent. But in the narrow bands of sediment that corresponded to periods of sudden warming, there were pronounced spikes in the concentration of diplopterol, largely absent from the rest of the section, with δ^{13}C values between -55 and -73 per mil. It had to derive from bacteria that were recycling ^{13}C-depleted organic matter and CO_2 from methanotrophic organisms, or from the methanotrophic bacteria themselves, and because the bulk organic matter was not significantly depleted in ^{13}C compared to that from primary producers, it probably came from the latter— though the 3-methyl hopanoids often associated with methanotrophic bacteria were absent.

Noting that the only known hopanoid-synthesizing methanotrophs are aerobic bacteria and the sediments in these intervals had clearly been anoxic all the way to the sediment–water interface; that the overall organic matter was not unusually depleted in ^{13}C; and that there were no ^{13}C-depleted carbonate formations like those at methane seeps, Hinrichs concluded that there had been no methane production *within* the sediments at this site in the middle of the basin— and that the ^{13}C depleted diplopterol was evidence that methane was suddenly

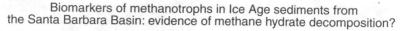

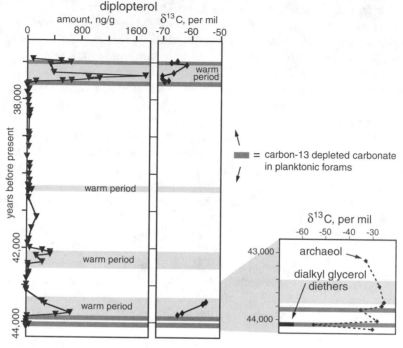

Biomarkers of methanotrophs in Ice Age sediments from the Santa Barbara Basin: evidence of methane hydrate decomposition?

released into the water of the basin at some time just prior to these warm spells, in keeping with Kennett's hypothesis. The ¹³C-depleted archaeal biomarkers indicative of anaerobic methane munchers were absent throughout most of the sediment section—with one notable exception. In a narrow band of sediments deposited about 44,100 years ago, the $\delta^{13}C$ value of archaeol dropped to –55 per mil, and several nonisoprenoid glycerol ethers that have been associated with sulfate-reducing bacteria in methane-oxidizing consortia were present, with $\delta^{13}C$ values of –70 to –65 per mil. Hinrichs postulates that the release of methane into the water at this time, about 400 years before the longest of the warm periods, was so massive and fueled so much methanotrophic bacterial activity in the water column that the oxygen in the poorly circulating deep water of the basin became depleted and anaerobic methane munchers flourished briefly at the bottom of the basin.

Though Hinrichs's results provide some evidence of sudden emissions of methane into the Santa Barbara Basin at times in the recent geologic past, and the breakdown of methane hydrates would seem the most plausible source, the actual extent, timing, frequency, and effects of such events even during the last ice age are far from settled. How much methane was released, and what proportion made it past the microbial control valve into the atmosphere? Did the decomposition of methane hydrate occur in other areas around the world, and to what degree were such events responsible for the fluctuations in atmospheric methane recorded in the ice core records, as Kennett hypothesized? How would this process play out under the current, human-induced warming regime? Answering such questions will require more extensive studies of other basins and continental shelf sediments over longer periods of time. But as Hinrichs found with his Santa Barbara Basin study, interpretation of the biomarker data in such environments is not always straightforward.

In methane seep environments, where methane consumption constitutes the primary means of productivity and is the dominant microbial process, lipids from methanotrophic archaea are prominent constituents of the organic matter and clearly marked for posterity with $\delta^{13}C$ values of –130 to –100 per mil. As fossil molecules in ancient sediments and rocks, they document the waxing and waning of methane seeps and mud volcanoes over the course of earth history. But in other environments, where methane is just one of many sources of carbon available to microorganisms, interpretation of biomarker structures and $\delta^{13}C$ values requires careful consideration. Freeman's and Hayes's early work on the ancient Messel Lake and Green River Basin shales indicate that this is the case even in restricted anoxic basins where methane concentrations are relatively high and methane oxidation is an important microbial process. And recent studies of methane oxidation and methanotroph lipids in the organic matter floating in the Cariaco Basin and Black Sea have raised the concern that even though methane oxidation is occurring in the water column, the organisms responsible may not leave a recognizable signature in the sediments. In the Black Sea, for example, Stu Wakeham and colleagues in the NIOZ group discovered that ^{13}C-depleted biomarkers of methanotrophic archaea in the suspended organic matter in the water either never reached the sediments, or their isotopic signal in the sediments was masked by relatively large quantities of structurally identical biomarkers from other types of archaea. In contemporary environments, the coupling of genetic investigations with lipid analyses and compound-specific isotope measurements provides a powerful means of determining possible sources for the lipids and of characterizing unknown organisms. But nucleic acids generally break down too quickly to be of much use in interpreting the geologic record beyond a few thousand years into the past, and interpretation of isotopic values that fall in the mid-range between those of photosynthetic and methanotrophic products is problematic for all but the most specific of biomarkers.

Does diploptene with a $\delta^{13}C$ of –45 per mil indicate that aerobic methanotrophic bacteria were present and their ^{13}C-depleted diploptene mixed with that from other diplopterol-containing bacteria? Or does it mean that the methane concentration was very low? Or that high productivity, rapid burial rate, and restricted water circulation resulted in ^{13}C-depleted carbon dioxide, which produced unusually depleted photosynthetic organic matter—and, accordingly, ^{13}C depletion in diploptene from the heterotrophs that had consumed it? Does archaeol or PMI with a $\delta^{13}C$ in the –50 per mil range come from methanotrophic archaea that have consumed thermally generated methane, or does it come from methanogens that have consumed ^{13}C-depleted CO_2, or from a mix of the two? Even now, we can hear John Hayes intoning, *there are no magic numbers*. It takes the compound-specific isotope analysis of an array of fossil molecules to untangle these various influences and gain insight into methane's role in climate change and earth history: the full spectrum of archaeal biomarkers, biomarkers from representative members of the food chain—phytane or sterols from photosynthetic algae, extended hopanoids for heterotrophic bacteria, and so on—and the isotopic makeup of the total organic matter and carbonate. Even then, interpretation of the microbial fossil record is all too often thwarted by our perennially incomplete understanding of an ecology wherein the "simplest" organisms perform in complex, multifarious ensembles.

The microbial role in controlling the chemistry of the rocks, oceans, and atmosphere has, over the long course of this book's creation, finally moved to center stage. Over the past few years, the discovery of methane-oxidizing microbial consortia and the meeting of molecular biology and geochemistry that it inspired have had a profound effect on the laboratories of organic geochemists, both in the way they operate and in the questions they address. Kai Hinrichs and Marcus Elvert joined forces at the University of Bremen and assembled an organic geochemistry laboratory whose focus is microbial ecology. Their group collaborates closely with the microbiologists and molecular biologists at the MPI, housed on the same campus. At NIOZ, molecular biologists and microbiologists have become a permanent feature in Sinninghe Damsté's biogeochemistry lab, and certain independent types like Stu Wakeham have spent their sabbaticals learning the genetic techniques themselves. As Geoff found out more than 50 years ago, it's all very well to work on the premise that "the present is the key to the past," like geologists did for over a century with their Uniformitarianism— but only if one can understand the present. Now, finally, the combined tools of gene and lipid analyses are providing the means to understand both present and past—just as Geoff and Dave Ward had hoped they would when the techniques first burst on the scene in the late 1980s.

8

Weird Molecules, Inconceivable Microbes, and Unlikely Environmental Proxies

Marine Ecology Revised

When a distinguished but elderly scientist states that something is possible, he is almost certainly right. When he states that something is impossible, he is very possibly wrong.
—Arthur C. Clarke, 1917–2008, English science fiction writer
  From *Profiles of the Future* (1962)

Anaerobic methanotrophs are not the only ecologically important archaea to surprise microbiologists in the last decade. And their isoprenoid ethers are not the only useful lipids—and certainly not the strangest—to have joined the lexicon of microbial biomarkers. Though much of that lexicon is still too generic to be of much use in understanding geologic history, some of these structures have allowed geochemists to transcend biological complexity and garner clues to past climates and environments.

In the 1990s, when Stefan Schouten first started finding ring-containing biphytanyl ethers in his sediment samples, he was still working on his doctorate at NIOZ. Like everyone else at the time, he assumed that they derived from the lipids of methanogenic archaea and that it was only a matter of time before ring-containing biphytanyl tetraethers would be identified among the lipids of some newly isolated culture of methanogens, as Guy Ourisson had predicted. Schouten was studying oxygen- and sulfur-bound biomarkers, which meant he treated his sediment extracts chemically to cleave the ether and sulfur bonds, and the treatments often turned up biphytanes. But then, he says, he and another student started finding the ring-containing compounds in some really unlikely places, such as the oxic surface layer of marine sediments where neither methanogens nor extreme thermophilic and halophilic archaea were likely to make a home. The only thing they could think of at the time was that the tetraethers had come from methanogens that lived in the oxygen minimum zone, the layer of water beneath the photic zone where heterotrophic bacteria are active, sometimes to the point of using up all of the oxygen.

When Schouten presented these ideas at the 1995 organic geochemistry meeting, Stuart Wakeham immediately piped up with the suggestion that they look for the lipids in the water column—and offered the perfect samples for the enterprise. He had collected particulate matter at different depths in the Black Sea and Cariaco Basin, just the sort of anoxic environments where one might expect to find methanogens in the water column....But to everyone's surprise, when they followed through with the analyses, the highest concentrations of biphytanyl compounds weren't in the samples collected from the anoxic deep water at all, but rather from the plankton and detritus collected in the completely oxic water near the sea surface! Not only that, but the $\delta^{13}C$ values of the tetraethers were generally *enriched* in ^{13}C compared to representative lipids from photosynthetic organisms, whereas methanogen lipids should have been slightly depleted. These biphytanyl ethers apparently came from planktonic archaea that had nothing to do with methane, and certainly nothing to do with extremely hot or acidic water. The isotopic composition and depth distribution of the lipids implied that they might be autotrophic organisms that derived their carbon from dissolved CO_2 or bicarbonate—but they clearly weren't photosynthetic, and what they used for energy was anybody's guess.

Around this time, Schouten says, he heard about some new species of archaea that marine biologists in California had discovered several years before. What Dave Ward had been doing with cyanobacteria in Yellowstone's hot springs, Ed DeLong, the leader of the California group, had done with the natural populations of archaea in the ocean. He'd extracted nucleic acids from marine plankton growing in all sorts of decidedly moderate, oxic environments—and found the rRNA sequences of organisms whose closest known relatives lived in hot springs and hydrothermal vents, a subgroup of archaea known as crenarchaea and hitherto comprised exclusively of extreme thermophiles. These new crenarchaea, however, appeared to be mainstream members of the marine plankton, present wherever one looked for them, even in freezing cold Antarctic waters. Were these the producers of the ring-containing biphytanyl compounds that had been puzzling the geochemists? Wakeham, who had also heard about the work in California, teamed up with DeLong to find out.

The planktonic crenarchaea had defied all attempts to grow them in laboratory cultures, but one strain was discovered growing in a sort of natural isolation that made it amenable to study. It lived in symbiotic relationship with a marine sponge that had been collected off the coast of California, and could be maintained in an aquarium, replete with its resident archaea. DeLong's and Wakeham's groups analyzed the rRNA, DNA, and lipid content of extracts from the aquarium sponges and compared the results to those from plankton samples they obtained from surface waters in two very different environments—off the coast of Southern California, and in the ocean around Antarctica. The only archaeal genes detected in any of the extracts were those of nonthermophilic crenarchaea, and in each case their presence was correlated with significant amounts of biphytanyl compounds, including ones with two and three rings.

When he saw the results of Wakeham's and DeLong's studies, Schouten says, things began to make sense—sort of. The existence of planktonic crenarchaea resolved one paradox—that of finding biphytanyl tetraethers in oxic sediments from cool water environments—and introduced another. Molecular models of archaeal membranes showed that the rings in the biphytanyl components of the tetraethers produced a densely packed and thermally stable membrane that was especially well-suited to extremely high temperatures. But how did these crenarchaea, which apparently had lots of ring-containing tetraethers, get along in the relatively cold ocean? Wouldn't their membranes be too stiff and impermeable to function? The genetic analyses allowed one to detect new species and place them with respect to known species, but one could only make inferences about their biochemistry, physiology, and biology based on those relationships—and it's a long haul from an organism that lives at temperatures above 60°C to one that thrives in the frigid waters around Antarctica. DeLong's surveys of rRNA were beginning to hint that the crenarchaea might hold an important place in marine food chains, but virtually nothing was known about how they lived, what they consumed and produced, or how plentiful they were. Schouten thought that a survey and comparison of the distributions of tetraether structures and isotopic compositions in plankton and sediments from different environments might be revealing—but such a survey was no easy task with the existing techniques. The tetraethers were too big and polar to analyze directly by GC-MS, and treating the extracts to cleave the ether bonds was a tedious multistep process. It took two full days just to analyze one sample, Schouten says, and then one could only determine the amounts and structures of the biphytanyl components, not of the actual tetraethers. The biochemists who had isolated tetraethers from thermophilic archaea had used thin-layer chromatography to separate and purify them, but that was untenable with the complex mixtures and relatively small concentrations in the sediments.

There happened to be a new instrument in the NIOZ lab around this time that was generally good for separating large polar molecules that couldn't be vaporized, but it had never been used to analyze anything like the tetraethers. It was an HPLC, like what James Maxwell and Kliti Grice used to analyze plant pigments, except the new instrument in the NIOZ lab had an interface to a mass spectrometer. In 1999, when Sinninghe Damsté got the idea that the machine might be used to separate and determine the structures of intact tetraethers, it was being operated by Ellen Hopmans, a postdoc who was an expert in toxicology and food science and had lots of analytical experience with HPLC. Hopmans knew it was a bit of a long shot—indeed, when she told the analytical chemists at the instrument company that she wanted to analyze dibiphytanyl glycerol tetraethers, they told her in no uncertain terms that the column she wanted to try wouldn't work and it would all be an exercise in futility. New to the lab, with no experience whatsoever in geochemistry, she was eager to please, and just rash and naive enough to try anyway.

As Schouten tells the story, the reason they happened to have a toxicologist and food scientist in the NIOZ lab at this particular juncture was that he'd gotten the urge to track down his old high school buddies on the Internet and found one of his former fellow nerds working in the food science and technology department at the University of California in Davis. The details may not be relevant to science, but somehow an idle electronic hello from an old school buddy halfway around the world led to the proverbial endless date, and since it was rather diffi-cult for a toxicologist in California and a geochemist in the Netherlands to have much of a marriage, the toxicologist had developed a sudden interest in geo-chemistry and the NIOZ group had decided it might be able to make use of a toxicologist's analytical talents.

The group obtained or prepared purified tetraethers from two cultures of thermophilic crenarchaea, and though Hopmans says that her first chromato-gram would best be described as a "humpogram," it wasn't long before she had managed to separate a standard mixture of nine compounds into nine sharp peaks with nine characteristic mass spectra. Now they could detect the entire range of tetraethers, with all of their structural information intact, in one fell swoop—the group immediately began running samples from its extensive collec-tion of sediment samples, from ancient to recent, open ocean to coastal, inland seas, lakes, and peat bogs.

All of the samples contained tetraethers of one sort or another, including many of the same structures that had been isolated from thermophilic archaea. But the tetraether structures present in the sediments were far more diverse than those in the membranes of the thermophilic crenarchaea: their distributions var-ied across different environments, and it was soon apparent that they came from large and diverse populations of archaea, adapted to all sorts of environments— from oxic to anoxic, surface waters to sediment, freezing to near-boiling—and deriving their carbon and energy from a variety of different sources. This was when Richard Pancost was at NIOZ as a postdoc, finding severely ^{13}C-depleted tetraethers in samples from the Mediterranean mud volcanoes—indicating that at least one of their source organisms was able to oxidize methane. The other thing that became apparent in the group's onslaught of intact tetraether analyses was that the most common, abundant compound in the sediments had a com-pletely unknown and difficult-to-determine structure.

Though the HPLC-MS system provided only basic mass spectral informa-tion, most of the unknown tetraether structures were pretty easy to figure out— except for the most prevalent one. Hopmans had to collect peaks from the end of the HPLC column during repeat runs until she had enough of it for other forms of analysis, and then the group went at it with every trick in the organic chemist's arsenal. Schouten says the structure had them fooled for almost two years, and in the meantime they obtained a sample from DeLong's sponge cren-archaea and found that the same compound comprised its dominant membrane lipid. More sediment analyses confirmed that it was the core lipid of the plank-tonic crenarchaea, a compound that the NIOZ group took to be a unique fea-ture, because to date it had never been detected in methanogens or thermophiles and was not among the ^{13}C-depleted tetraethers that seemed to derive from methanotrophs. Finally, in late 2001, Sinninghe Damsté cracked the structure using two-dimensional nuclear magnetic resonance spectroscopy, a favorite tool of natural products chemists that was generally out of bounds for geochemists because of the amount of pure compound required. The main problem in this case, he says, was not isolating the compound, but rather the incredibly complex spectra it yielded and the endless interpretations required to determine the relative positions of each of its atoms. It was distinguished from the other ring-containing tetraethers by a biphytanyl chain that had a *six*-carbon ring in addition to two of the usual five-carbon rings. Damsté christened the compound "crenarchaeol" and, noting that the six-carbon ring created a sort of bulge in the middle of the tetraether, teamed up with theoretical chemists from the University of Amsterdam to create a molecular model of an archaeal membrane structure with the new compound as a core lipid. The result offered one possible solution to the apparent paradox of cold water crenarchaea: whereas the five-carbon rings made the biphytanyl chains fit together more snugly in a dense structure, addition of crenarchaeol with its bulging six-carbon ring kept them from packing together too closely and would, conceivably, allow the membrane to maintain its fluidity at low temperatures.

crenarchaeol

For his part, Schouten says, the ongoing analyses of tetraethers in sediments and plankton had almost gotten boring, because the compounds were so widespread. It seemed they were everywhere, all the time—in plankton at all depths and seasons, recent sediments, ice age sediments, even Cretaceous sediments, from all around the world. The group did another study with Wakeham, examining particulate matter in the Arabian Sea and comparing the concentrations of tetraethers with those of algal steroids and fatty acids. The results indicated that these planktonic crenarchaea could live not only at all depths, but also at very low oxygen levels. They were certainly not photosynthetic, and they weren't particularly abundant in the zones where organic matter accumulates and heterotrophic bacteria prosper. There was some correlation between the concentration of crenarchaeol and the concentrations of nitrate and nitrite, as if the organisms might be using the former and producing the latter. But the ecology and metabolic strategy of these tiny but ubiquitous organisms were still a mystery when Schouten and Hopmans noticed another pattern, this time in the overall distributions of the tetraethers.

Schouten says that he was standing by the HPLC-MS with Sinninghe Damsté and Hopmans late one afternoon, contemplating the chromatograms from the day's last analyses and getting ready to go home, when he wondered idly why the relative amounts of tetraethers in the two chromatograms were so different. "Oh, it's probably just temperature," Hopmans said. She was too new to geochemistry to know about the alkenone U^K_{37} index or to realize the importance of her insight, but Schouten and Sinninghe Damsté turned and stared at each other. Of course.

One of the samples was from North Sea sediments, and the other from the Arabian Sea. The latter contained more tetraethers with lots of rings in them, and fewer of the ones with zero or one ring: warm water, more rings. This even made some biochemical sense, more so than the alkenones' much-studied decrease in unsaturation with increasing temperature. There was overwhelming evidence that ring-containing tetraethers were components of cell membranes in the phylogenetically related thermophilic crenarchaea. And experiments with cultures of the thermophiles had shown that they biosynthesized tetraethers with more rings when grown at higher temperatures, in keeping with the observation that the five-membered rings caused the biphytane chains to pack together more closely and make a denser, more protective membrane.

It was a quarter to five in the afternoon when the group had this epiphany, and the problem with working at NIOZ—or advantage, if one happens to be a workaholism-prone scientist with a family—is that it is on an island, and the last convenient ferry and train connection to Amsterdam departs at 5:00 P.M. sharp. But the following morning Schouten set to work trying to quantify the

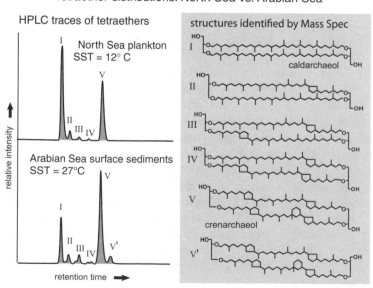

Tetraether distributions: North Sea vs. Arabian Sea

relationship between temperature and the relative amounts of the various tet-raethers. His first attempts were, Schouten says, a bit discouraging. He used a weighted average of the number of rings in the compounds, which gave a positive correlation with temperature, but the data were too scattered to make a good temperature proxy. Then he tried using the ratio of isomers with more rings to isomers with fewer, and finally, when he left caldarchaeol out of his equation, he got a clean correlation. Precisely why was not entirely clear, but it now appears that caldarchaeol is a core lipid in many groups of archaea, and large contributions from other organisms in the sediments may have been skewing the data. Whatever the case, the correlation between the new index and temperature, based on 40 surface sediment samples from around the world, was worthy of the best proxy. Following suit with the Bristol group's U_{37}^K, they called their proxy the "Tetraether Index for tetraethers with 86 carbon atoms" and gave it the acronym TEX_{86}, a small tribute to local pride, as the Dutch island where NIOZ is located is called Texel.

The possibilities for the new proxy were exciting. The tetraethers were found in significant quantities in sediments dating back to the Cretaceous, in Arctic and Antarctic waters, and in lake sediments—times and places where use of the U_{37}^K temperature proxy was proving problematic. In 1999, an international conference on the alkenones had convened in Woods Hole, examining the U_{37}^K proxy from all angles, reinforcing its success, highlighting its limitations, and, in the process, providing a recipe book for development and validation of new

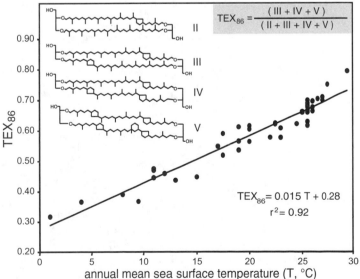

Calibration of the Tetraether Index (TEX_{86}) in surface sediments, 2002

$$TEX_{86} = \frac{(\,III + IV + V\,)}{(\,II + III + IV + V\,)}$$

$$TEX_{86} = 0.015\,T + 0.28$$
$$r^2 = 0.92$$

annual mean sea surface temperature (T, °C)

temperature proxies. Does the calibration hold up under controlled conditions in the laboratory? Is it affected by changes in salinity and growth rate? Are there different strains of crenarchaea with different temperature dependences? For that matter, are the tetraethers used in TEX_{86} found in other archaea that might confuse the signal in the sediments? Do some of the compounds break down more quickly than others, such that the ratios are affected by diagenesis? And so on.

The NIOZ group had one difficulty the Bristol contingent hadn't had: whereas *Emiliania huxleyi* was already happily growing and reproducing in the laboratories of the Marine Biological Association in Plymouth, the crenarchaea had defied all attempts at domestication. Sinninghe Damsté's lab had recently acquired both a microbiologist and a molecular biologist—another affair of the heart, which NIOZ seems to have had good luck with—but how could the two do experiments without a laboratory culture of the organisms in question? They didn't necessarily need a *pure* culture.... Could they just obtain a large sample of water where crenarchaea were growing naturally, put some in a tank where the temperature could be adjusted, and see if they'd keep growing? The very idea was anathema to most microbiologists, and Schouten says everyone he talked to thought such an experiment was preposterous, doomed to failure. But his student Cornelia Wuchter took up the challenge: she used water from the North Sea and a series of simple 20-liter tanks, and then monitored the total concentration of tetraethers to see if and how fast the crenarchaea were growing.

Remarkably, the organisms no one could get to grow in a carefully controlled bacterial medium were off and running within a week. Molecular biologist Marco Coolen analyzed the rRNA and DNA and found that the North Sea water contained a simple mixture of two strains, one of which eventually died out. The "preposterous" improvisation would provide Wuchter with results for an entire doctoral thesis. She could heat the tanks at different temperatures and calibrate TEX_{86}, and she could track the chemistry of the water and learn something about what the crenarchaea were consuming and producing.

The first round of studies with the captive crenarchaea not only verified that the lipid composition within a single strain varied with temperature, but also provided a calibration of TEX_{86} that was gratifyingly close to that from the sediments.... But if, as DeLong's rRNA studies and the NIOZ team's lipid surveys indicated, the crenarchaea thrived throughout the water column, *what* temperatures would they be recording? It was all very nice that the values of TEX_{86} in the sediments correlated so well with sea surface temperatures—but *why* would they, when temperatures at 500 meters depth were some 10°C colder? More studies of water column particulate matter with Wakeham soon made it apparent that this was a question not only of where in the water column the crenarchaea *live*, but also of how their remains get to the sediments once they're dead.

Crenarchaea are less than one-fifth the size of coccolithophores, and with no heavy coccolith shields for ballast, they apparently remained suspended in the water—unless they were living in the upper 100 meters or so, where they were likely to be consumed by larger grazers. In this case, some undigested crenarchaea lipids might well end up in zooplankton fecal pellets and sink to the seafloor. What exactly happens to the biomass of such tiny, free-living organisms in the deep water, where zooplankton activity is limited, remains a matter of conjecture. But Wuchter's work with Wakeham indicated that the dominant portion of the crenarchaeal lipids in the sediments came from organisms living in the upper 100 meters of the water, where predators are most active. This might explain the group's observations from the Black Sea, and those of Kai Hinrichs from the Santa Barbara Basin, where the lipids of organisms that had presumably lived in the oxic surface water—planktonic crenarchaea in the one case and aerobic methanotrophic bacteria in the other—were found in the sediments, and the ^{13}C-depleted lipids of methanotrophic archaea living in the deep anoxic water were not. And it is the best explanation to date for why TEX$_{86}$ values in surface sediments at most sites have recorded sea surface temperatures, despite thriving populations of planktonic crenarchaea in deeper waters.

Even as the Dutch group worked through the recipes for calibrating and validating its new paleothermometer, first applications of TEX$_{86}$ began yielding some intriguing results. In sediment cores from the western Arabian Sea, TEX$_{86}$-derived temperatures are in good agreement with the U^K_{37} temperatures for most of the current interglacial period, but during the last glacial to interglacial transition, the two proxies tell significantly different stories. Whereas the U^K_{37} temperatures are correlated with the $\delta^{18}O$ values in the ice of the Greenland ice cores, the TEX$_{86}$ temperatures record a larger temperature shift and follow the rather distinct trend recorded in the Antarctic ice core. The reasons for this discrepancy are not at all clear, but sea surface temperatures in the region tend to exhibit marked seasonal variations due to upwelling of cool water caused by the seasonal monsoon winds. The Dutch group speculated that whereas TEX$_{86}$ was registering average annual sea surface temperatures, exceptionally large pulses in coccolithophore productivity during upwelling may have skewed the temperatures registered by U^K_{37} toward those of the cool deep upwelled water, thus partially obscuring the glacial to interglacial sea surface warming. This raises the possibility that the two proxies can provide complementary climate information, maybe even a record of how seasonal differences varied during the ice ages—but it poses the conundrum that both proxies have been calibrated and correlate well with *average annual* sea surface temperatures, even in the western Arabian Sea. It also highlights the need for careful interpretation of the temperatures obtained with such proxies, because both can, under some circumstances, be affected by the dynamics of seasonal biological cycles and structure of the water column.

The new proxy is also proving useful for estimating the temperature history in regions or time periods where $\delta^{18}O$ and U^K_{37} measurements are not possible. Planktonic crenarchaea are not as ubiquitous in lakes as they are in the marine environment, but they do seem to be abundant in the largest lakes, and unlike U^K_{37} measurements in lakes, calibrations of TEX$_{86}$ in these environments fit smoothly into the calibration from marine sediments. In a sediment core from southeast Africa's Lake Malawi, temperatures derived from TEX$_{86}$ provided new insights into glacial to interglacial temperature changes in Africa, raising hopes that similar studies might fill in missing information about how temperatures on the continents were affected by changes in the earth's climate over the past few hundred thousand years.

The calibration of TEX_{86} for temperatures above 27°C is incomplete and appears to deviate somewhat from that at lower temperatures, but with this uncertainty in mind, the tetraethers in sediments from Cretaceous black shales indicate that sea surface temperatures were between 33°C and 36°C during the early to mid Cretaceous, consistent with the upper estimates from [18]O measurements— and warmer than anywhere in the contemporary ocean. More provocative is an analysis of tetraethers in Arctic Ocean sediments from the late Cretaceous, which provides one of the first direct estimates of polar sea surface temperatures during the period. The TEX_{86}-derived value of 15°C adds to evidence that the planet was entirely free of ice, on the one hand, but it also implies a steeper pole-to-tropics temperature gradient than previously hypothesized. A recent study of Arctic Ocean sediments from the relatively short, putatively methane-induced heat wave at the end of the Paleocene indicates that sea surface temperatures near the North Pole at this time were even higher, ranging from 18°C to 23°C. Of course, these TEX_{86}-determined temperatures rely rather heavily on the principle of biochemical Uniformitarianism, based as they are on the assumption that lipid biosynthesis in microbes living tens of millions of years ago—for which there is no record but the tetraethers themselves—had the same temperature dependence as it does in contemporary crenarchaea.

Whether or not TEX_{86} will hold up to scrutiny as well as the U^K_{37} has, and precisely what caveats should accompany its interpretation, remains to be seen. With so much of the microbial world yet to be discovered, the specificity of bio-markers like the ring-containing tetraethers is likely to remain in question for some time. Wuchter's and Coolen's combined genetic and lipid analyses of plankton indicate that other planktonic archaea in the surface waters of the ocean do not make them, and tetraethers produced by methanotrophic archaea can usually be distinguished by their pronounced negative $\delta^{13}C$ values. But explorations with genetic probes have identified a number of crenarchaeal strains living in soils, and the NIOZ group recently discovered that ring-containing tetraethers, including crenarchaeol, are common components of the organic matter in soils and may complicate the use of TEX_{86} in areas where there have been relatively large inputs of material from rivers and continental runoff. And Kai Hinrichs's Bremen group recently spearheaded a big multidisciplinary study of ODP sediments from off the coast of Peru that revealed a whole range of apparently heterotrophic crenarchaea living buried deep in marine sediments and producing crenarchaeol and other ring-containing tetraethers with $\delta^{13}C$ values similar to those of lipids from planktonic crenarchaea. Schouten is guessing that the contribution of lipids from these sediment organisms is quantitatively minimal compared to inputs from the autotrophic organisms in surface waters, and thus has little effect on TEX_{86} values and the use of tetraethers to gauge the productivity of planktonic crenarchaea, but Hinrichs is not so sure this is the case.

Several studies, including Wuchter's tank-based experiments, have now confirmed that at least some of the planktonic crenarchaea are in fact autotrophic.

Their carbon comes from bicarbonate, which is typically ^{13}C-enriched compared to CO_2, accounting for the observation that crenarchaeal lipids are consistently enriched in ^{13}C compared to lipids from photosynthetic organisms or their consumers. And their energy comes from the oxidation of ammonium to nitrite. The discovery of a cosmopolitan marine primary producer that doesn't require light; is not constrained to nutrient-limited surface waters or to the neighborhood of hydrothermal vents; is adapted to a wide range of climates; converts ammonium from decomposing organic matter in mid-depth waters to nitrite that phytoplankton can more readily use, a process previously thought to be the exclusive domain of a handful of specialized and elusive bacteria; and comprises a major portion of the microbial biomass in the huge expanse of the deep sea raises a lot of questions, to say the least. What does this signify for estimates of primary productivity and the cycles of carbon and nitrogen, both present and past? How did these planktonic crenarchaea affect the ocean's biological pump for CO_2—what role did they play in past climates and environments? When did they evolve...? And what other important microbial players have we missed?

Another previously unimagined microbe recently found its way to center stage, thanks in part to the NIOZ group's discovery of yet another weird lipid molecule—not, initially, in marine plankton or sediments or rocks, but in bacteria from a Dutch wastewater treatment plant. Microbiologists were trying to find a way to get rid of ammonium in the wastewater and had isolated a bacterium that could convert ammonium to nitrogen gas in the anoxic environment of the sewage sludge. Sinninghe Damsté says he got a request from someone at Delft University who had seen one of the NIOZ group's publications on the archaeal lipids. The microbiologists wanted to know the lipid composition of their bacterium's cell membrane and wondered if he would do the analysis. It was a routine task that Sinninghe Damsté agreed to do more out of a sense of civic duty than out of scientific curiosity... or so he thought, until he saw the huge unidentified peak in the chromatograms from the extracts.

If the ring-containing biphytanyl tetraethers seemed strange to biochemists and organic geochemists when they first started finding them in the 1970s, the structures that Sinninghe Damsté and his colleagues determined among the lipids of the anaerobic ammonium-oxidizing bacteria were so bizarre it was hard to believe they could possibly exist in nature. Indeed, Sinninghe Damsté says that even the chemists he worked with at the University of Amsterdam couldn't believe it at first. The ladder-shaped hydrocarbon extensions were made up of fused four-carbon rings, a construction that was so strained it looked like an organic chemist's joke, something one might try to synthesize on a dare, simply because it looked impossible.

The ladderane lipids made their debut in a poster at the 2002 Gordon Research Conference, and Geoff says he laughed out loud when he saw it. Jan de Leeuw had explicitly led him past the other presentations to show him the poster, which depicted the lipid structures and described their extraction from a wastewater bacterium sporting the unwieldy acronym "anammox," apparently because they engaged in anaerobic ammonium oxidation. Geoff says he stood there looking at

Examples of ladderane lipids from anammox bacteria

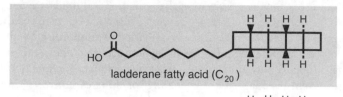

ladderane fatty acid (C$_{20}$)

ladderane glycerol ester/ether

it for several minutes before he decided that his old friend and colleague must be having him on. Ladderane reminded him of his early career, when he considered synthetic organic chemistry a form of modern art in which one tried to outdo one's cohorts by creating the weirdest shapes. But de Leeuw was indignant at the suggestion that the poster was a prank. "I do not joke about science," he said. Even then, Geoff wasn't convinced. Sinninghe Damsté says that on the last night of the conference Geoff took him aside and asked again if the presentation wasn't, in fact, a joke. For his part, Jürgen refused to believe the structures were correct until he heard that they had been proven by synthesis—which they were in 2004, by one of the world's foremost synthetic organic chemists.

Ladderanes had, in fact, been synthesized by chemists as a novelty and object of theoretical interest. But finding such a thing as a major component in a bacterial membrane was another matter altogether. Sinninghe Damsté, now thoroughly interested in the project, discovered that the ladderane lipids weren't in the cell membrane, but rather formed part of a dense, relatively impermeable internal membrane. This unusual membrane enclosed the bacterium's energy-producing reactions and was apparently the solution to one of the enigmas of the anammox, protecting the cell from the indisputably toxic reaction intermediates—which include the common rocket fuel hydrazine—while maintaining the necessary concentration gradients for energy production. The ladderane lipids take a variety of forms: in addition to fatty acids, there are glycerol ethers, esters, alcohols, and dialkyl glycerol diethers, all with ladderane groups attached. Indeed, the diverse anammox lipids even include hopanoids, which were previously considered the exclusive dominion of aerobic bacteria. But for all the novel chemistry and apparatus contained in these tiny organisms, they are, as it turns out, no more a novelty in nature than the anaerobic methanotrophs or the planktonic crenarchaea.

Like methane, ammonium had long been considered biologically inert under anoxic conditions. Likewise, it was oceanographers who first observed that large quantities of the ammonium produced by the decomposition of organic matter disappeared in anoxic waters and sediments, despite the fact that the only bacteria known to oxidize ammonium required oxygen. Unlike methane, ammonium could, in theory, be oxidized by nitrite or nitrate in the absence of oxygen and produce enough energy to be of use to some microorganism. But no such organism could be found, and so the oceanographers' observations were largely ignored and biologists generally thought they had the marine nitrogen cycle figured out, until the late 1980s when microbiologists in Delft first demonstrated that the process was occurring in an experimental waste-processing plant. The organisms, which they isolated in the mid-1990s, were autotrophic, using CO_2 as their carbon source and obtaining their energy by converting ammonium and nitrite—two of the three forms of nitrogen that phytoplankton use as nutrients—to nitrogen gas, which is plentiful in the air and sea but useless to most organisms.

In wastewater facilities and cultures, the anammox bacteria grow exceptionally slowly, and initially microbiologists assumed they wouldn't make much of a contribution to the cycling of nitrogen in natural ecosystems. But once again, the microbial communities in natural environments exceeded expectations. Explorations with genetic probes have identified several phylogenetically related anammox species in coastal basins and freshwater lakes. Marcel Kuypers, who had been working on his doctoral thesis at NIOZ when the ladderane lipids joined the biomarker lexicon, teamed up with microbiologists at his new post at the Bremen MPI to study the activities of anammox in Europe's best anoxic laboratory, the Black Sea, where there were high concentrations of ammonium. Together with Sinninghe Damsté and the Delft microbiologists, they used lipid analyses, genetic probes, and water sample incubation experiments and found several of the unusual ladderane lipids and rRNA sequences closely related to those of the cultured anammox, along with nitrite, in a thin layer of water between 90 and 100 meters depth. Ammonium suddenly disappeared at the same depth, though the water was still anoxic. Fluorescent labeling of the genetic probes allowed the group to count anammox cells: even at the exceptionally slow metabolic rates that had been observed for the organism in the laboratory, the anammox in this layer of water were plentiful enough to account for the disappearance of most of the ammonium diffusing up from the deep waters of the Black Sea.

Studies of the oxidation of ammonia by nitrite in oxygen minimum zones in the ocean and in anoxic sediments, now tuned to recognize the anammox process, indicate that anammox may actually be responsible for conversion of up to 50% of the ocean's nutrient nitrogen to inert nitrogen gas—in other words, for limiting the amounts of ammonium, nitrite, and, indirectly, nitrate that is ultimately available to phytoplankton in the surface waters. Accordingly, these strange organisms, with their rocket-fueled innards, may actually have played a key role in determining the amount and nature of primary productivity in the world's oceans, particularly at times in the past when oxygen was more limited, such as during the episodes of anoxia and enhanced productivity that produced the Cretaceous period's ever-enigmatic black shales. One might, naturally, look to the distinctive lipids or, rather, their bizarre ladderane hydrocarbons for evidence—but it seems unlikely that such an incredibly strained structure would be stable for long, and one is hard-put to predict what sort of fossil molecules it might leave behind. Degradation and artificial maturation experiments may provide some clues as to possible fossil structures, and the ladderane lipids do have a distinctive ^{13}C signature that might help, some 45 per mil more depleted than the CO_2 the organism consumed. In the meantime, the NIOZ group has hypothesized that the anammox played a pivotal role during two of the Cretaceous oceanic anoxic events, relying on indirect evidence from a more mundane series of microbial biomarkers—the hopanoids produced by organisms at the opposite end of the nitrogen cycle.

Cyanobacteria are among the few known photosynthetic organisms in the ocean that can make use of the abundant nitrogen gas dissolved in the water. They thrive in areas where other phytoplankton are limited by a lack of nutrient nitrogen, and in the contemporary ocean they dominate phytoplankton populations in low-productivity central ocean areas. According to Roger Summons's and Linda Jahnke's surveys of bacterial hopanoids, one of the few hopanoid innovations to be incorporated into the sturdy ring structure and preserved for posterity is a methyl group on the second carbon atom of the A-ring, a structure that appeared to be synthesized almost exclusively by species of cyanobacteria. Summons also found that these 2-methyl hopanoids were only prominent in rocks and sediments formed in environments where cyanobacteria were, for one reason or another, prominent. When the NIOZ researchers detected exceptionally large amounts of 2-methyl hopanoids in Cretaceous black shales, they knew they weren't looking at sediments from an ocean where nutrients were perennially limited and productivity was generally low, as in the contemporary central oceans. Rather, they speculated, something must have specifically depleted the available nitrogen in the photic zone, giving the cyanobacteria a competitive edge, while other key nutrients such as phosphate and iron remained in good supply. In black shales from 93 and 120 million years ago at sites in both the Atlantic and Pacific oceans, the concentration of 2-methyl hopanoids relative to other extended hopanoids increased from less than 2% before shale deposition to more than 20%. The proportion of bacterial hopanoids relative to algal steroids also increased, suggesting that cyanobacteria may have dominated phytoplankton communities in large areas of the world's oceans during these periods.

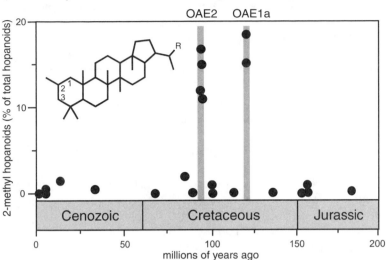

Elevated levels of 2-methyl hopanoids in Cretaceous black shales: Did cyanobacteria take over during oceanic anoxic events?

Drawing an analogy to what they'd observed in the Black Sea, the NIOZ group speculated that anammox bacteria could have been active in the deep ocean and even into the photic zone during these periods of ocean anoxia, and that they would have used up the ammonium and nitrite, and indirectly the nitrate, that most phytoplankton require. This would have given the cyanobacteria, with their ability to use the always abundant nitrogen gas, a decided advantage over the coccolithophores, dinoflagellates, and diatoms that generally dominate marine phytoplankton communities. Cyanobacteria not only use the inert nitrogen gas, but also incorporate it into nitrogen-containing organic compounds and ultimately resupply the water with nitrogen in nutrient form, and the NIOZ group suggests that this may have been one of the mechanisms for sustaining the long periods of high primary productivity that produced the black shales.

Surpassing even Guy Ourisson's enthusiastic prognoses in the 1980s, fossil lipid molecules in sediments and rocks have led to or accompanied the discovery not just of new species, but of energy-harvesting systems that redefine what is "biochemically feasible." As the anammox bacteria and methanotrophic archaea illustrate, if a reaction is chemically possible and releases some modicum of energy, then it's quite likely that some microbe somewhere has figured out how to make use of it: offer a free lunch, no matter how humble, and someone, or some few, will find a way to partake.

Recent work verifies that there are microbes living more than half a kilometer beneath the surface of the seafloor, somehow making a home in sediments where the sequence of good oxidants and readily oxidized fuels has long been exhausted and it would seem that not even the most minimal source of energy is available. Microbes have even been found living within the underlying basalt of the oceanic crust and deep within the granite bedrock on the continents. The microbial sector of life is not only vaster and more varied, but also more resourceful and enterprising than even the most forward-thinking microbiologists imagined. Its exploration requires the tools and expertise of chemists, biologists, and geologists, in every combination and permutation. Ironically, organic geochemists, for whom the perfect biomarker is one that survives the ravages of time, now find themselves developing HPLC techniques that allow them to detect the most fragile biological molecules—glycerol ethers, esters, and fatty acids with the hydrophilic phosphate groups still attached—and thus distinguish the lipids of live cells. Used together with fluorescently labeled genetic probes, these biomarkers allow both identification *and* quantification of the active microbes in a sedimentary environment.

Current hypotheses hold that the organisms in the deep subsurface microbial communities grow in slow motion, so to speak, on a geologic timescale, and that their primary producers derive their energy by exploiting reactions between the reduced minerals in the rocks and the oxidized elements in seawater that circulates through cracks in the igneous rocks beneath the sediments—similar to some of the reactions exploited by hydrothermal vent microorganisms, except at much, much lower temperatures. In some areas, oxidants such as sulfate seem to be introduced from below by seawater that has made its way deep into the basement rocks and is now circulating back toward the surface. Another potential source of energy for these sleepy microbial communities may be the hydrogen generated when water reacts with the reduced metals in the basalt. There is some concern that the presence of live microbes in even the most ancient sediments may hinder interpretation of the molecular fossil record of life and environments in the past, but, so far, it appears that the deep subsurface microorganisms grow so slowly that their lipid biomass is minimal compared to that of the fossil lipids in most rocks.

Exploration of the microbial realm is only just begun, but the repercussions are already sounding in almost every field of the natural sciences. The recognition that microbial photosynthesis is the norm in the surface waters of the ocean, along with the realization that the capture and storage of energy in reduced organic compounds are not confined to the sun-graced surface of the planet but are prevalent in the deep sea and sediments—the two largest habitats on Earth in terms of sheer volume—requires reassessment of how key elements such as carbon and nitrogen cycle through the atmosphere, ocean, and rocks. Organic geochemists are now struggling to understand which and how much of the microbial organic matter formed in the ocean is transported to the sediments, while geologists scramble to understand the movement of nutrient-rich fluids deep within the earth, biologists expand their ideas of what nutrients and energy sources determine primary productivity, climate scientists reconfigure their models for the movement of carbon dioxide and methane into and out of the atmosphere, and paleontologists shift their views of life's history on the planet— all in an attempt to accommodate the flood of new information from life's tiniest, most inventive, and most primordial representatives.

9

Molecular Paleontology
and Biochemical Evolution

Men do generally too much slight and pass over without regard
these Records of Antiquity which nature have left as Monuments
and hieroglyphick characters of preceding Transactions in the like
duration or transactions of the Body of the Earth.
—Robert Hooke, 1635–1703, British natural philosopher, sometimes referred to as
  "London's Leonardo"
  From *The Posthumous Works of Dr. Robert Hooke* (1688)

The philosophical study of nature rises above the requirements of
mere delineation, and does not consist in the sterile accumulation of
isolated facts. The active and inquiring spirit of man may therefore
be occasionally permitted to escape from the present into the
domain of the past, to conjecture that which cannot yet be clearly
determined, and thus to revel amid the ancient and ever-recurring
myths of geology.
—Alexander von Humboldt, 1769–1859, German naturalist and explorer
  From *Views of Nature* (1850 translation)

Carl Woese's drive for a unified system of biological classification didn't just open the microbial world to exploration: it reshuffled the entire taxonomic system and revolutionized the way that biologists study evolution, reigniting interest in pre-animal evolution. Studies of evolution from the mid-nineteenth through most of the twentieth century relied on the comparison of forms in living and fossil organisms and were limited to the complex multicellular organisms that developed over the past 550 million years. In other words, much was known about the evolution of animals and land plants that left distinctive hard fossils, and very little was known about the unicellular algae and microorganisms that occupied the seas for most of the earth's history. Woese's Tree of Life, derived from nucleic acid sequences in ribosomal RNA, has revealed ancestral relationships that form and function don't even hint at, allowing biologists to look beyond the rise of multicellular life and link it with less differentiated, more primal forms—which was precisely Woese's intention. But evolution is a history, not just a family tree of relationships. If the information stored in the genes of extant organisms is to provide true insight into that history, it needs to be anchored in time, linked to extinct organisms and to past environments. Ultimately, we must look to the record in the rocks and sediments, just as paleontologists and biologists have been doing for the past two centuries.

In Darwin's time, that record comprised rocks from the past 550 million years, a span of time that geologists now call the Phanerozoic eon, based on Greek words meaning visible or evident life. The eon began with the rocks of the Cambrian period, in which nineteenth- and early-twentieth-century paleontologists discovered a fabulous assortment of fossils—traces of trilobites, anemones, shrimp, and other multicellular animals that were completely missing from any of the earlier strata. Thousands of new animals and plants, including representatives of almost all contemporary groups, as well as hundreds of now-extinct ones, appeared so suddenly between 542 and 530 million years ago that paleontologists refer to the phenomenon as the Cambrian "explosion." Darwin speculated that the progenitors of these animals had been evolving for vast periods of time prior to the explosion and that their remains were still hidden somewhere in the earth's rocks, yet to be discovered. "We should not forget," he noted in *The Origin of Species*, "that only a small portion of the world is known."

Even in the twenty-first century, it's humbling to consider how dependent our knowledge is on luck and chance—on fortuitous geologic events and the chance of their discovery. Trillions of cubic kilometers of rock strata lie buried and hidden from view unless an oil company decides to drill a core to hunt for oil or a miner excavates a quarry, an engineer cuts a road...or an earthquake creates a rift or a river erodes a canyon. The famous Burgess shale fossil assemblage in the Canadian Rockies owes its preservation to a massive primordial mudslide that buried a complete marine ecosystem in the peak of its Cambrian glory, and its bizarre fossil creatures were discovered by chance when a group of adventuresome nineteenth-century geologists were returning home from their explorations and had to move a stray boulder out of their donkeys' path. Our understanding of the earliest land plants owes in large part to a volcanic eruption or similar disturbance that covered an ancient moor with a thick layer of dust that formed the chert that some farmer used for his boundary wall 400 million years later—unusual stones that a doctor *cum* geologist happened to notice as he was out walking near the Scottish village of Rhynie at the beginning of the twentieth century.

Just as Darwin suspected, the discovery of larger portions of the rock record, combined with advances in microscopic techniques and a better understanding of the mineral and chemical consequences of microbial life, has revealed rocks formed during vast periods of Precambrian time that were teeming with life. Three of the nearly four billion years of earth history that predate the Phanerozoic have now earned division into eons, if not eras and periods: the ill-defined Archean, which includes the first sedimentary rocks and the first putative stromatolites and microfossils, as well as organic carbon that is depleted in ^{13}C compared to carbonates; followed by the long Proterozoic, which extends from 2.5 billion to

542 million years ago and includes evidence of the first continents, the first ice ages, and the first definitive fossils of bacteria and microalgae. Darwin was, however, only partially correct in his expectation. The detailed fossil imprints of a rich variety of relatively large, soft-bodied, apparently multicellular creatures have indeed been discovered in Precambrian rock formations around the world, most notably in Australia's Ediacaran hills—but only in rocks formed during the 40 million years directly preceding the Cambrian explosion. The fossils are plentiful enough to have earned the time from 620 to 542 million years ago its own name, the Ediacaran period, but they are completely enigmatic with respect to anything that came before or after. In other words, though there is now plenty of evidence to document life's presence on Earth for billions of years before the Cambrian explosion, we require another echelon of information, beyond the morphological, if we are to understand how, when, and which of the microscopic spheroidal marks and bloblike polymeric cysts that comprise its fossil record gave way to the plethora of complex forms that appeared in the Cambrian.

In fact, the phylogenetic work of the past few decades has shown that a lot of genetic variation and, presumably, evolutionary diversification can occur before it is expressed in morphological traits—not only in simple unicellular microorganisms, but also in complex multicellular plants and animals. Of course, botanists and natural products chemists had been noting this indirectly since the 1960s, when they started using the structures of biochemicals—proteins, or the triterpenoids that Guy Ourisson studied, or Geoff's distributions of leaf waxes—to distinguish between related groups of plants that morphology failed to clearly differentiate. So it makes sense that fossil lipids, in addition to being the *only* fossils preserved in many environments, may reflect genetic change that morphology has yet to register, even in highly differentiated, well-characterized, multicellular forms of life such as the flowering plants.

Though biologists and paleontologists have been studying the morphology of flowering plants for almost two centuries and their fossils show up clearly during a relatively well-documented period of earth history—though they are the most abundant plants on the contemporary earth and our agriculture depends on them—we know as little about their early evolution as we do about the unicellular algae that left their amorphous imprints and blobs in Precambrian rocks. Fossils of flowering plants appear in rocks from the mid-Cretaceous period, but paleontologists could find no transitional forms, no morphological characteristics that might indicate when, where, or how the flowering plants diverged from the other groups of seed plants, such as conifers, ginkgoes, and mosses, whose fossils predate them in the rock record. It was, Darwin noted in a letter to botanist James Hooker in 1879, an "abominable mystery," and the situation wasn't much improved in the 1990s, when Mike Moldowan left Chevron for an academic post at Stanford and turned his attention to the problem. Even the advent of modern phylogenetic analyses had not shed much light on the origin of the flowering plants: they seemed to have no relatives.

Moldowan's geochemical approach was based on a phenomenon that he had been seeing in oil and rock samples for years, the chemistry of which Jürgen's group had recently clarified: the various β-amyrin–type pentacyclic triterpe-noids that flowering plants use to ward off insects and other pests are converted to oleanene and oleanane during diagenesis. If, as appeared to be the case, the β-amyrin structure is made by a biosynthetic pathway that is limited to flowering plants, and there are no other sources of oleanane in the sediments; and if one looked for oleanane in ancient marine sediments, rocks, and oils that had formed along continental margins, where runoff and rivers had delivered sediments eroded from large areas of land—then, Moldowan reasoned, one could extend the search for clues to the origins of flowering plants far beyond the paleontolo-gists' quarries and explorations.

The incidence of oleanane in the large group of samples that Moldowan and his colleagues analyzed was in rough accord with the fossil evidence, generally increasing from the late Cretaceous onward. The additional chemical informa-tion invites speculation about why the flowering plants diversified when they did: is this expansion of triterpenoid-producing flowering plants—presumably repre-sentative of the success of organisms with the β-amyrin biosynthetic capability—linked to the evolution, diversification, and expansion of insect pests? We can, to date, only speculate. The presence of oleanane in early Cretaceous and older rocks and fossils has, however, finally revealed a connection between flowering plants and older forms of seed plants—a connection based not on common mor-phological characteristics or on genetic information in extant species, but on a common biosynthetic pathway for a particular group of secondary metabolites. Oleanane was detected in rocks that formed during the Permian period, some 100 million years before unequivocal fossils of flowering plants or their pollen begin to show up in the rocks. In an attempt to pin down its source, Moldowan's group analyzed samples that contained recognizable fossil leaves or fragments, and even managed to extract the lipids from some of the fossils themselves. The results were provocative. Oleanane was absent from most of the fossil conifers and seed ferns the researchers analyzed, just as it and its precursor triterpenoids are absent from living representatives of these groups of seed plants. An extinct group of cycad-like seed plants were an exception: fossils of three species of ben-nettitaleans, found in rocks from the early Cretaceous period, contained signifi-cant amounts of oleanane. More intriguing still, oleanane was present in an even older group of extinct fossil plants, the gigantopterids.

About 65 million years ago, at the end of the Cretaceous period, a giant aster-oid apparently crashed into Mexico, frying a good part of North America and wreaking havoc with the earth's climate system. The extent to which the asteroid was responsible is still a matter of hypothesis, but massive die-outs and extinc-tions of entire groups of Mesozoic era organisms, including dinosaurs, are appar-ent in the narrow strata of rocks and sediments deposited at this time. According to the fossil record, the bennettitaleans were among the victims of this holocaust, but unless the ability to make β-amyrin–type pentacyclic triterpenoids evolved more than once, some bennettitalean offshoot that paleontologists haven't found must have survived, eventually giving rise to the flowering plants. And, given the presence of oleanane in gigantopterid fossils from the Permian period,

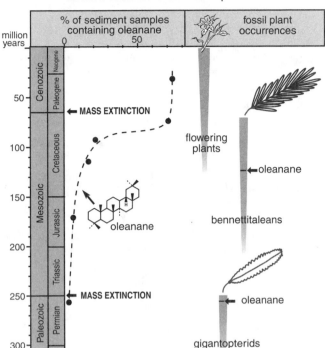

Geologic records of oleanane in
ancient sediments and fossil plants

it may not have been the first time that the oleanoid producers slipped by: the fossil record indicates that the gigantopterids disappeared during an even more catastrophic mass extinction of species at the beginning of the Triassic, some 251 million years ago.

Botanists have found various reasons beyond oleanane to believe that the bennettitaleans and gigantopterids were related to the group of plants that eventually took over the continental landscape and won the hearts, not to mention the stomachs, of its human inhabitants—in which case the flowering plants may indeed be the sole survivors of a larger group of seed plants that diverted onto its own evolutionary pathway during the Permian period. But another possible explanation for the oleanane data is that the ability to produce β-amyrin evolved more than once in different, unrelated groups of plants at various times in the earth's history. John Volkman has become increasingly interested in the evolution of biosynthetic pathways in recent years, inspired by the writings of Guy Ourisson and Michel Rohmer, on the one hand, and by the dynamic group of molecular biologists at his institute in Australia on the other. He points out that, like sterol biosynthesis, the biosynthesis of β-amyrin begins with squalene epoxide and is catalyzed by an enzyme that is similar to the enzymes that catalyze sterol biosynthesis, so perhaps chance mutations leading to the ability to make it were not such a rare event. Triterpenoids with the oleanoid skeleton have, in fact, been found in a species of marine fungus, which might lend some support to the hypothesis that this biosynthetic capability evolved repeatedly in different groups of organisms. Or perhaps, Volkman notes, rather than a biochemical déjà vu, the ability to make oleanoids was a common property of early eukaryotic cells that was subsequently lost in most lineages—with the exception of those leading to the flowering plants and, apparently, certain fungi. If this were the case, however, the fossil oleanane trail, which seems to begin with the gigantopterids in the Permian period, might be expected to extend back even beyond the first fossils of land plants, which are about 425 million years old, to some common ancestor among the algae and the early eukaryotes that emerged during the Proterozoic eon.

Similarities in cell structure and pigment composition imply that the precursors to land plants may have been members of the green algae, one of the three broad groups of algae—red, brown, and green—that can be distinguished by their different pigment compositions and types of photosynthetic apparatus. Phylogenetic analyses also indicate that the green algae are related to liverworts and mosses, and the oldest fossils of land plants are very similar to contemporary members of these two groups. Jan de Leeuw has speculated that the cross-linked alkyl chains of algaenan may have acted as a sort of plastic coating that provided protection against dehydration for early land-colonizing algae and allowed them to survive dry spells in shallow-water environments or during transport through the air by wind. But, of course, we don't know when or even which green algae first started producing algaenan, and attempts to distinguish it or similar biological macromolecules in ancient kerogen and use it as a biomarker have, to date, been unsuccessful. On the contemporary earth, green algae are a minority group among marine phytoplankton, generally prominent only in freshwater environments, where they are often the dominant group. But microfossils in rocks formed prior to the Mesozoic era indicate that they may have been more widespread and prevalent in marine environments in times past. Paleontologists are at a loss to apply specific taxonomic assignments to many of these fossils, which are composed of insoluble, acid-resistant organic material in the form of a cell or cyst. They have been lumped under the ignominious heading "acritarch," meaning "of uncertain origin" in Greek, but many of those found in marine rocks from the Paleozoic era exhibit just enough detail to be broadly identified as deriving from some species of green algae.

The distribution of steranes in ancient oils and rocks indicates that green algae may actually have dominated marine systems throughout the Paleozoic, and that the triumvirate of diatoms, photosynthetic dinoflagellates, and coccolithophores that currently rules ocean surface waters didn't begin its rise to power until the Mesozoic era. Geochemists have been noting since the 1980s that the steranes in pre-Mesozoic crude oils and source rocks were predominantly C_{29} compounds, whereas those formed later were predominantly C_{28} compounds. By the 1990s, Volkman's and others' analyses of algal sterols were showing that, like most land plants, green algae typically produce relatively large amounts of stigmasterol and sitosterol, both of which are reduced in the sediments to some stereoisomer of the C_{29} sterane 24-ethylcholestane. The most abundant sterols in diatoms, dinoflagellates, and coccolithophores—all phylogenetically related to the red lineage of algae—are compounds such as brassicasterol, ergosterol, and 24-methylenecholesterol, which are reduced in the sediments to the C_{28} sterane 24-methylcholestane. Recently, Roger Summons has been working with the geochemist John Zumberge, whose private consulting company, GeoTech Oils, has assembled a database of results from careful, standardized biomarker analyses of hundreds of well-characterized and dated crude oils and source rocks: here the ratio of C_{28} to C_{29} steranes is consistently less than 0.6 from the late Proterozoic eon to the beginning of the Mesozoic era, but increases to more than 1.0 by the end of the Cretaceous period.

Silica frustules, calcium carbonate coccoliths, and elaborate organic cysts document the general expansion and diversification of diatoms, coccolithophores, and dinoflagellates, as well as the fluctuations in the relative status of their populations, since the Cretaceous. Diatoms appear to be the most recent of the three to gain prominence, with the largest increase in diversity having occurred within the last 50 million years. But like the fossil record of land plants and animals, the preservation of all these tiny microfossils is dependent on the whims of nature, in this case on the depth and chemical composition of the ocean water at a given time and place—and, of course, there is only so much one can learn about the evolution of an organism from its mineral encasement or, in the case of dinoflagellates, the cysts formed during its dormant periods. Here again, the molecular fossils tell a longer and potentially enlightening story.

According to their microfossil record, the diatom family has been around for about 130 million years. Fossils of the genera known to produce HBIs aren't present until about 70 million years ago, but in a massive survey of sediments and oils formed over the past 500 million years, Sinninghe Damsté's NIOZ group discovered that the HBIs had been around for much longer. According to this biomarker evidence, HBI-producing genera appeared on the scene about 91 million years ago—just after one of the most pronounced Cretaceous oceanic anoxic events, so it seems that the reorganization of marine ecology and nutrient cycles may have paved the way for their rise to dominance. The physiological role of HBIs in contemporary species of diatoms remains a mystery, but the Plymouth group's investigations of how the organisms make them have raised some interesting questions. The two most prominent groups of HBI-producing marine diatoms, *Haslea* and *Rhizosolenia*, use two *different* biosynthetic mechanisms to produce the isoprene building blocks for their HBIs. Furthermore, phylogenetic analyses of more than a hundred species from all four HBI genera indicate that the *Haslea* and the *Rhizosolenia* fall into two very distinct genetic groups. It seems, then, that the ability to make HBIs evolved independently in different groups of organisms during at least two different episodes in earth history, implying that the compounds provided an important evolutionary advantage— but what that advantage was and how and whether it was related to the transformation of marine ecosystems that occurred during the oceanic anoxic event 93 million years ago are still anybody's guess.

Another group of organisms that may play an important role in contemporary marine ecosystems, but is unrelated to any algal group and left no hard fossils whatsoever to attest its passage, may have made an even more decisive evolutionary leap during another of the Cretaceous anoxic events. The preponderance of 2-methyl hopanoids and relative paucity of algal steroids in black shales deposited during the oceanic anoxic events at 120 and 93 million years implies that nitrogen-fixing cyanobacteria were the primary producers of choice for these stratified, low-oxygen ocean regimes. But all Cretaceous black shales were not, apparently, created equal, despite their similar appearance. Black shales deposited in the newborn Atlantic and shrinking Tethys oceans about 112 million years ago contain an entirely different contingent of organic matter than anything that came before or after. Algal steroids are scarce, as in other black shales, but 2-methyl hopanoids are entirely absent, and the extractable organic matter consists mostly of archaeal tetraethers, with crenarchaeol being the most abundant compound in the polar fraction, and of acyclic isoprenoid alkanes, including pentamethylicosane and a similar compound, tetramethylicosane. Stranger still, the kerogen consists of a macromolecular material composed almost entirely of the latter, laced together by a network of ether linkages. In both the extract and the kerogen, these lipids are significantly enriched in ^{13}C compared to the algal steroids—as is typical for crenarchaeal lipids. Faced with this evidence, the NIOZ group hypothesized that during this particular Cretaceous anoxic event, it was the planktonic crenarchaea, rather than cyanobacteria, that gained an upper hand. Perhaps the event was even a major turning point in crenarchaeal evolution, marking the expansion of cool-water, crenarchaeol-producing planktonic forms.

Phylogenetic analysis does indeed indicate that the marine planktonic crenarchaea diverged relatively recently from their hyperthermophilic antecedents, though just how recently is unclear. Hyperthermophilic crenarchaea have been discovered living within hydrothermal vents, and perhaps more moderate forms developed in the warm outflow waters around the edges of the vents, which were evidently quite plentiful during the mid-Cretaceous. According to Sinninghe Damsté's hypothesis about membrane density, these more moderate thermophiles would have begun synthesizing crenarchaeol, with its six-membered bulge-producing ring...and then what better time for heat-loving, ammonium-using, low-oxygen-tolerant organisms to venture out into the wider world than during an oceanic anoxic event when, according to the TEX$_{86}$ index and other proxies, surface water temperatures were as high as 35°C? Crenarchaeol or its distinctive biphytanyl component has been found in rocks from the late Jurassic, however, some 50 million years before the mid-Cretaceous expansion. Some researchers think that cold-adapted crenarchaea evolved much earlier, whereas others doubt that crenarchaeol had anything to do with cold adaptation and maintain that it can be made by thermophilic crenarchaea as well. A group of microbiologists and organic geochemists at Harvard and the University of Georgia recently detected crenarchaeol in hot spring microbial mats with temperatures ranging from 40°C to 85°C. They proposed that the ability to biosynthesize crenarchaeol may be

more related to metabolic function and the permeability of the membrane to specific compounds than it is to temperature. However, the amount of crenarchaeol
in the hottest mats they studied was, in fact, negligible, whereas it comprised
a significant fraction of the archaeal tetraethers in the lower temperature ones,
and completely dominated them at 40°C—just as one might expect for a lipid
component that facilitates membrane function at less extreme temperatures. The
presence of crenarchaeol in continental hot spring mats, like its association with
heterotrophic organisms in sediments and in soils, indicates that the compound
is synthesized by organisms occupying a variety of ecological niches and is not
limited to planktonic marine crenarchaea, as initially supposed. Whether it is
produced in the hot spring mats by the crenarchaea whose genes the Georgia
group found, which are closely related to thermoacidophilic crenarchaea, or by
an undetected strain of "cold-adapted" crenarchaea that can survive at higher
temperatures, or by some yet-undiscovered evolutionary intermediate is, as of
this writing, unclear.

Another story that brings us tantalizingly close to linking the evolution of
a particular biochemical trait with a particular set of environmental conditions
and juncture in earth history—and is likewise missing an important chapter, or
two or three—is that of the alkenone-producing coccolithophores. The incidence
of long-chain alkenones and related esters in marine sediments can be loosely
correlated with the different forms of coccoliths deposited over the past 50 million years. The most recent of these are, of course, the coccoliths of *Emiliania huxleyi* and members of the genus *Gephyrocapsa*. The latter first appear in Pliocene
sediments and dominate the coccolith record for about three million years, until
Emily gained the upper hand around 75,000 years ago. During the Miocene and
Oligocene epochs, the incidence of alkenones correlates roughly with the presence of the now-extinct genus *Reticulofenestra*, which, based on coccolith morphology, appears to be closely related to *Gephyrocapsa*. During the Eocene, the
record of alkenones becomes sketchier, either because they have been degraded
or reduced to alkanes, or because the organisms producing them were scarcer.
Prior to the Eocene, the taxonomic association breaks down altogether, as the
coccoliths are too morphologically indistinct to classify, and there are only a few
reports of alkenones.

Alkenones have, in fact, been identified in Cretaceous sediments as old as
120 million years, but the distribution of chain lengths and saturation patterns
is different from the suite of eight made by Emily, most notably in the absence of
compounds with more than two double bonds. The triunsaturated compounds
are slightly more reactive than their more saturated counterparts and could have
simply disappeared from some older sediments—but they are also absent from
black shales where much more labile compounds persist, and it seems likely that
compounds with more than two double bonds simply didn't exist yet. The mod-
ern-day distribution of homologues, including the triunsaturated 37-carbon alkenones, makes its debut in sediments deposited at high latitudes about 50 million
years ago, during the first round of cooling and variable climate regimes that followed the long heat waves of the Cretaceous and early Paleogene periods. Simon

Brassell has proposed that this was when the temperature-dependent regulation of unsaturation in alkenone biosynthesis emerged, though whether the algae gained some advantage from storing a larger proportion of unsaturated compounds when the water cooled, or whether this temperature "adaptation" is just the direct effect of temperature on the biosynthetic reactions, is still the question of the day when it comes to alkenones. We remain nearly clueless as to what adaptive benefits these unusual lipids provided for the organisms that started making them to begin with. According to the hard fossil record, phytoplankton diversity, and coccolithophore diversity in particular, was on the rise during the early Cretaceous, implying that competition was also increasing. *Trans* double bonds make the alkenones difficult for zooplankton to digest and relatively resistant to degradation by sunlight, so perhaps the coccolithophores that invented them were less vulnerable to grazing, or perhaps the alkenones provided a particularly long-lasting energy supply that put them at an advantage when nutrients became depleted in the surface waters.

Among the phytoplankton that may have been competing with the coccolithophores in the oceans of the early Cretaceous, the dinoflagellates appear to have the longest and most puzzling history. Their organic cysts, replete with telltale details and ornamentation that allow paleontologists to classify them, are indisputably present in sediments dating back to the beginning of the Triassic period and reflect a phenomenal degree of evolutionary innovation and diversification during the Jurassic and Cretaceous. Summons and his cohorts found that an increase in the relative abundance of dinosterane is correlated with the incidence of cysts in rocks and sediments, and Moldowan's group obtained a similar result from a more extensive analysis of triaromatic dinosteroids—presumably diagenetic end members of dinosterol—in mature sediments and oils. The appearance of the dinoflagellates at the beginning of the Triassic is preceded by a mass extinction of invertebrates and land plants that was so pronounced it almost seems like retribution for the earlier excesses of the Cambrian—a severe retrenchment that cleared the earth of more than 75% of its fossil-producing species, including the trilobites that had reigned in the ocean for hundreds of millions of years. Few of the phytoplankton that lived in pre-Triassic times left hard fossils, and little is known about what happened at the base of the marine food chain during the extinction. The Bristol group—now under Richard Evershed's leadership, with Richard Pancost among its permanent members—recently analyzed rocks spanning this period and found sudden peaks in the abundance of 2-methyl hopanoids immediately following the two stages of invertebrate extinction indicated by the fossils. The cyanobacteria, it seems, were just waiting in the wings to take immediate, though transitory, advantage of the lack of grazing pressure and competition, much as they apparently did during some of the Cretaceous oceanic anoxic events. Kliti Grice, now leader of her own research group at Australia's Curtin University of Technology, has found isorenieratane and related compounds in rocks from this same period, indicating that the oceans were similarly oxygen depleted. Both the fossil and the biomarker records suggest that the dinoflagellates rapidly evolved and diversified in the wake of

Evolution and diversification of marine algae:
microfossils and molecular fossils

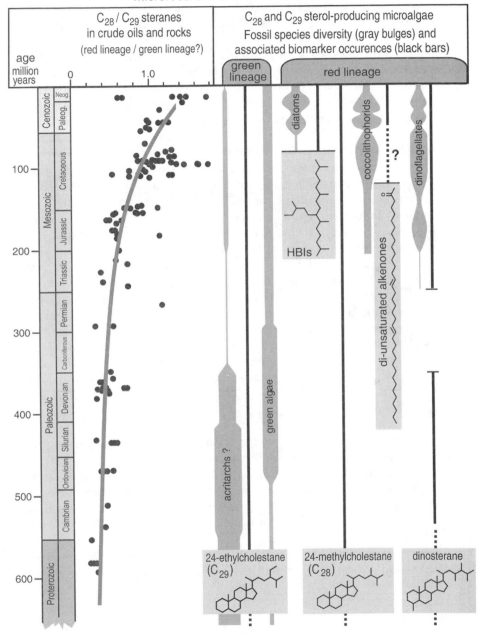

the Permian-Triassic holocaust, and there is no sign, either fossil or biomarker, of their existence during the 100 million years prior....And yet the story is not, apparently, so simple: there are hints in the geological record that this might have been a repeat performance for the dinoflagellates, either a comeback after a long period of reclusive inactivity, or a reinvention.

Though there are no signs of dinoflagellates in rocks formed during the Permian and Carboniferous periods, acritarchs that bear a marginal resemblance to dinoflagellate cysts are frequently encountered in rocks that formed *before* 350 million years ago, and independent studies from both Summons's and Moldowan's labs have found relatively large amounts of dinosterane and triaromatic dinosteroids in the rocks associated with them. Did these come from early dinoflagellates or their ancestors? And did they simply recede into obscurity 350 million years ago, eking out a living in some ecological microniche where they survived through the great apocalypse at the end of the Permian—lying low, so to speak, until they could expand into an empty world where their every evolutionary innovation would be accommodated? Many dinosterol-producing dinoflagellates are heterotrophic and would have been less productive than their photosynthetic relatives—perhaps their fossils and biomarkers are simply not plentiful enough for detection. Or perhaps the early dinosterol-producing, cyst-forming organisms had nothing to do with dinoflagellates at all and went extinct 350 million years ago—perhaps the dinoflagellates did evolve from scratch after the Permian-Triassic mass extinction, and simply reinvented some of the same characteristics, including dinosterol....But there is another clue, this time from the phylogenetic tree of life: the "primitive" morphology and genetic makeup of contemporary photosynthetic dinoflagellates imply that they diverged onto their own idiosyncratic evolutionary pathway early in the evolution of the eukaryotic cell structure, indeed, not long after the divergence of the red lineage of algae....And when was that?

The oldest occurrence of fossils that can be confidently identified as red algae have been found in a rock formation in Canada and are about 1.2 billion years old. The consensus among paleontologists is that acritarchs found in Australian shales that formed between 1.49 and 1.43 billion years ago are definitely the remains of some sort of eukaryotic algae. And most paleontologists agree that microfossils found in 1.9- to 1.8-billion-year-old rocks correspond to organisms that boasted a eukaryotic cell structure. But once again, the molecular fossils may have a much longer story to tell, a story that is directly tied to the accumulation of free oxygen on the early earth and the evolution of oxygenic photosynthetic bacteria—and perhaps, counterintuitively, methanogenic archaea—among other things.

10

Early Life Revisited

The surface of the earth is not simply a stage on which the
thousands of present and past inhabitants played their parts in
turn. There are much more intimate relations between the earth
and the living organisms which populated it, and it may even be
demonstrated that the earth was developed because of them.
—Jean Louis Rodolphe Agassiz, 1807–1873, Swiss-American geologist and natural
   historian, the first to propose that the earth had experienced an ice age
   From *Études sur Les Glaciers* (1840)

False facts are highly injurious to the progress of science for they
often long endure: but false views, if supported by some evidence,
do little harm, for everyone takes a salutary pleasure in proving their
falseness; and when this is done, one path towards error is closed
and the road to truth is often at the same time opened.
—Charles Robert Darwin, 1809–1882
   From *The Descent of Man, and Selection in Relation to Sex,* Vol. 2 (1871)

The evolution of microbes appears to have been characterized by a sort
of "Volkswagen syndrome," a lack of change in external body form that
has served to mask the evolution of internal biochemical machinery.
—J. William Schopf, 1941–, American paleontologist known for work on Precambrian
   microfossils; John M. Hayes, 1940–, American organic geochemist; and Malcolm R.
   Walter, 1944–, Australian geologist
   From J. William Schopf, ed., *The Earth's Earliest Biosphere,* ch. 15 (1983)

That the evolution of organisms depends in large part on the evolution of their
environment is something paleontologists have been noting since the early nineteenth century, and indeed, it is so inherent in Darwinian theory as to seem almost
banal. That this dependency might have been two-way—that the earth's minerals, atmosphere, oceans, and climate have been in large measure determined by
the evolution of different life-forms—was somewhat harder to document and
accept, partly because the most dramatic evidence was hidden, at the molecular level, in the elusive Precambrian rocks. The concept of the coevolution of

Earth and life saw its first cohesive and most provocative expression when James Lovelock presented his Gaia hypothesis in the early 1970s, but not until the end of the twentieth century were the basic tenets of the hypothesis accepted as a valid theory. Lovelock began conceiving the Gaia hypothesis when he was designing instruments for NASA's first extraterrestrial explorations and it occurred to him that, unlike the moon and Mars, the earth had an atmosphere composed of gases that couldn't and wouldn't coexist without life's intervention. At the same time, a handful of paleontologists and geochemists had been conceiving similar if less provocatively formulated hypotheses based on their studies of the earth's most ancient rocks and sediments. In 1979, a decade after Geoff, Thomas Hoering, and Keith Kvenvolden had more or less given up on the prospect of garnering clues about early life-forms from the fossil molecules in Archean and early Proterozoic rocks, one of those paleontologists inadvertently inspired a certain Australian chemist to give it another go.

Roger Summons met the paleontologist Preston Cloud when Cloud was on sabbatical at the Australian Institute of Marine Science. Summons was working in the biology department at Australian National University and had been assigned to play guide and chauffeur for Andrew Benson, a visiting American plant physiologist who was staying out at the marine institute. "There was a couple living in the guesthouse next to us," Summons tells me. "And this guy was a jogger. He'd leave every morning at 5:00 A.M. and run past the house, clump clump clump clump, and I'd wake up." Summons didn't know that the neighbor who was disturbing his sleep was a famous paleontologist, as accomplished as the guest charged to his care—who, together with Melvin Calvin, had unraveled the biochemical reactions of photosynthesis. Having spent his early career studying plant hormones and biochemistry, Summons says he was completely ignorant of paleontology or geochemistry—until Cloud gave his requisite lecture for the institute's weekly seminar series. "He came in wearing jogging shorts and sandals, and announced that this was the first time he'd ever been able to give a talk in his underwear." And then, apparently, Cloud proceeded to give one of the most exciting, captivating lectures that Summons had ever heard.

"It was a fabulous story about the history of the earth," Summons recalls, 25 years later. Cloud talked about how reduced minerals in the earth's oldest rocks indicated that free oxygen was largely absent from the atmosphere and ocean until at least 3.7 billion years ago, when the oldest known sedimentary rocks begin to bear witness. He talked about the reputed microfossils that had been found in those sedimentary rocks and the doubts regarding their nature and veracity, and he discussed how the Precambrian record of carbon-13 depletion in kerogen relative to carbonates might well chronicle the rise of CO_2-fixing, oxygen-producing photosynthetic organisms, but was also inconclusive. The question that Calvin and James Maxwell briefly dreamed of answering with their stereochemical analyses of phytane in the 1960s remained unanswered and, in Cloud's analysis, key to understanding the history of the earth and the development of its atmosphere, oceans, and rocks: when did oxygen-producing photosynthetic organisms first develop and spread? Some paleontologists considered the stromatolite-like structures in the oldest sedimentary rocks proof that cyanobacteria evolved more than 3.7 billion years ago. Others rejected or were ambivalent about this early evidence. Cloud maintained that the laminated calcium carbonate structure of stromatolites formed two billion years ago confirmed that they, at least, were the product of CO_2-fixing, oxygen-producing photosynthetic microorganisms like cyanobacteria. He discussed the microfossils that he and others had found in rocks from this period, which offered convincing evidence that microbial communities were thriving at least two billion years ago. But though techniques for validating such fossils had improved since the 1960s, the fossils could only show the microbes' external morphology and provided no information about their metabolic and energy-harvesting systems—information, Cloud said, that was key to understanding not only the evolution of life, but also the history of the earth itself. He went on to describe the beautiful red and black bands of iron oxides that began to appear in marine sediments about 2.6 billion years ago, while other minerals remained unoxidized. This, he argued, indicated that oxygen was being produced by *something* in the ocean, but that it didn't persist or accumulate in the atmosphere. His own hypothesis was that the very first photosynthetic organisms must have been anaerobic, like some contemporary photosynthetic bacteria that get their electrons from hydrogen sulfide instead of water and produce elemental sulfur instead of O_2. Free oxygen would have been highly poisonous for such organisms—indeed, for all forms of life at the time—and Cloud's idea was that as photosynthetic bacteria began to utilize the plentiful water in their environment and produce oxygen, iron oxides formed and removed the toxic O_2. In other words, the beautiful banded iron oxide formations found in rocks formed between 2.7 and 1.8 billion years ago were actually toxic waste dumps, keeping oxygen from building up in the environment until the organisms developed mechanisms for protecting against it!

The whole talk went on like this, Summons says, a grand melding of biology and geology unlike anything he'd ever heard. Cloud had assembled every scrap of information available from dated rocks of the Precambrian—the elemental makeup of sedimentary minerals and trends in their oxidation states; the quantity and nature of organic matter and igneous rocks; geologic signs of weathering, volcanism, and tectonic activity; and the paleontological record of early life. With it, he wove a tale of the earth's epic conversion from a cooling mass of matter headed for chemical and physical equilibrium, to a planet that seemed to defy the most essential laws of chemistry and could be recognized from halfway across the solar system, a planet ruled, no less, by one of the most destructive molecules of all: O_2. By 1.8 billion years ago, the banded iron formations ceased to accrue and the minerals shifted to more oxidized forms, signs that O_2 was finally accumulating in the atmosphere and oceans. But, Cloud explained, either the accumulation was slow, or life was slow to respond, because the first multicellular oxygen-respiring animals didn't appear until about 600 million years ago, according to their fossil record—which, in contrast to the microbial one, was relatively clear.

"I was fascinated," Summons says. "I thought this was the best thing I'd ever heard." He commissioned Cloud to give the same talk at the university and found it even more exciting the second time around—but the biologists in his department were decidedly ambivalent. It seemed that the only kindred spirits in Australian academe at the time were the scientists at the Baas Becking Geobiological Laboratory who had invited Cloud to Australia to begin with. Summons says that he gave up hope of working on such issues in his university post until, two years later, a couple of Baas Becking scientists showed up in his lab to ask for advice about buying a new mass spectrometer and he saw a chance to join their ranks. Summons had been using mass spectrometry extensively in his work with plant hormones and was something of an expert. After showing the Baas Becking researchers the ins and outs of the state-of-the art suite of instruments he maintained in the university lab, he blithely advised them that they would be foolish to purchase such an instrument unless they hired someone with a lot of experience to run it…and soon he had the job he'd been coveting, working alongside microbiologists, geochemists, geologists, and paleontologists who shared his newfound fascination.

When Summons joined the staff in 1983, the Baas Becking Geobiological Laboratory was already struggling to justify its basic research programs to pragmatic, budget-conscious government officials, and he was assigned the politic task of analyzing the kerogen in Australia's ancient Precambrian rock formations to determine if they might have generated oil. Like Jürgen when he started working for Welte at the Institute of Petroleum and Organic Geochemistry in the 1970s, Summons was trained as a chemist and didn't even know what kerogen was. "They gave me a job to find out," he says, laughing. "But twenty-three years later, I still don't know!" The assignment, however, stood him in good stead in the lifetime quest that Cloud's lecture had inspired: to understand the evolution of the earth and its earliest ecosystems.

Decades of work by petroleum geologists indicated that the big deposits of oil had formed from the lipids of algae and green plants, but the Precambrian world was a predominantly microbial world, and it was uncertain whether microbial organic matter could generate significant amounts of petroleum hydrocarbons even under ideal conditions. The question with which Summons began his geochemical career was, essentially, a question about microorganisms: what sorts were thriving when these ancient rocks formed, and what lipids did they make? He adopted a strategy that combined analysis of Precambrian rocks and oils with study of extant microorganisms. It was a strategy that seemed custom-designed for the institution where it was born, requiring close collaboration with geologists and microbiologists, but Summons sustained it for decades after the Baas Becking laboratory closed shop and he continues to use it to this day.

It soon became apparent, in Summons's lab and elsewhere, that both bacteria and archaea lack storage lipids like those that algae keep in reserve—a rich source of hydrocarbons for petroleum—and they lack the macromolecular cell wall constituents that seemed, by some analyses, to comprise such an important component of oil-generating kerogens. The Precambrian kerogens themselves are intrinsically different from those formed in the Phanerozoic: hydropy-rolysis releases aromatic compounds but few of the saturated hydrocarbons or isoprenoid biomarkers found in the associated bitumen. It now seems that this disparity between the kerogen and the bitumen, once considered a sure sign that the bitumen was contaminated, is an inherent quality of the organic matter in all Precambrian rocks, no matter how mature or immature, and can be attributed to the microbial biomass that formed it. Whatever the precise nature of the com-pounds or macromolecules that went into forming these ancient kerogens, they were clearly not the sort of stuff that would have generated commercially viable deposits of petroleum. But Summons and his collaborators weren't just concerned with ascertaining that Australia's extensive formations of Precambrian sedimen-tary rocks were not likely to gain it admission to OPEC. They were reopening the door to the study of molecular fossils in the earth's oldest rocks. With direct evidence of a prebiotic earth clearly out of reach, their goals were arguably more modest than Calvin's were in the 1960s, when he hoped to find the chemical imprint of the first life-forms: they just wanted to know what happened in the three billion or so years before life began leaving its unmistakable graffiti in the rocks of the Cambrian.

The problems that plagued the early studies haven't gone away, of course, but decades worth of research were not for naught. Better maturity parameters now allow comparison of the maturity of the kerogen with that of compounds identi-fied in the bitumen. More sophisticated understanding of how oil moves through different types of rock now makes it possible to ascertain whether hydrocarbons from more recent sediments or oil reservoirs have migrated into a particular Precambrian formation and contaminated it. Relatively gentle hydropyroly-sis and chemical degradation techniques for cleaving bonds in kerogen can, in principle, provide some access to molecular information from the kerogen itself. New understanding of kerogen formation and the relationship between it and the bitumen opens avenues of interpretation that were unavailable to Thomas Hoering in the 1960s, when he determined that the organic compounds detected in the bitumen of Precambrian rocks were not as old as the rocks. And, finally, compound-specific isotope analysis can, in principle, provide verification of the provenance of fossil molecules in the bitumen, as well as clues to ancient met-abolic processes. The result is that analyses of organic matter in Precambrian rocks now come replete with a system of qualifiers that might make the cynical laugh, but which are nevertheless based on a series of well-delineated criteria, many derived from Hoering's early efforts: molecular fossils can be "certainly syngenetic" or "probably syngenetic" or "not syngenetic," depending on how well one has been able to verify that they are indigenous to the rocks and time where they are found and not migrants from distant rocks and eras.

With the exception of the 1.1-billion-year-old Nonesuch shale and a couple of more recent Neoproterozoic formations, most of the Precambrian rocks that were available for study in the 1960s had been too thoroughly baked, squeezed, stretched, and squashed to yield anything that was even possibly syngenetic. But extensive mineral and petroleum exploration and prospecting in Australia during the 1980s and 1990s revealed a number of Precambrian formations with a much milder geologic history. And more than 40 years of development of analytical techniques and instrumentation have done more than expand the list of instrumental acronyms. Compound detection and identification are now vastly more sensitive and dependable than in the 1960s and 1970s. Ever narrower capillary columns and more sophisticated lining materials provide better separation of compounds. Better understanding of molecular fragmentation patterns in mass spectrometry allows for the detection of overlapping compounds. The use of two mass spectrometers in tandem allows one to decipher the structures of complex fragments. Improved detector technology can reveal the presence of even the tiniest traces of fossil molecules, and integrated data processing systems facilitate faster, more all-encompassing molecular surveys.

By the mid-1990s, Summons and the geologist Malcolm Walter, whom Summons followed to the Bureau of Mineral Resources when the Baas Becking laboratory closed, were well embarked on an exploration of the Australian Precambrian rocks. They spent over a decade juggling their quixotic investigations of life's incomprehensibly long prehistory with the somewhat more pedestrian petroleum research that the Bureau of Mineral Resources favored—until John Hayes convinced Summons to reverse their long-established trans-Pacific migration pattern and set up a dedicated "geobiology" lab at MIT, just up the road from Hayes's new home at Woods Hole. From this perch, Summons and his former student Jochen Brocks—who was then at nearby Harvard—went on the offensive and directly readdressed the questions about early life that organic geochemists had generally been shying away from since the early 1970s.

The relative importance of microbial ecosystems in the Precambrian rocks is now readily apparent in the overall distributions of biomarkers that Summons and crew have found. All traces of higher plant pentacyclic triterpenoids such as oleanane are absent, as are the distinct long-chain *n*-alkane distributions characteristic of plant waxes. There are notably high concentrations of the mid-chain branched alkanes that derive from bacterial fatty acids, and the 2-methyl extended hopanes associated with cyanobacteria are significantly more abundant than in Phanerozoic deposits, except, of course, for those formed during oceanic anoxic events. Steranes are present in most of the Precambrian rocks analyzed, but in much lower abundances relative to hopanes than in most Phanerozoic rocks and oils. The biomarkers thus paint a picture of Precambrian life that includes diverse microbial communities, with a major role for photosynthetic cyanobacteria, and a subdued community of eukaryotic algae.

Perhaps the biggest surprise of this work has been that there are no real surprises, no dinosaurs or trilobites, no long-extinct monsters discovered among the molecules. Rather, the fossil molecules in these ancient sediments—the certainly and probably syngenetic fossil molecules in the 550-million-year-old oil from Oman, or the 850-million-year-old Chuar Formation from Arizona's Grand Canyon, or Australia's 1.6-billion-year-old McArthur Basin or 2.7-billion-year-old Hamersley Basin—can also be found in sediments, rocks, and oils formed in the past 300 million years. Species and genera, even whole phyla of organisms, appear and disappear in the fossil record, succumbing to all manner of environmental catastrophe and mass extinction, ecological transformations, and genetic innovation—while their lipid inventions seem to appear and *persevere*. It is quite possible that numerous lipid structures did in fact evolve, reign successfully, and go extinct with the organisms that fabricated them, and that their fossil remains have simply not been found, or were not persistent or abundant enough to leave a detectable biomarker trail. But so far, we find no indication in the geologic record of lipids that have gone out of business altogether. It appears that once invented, the biochemistry of lipid construction is generally conserved, with old compounds being employed in new ways even as the organisms that invented them go extinct or, as may be the case for cold-adapted crenarchaea, evolve in new directions.

The other surprise in these ancient rocks is that the capacity to make the carbon skeletons of the principal groups of lipids used by all three domains of life—Bacteria, Archaea, and Eukarya—seems to have been in place for at least two-thirds of life's sojourn on Earth. Mid-chain branched alkane skeletons, like those produced by many heterotrophic bacteria and cyanobacteria, comprise a prominent fraction of the bitumen in most of the Precambrian rocks analyzed to date. But phytane and pristane are also in evidence, as are hopanes and even steranes with the complete range of branching patterns known in contemporary sediments and organisms. Brocks has done an exhaustive analysis of samples from cores drilled in western Australia's Hamersley Basin, including strata that formed 2.78–2.45 billion years ago. Though the rocks are thermally mature, extensive analysis and consideration earned the steranes and hopanes identified in the bitumen the designation "probably syngenetic." In an independent study of samples from the same period and basin, but formed in a slightly different depositional environment, Kate Freeman and her student Jennifer Eigenbrode at Penn State found a similar array of compounds and came to the same conclusion. The implications are sobering. Paleontologists have been arguing for decades about whether the microfossils and stromatolites they find in early Proterozoic and Archean rocks were left by cyanobacteria. Now the significant quantities of 2-methyl extended hopanes in the Hamersley Basin rocks, along with pristane and phytane with $\delta^{13}C$ values of –27 to –30 per mil, provide substantial evidence that oxygen-producing cyanobacteria were "probably" extant and thriving in the surface waters of what appears to have been a restricted marine basin or large lake 2.7 billion years ago. The presence of steranes in these rocks is even more telling, for two reasons: the biosynthesis of sterols is one of the defining traits of eukaryotes, and several of the enzymes that are essential for that synthesis cannot function without O_2.

Whatever primordial organisms were synthesizing sterols in the ancient Hamersley sea or lake must have had access to free oxygen: either they were living in direct association with the oxygenic cyanobacteria, or oxygen had accumulated in their immediate environment, perhaps in a thin layer of water near the surface. But what sort of organisms were they? There is now strong evidence that cyanobacteria do not have the genetic wherewithal to synthesize sterols, despite early reports to the contrary. In fact, with the exception of a few ciliates, all organisms with a eukaryotic cell structure have the capacity to synthesize sterols, but only three unrelated species of bacteria are known to do so—and none of the latter have the capacity to produce branched carbon skeletons like the 24-ethylcholestane found in the Hamersley Basin bitumen.

Finding "probable" molecular fossils of cyanobacteria in the Hamersley Basin rocks reinforces other geochemical and geologic evidence that oxygenic photosynthesis was already well established by 2.7 billion years ago, and that some minimal amount of oxygen had accumulated in the atmosphere by 2.3 billion years ago. But does the presence of steranes like 24-ethylcholestane mean that the essential elements of the eukaryotic lineage, which eventually gave rise to the fabulous array of plants and animals that populated Cambrian seas, were also in place by this time? If so, if the basic biochemical wherewithal was up and running 2.7 billion years ago, why did it take more than a billion and a half years for anything more than single-celled organisms to emerge? And why was so much of the action then crammed into this particular, relatively brief period at the beginning of the Cambrian? When nineteenth-century geologists began to understand just how unimaginably old the earth was, they thought this near infinity of time could account for the glory of creation and the majestic diversity of life. And yet now it seems that rather than taking advantage of all that time for the gradual evolution of progressively more refined life-forms like it has during the past 542 million years, life may have spent most of its tenure on Earth locked in microbial stasis. Did it simply take that long—more than two billion years— for key groups of enzymes to change? Was there an evolutionary hurdle to be overcome? Or was life stuck in some sort of couch-potato syndrome, waiting for the earth to challenge its comfortable existence and inspire the development of its latent eukaryotic potential? Or was there, in fact, some real environmental factor blocking further development?

Though none of the many hypotheses to explain the Cambrian explosion can be fully supported by the geologic and genetic evidence at hand, consensus is growing that both its timing and the long history of retarded eukaryotic evolution that preceded it were in large part contingent on oxygen—not just how much and *when*, but how much and *where*.

According to Preston Cloud's reconstructions, and the paradigm-setting model that marine chemist Heinrich Holland presented in his 1983 classic, *The Chemical Evolution of the Atmosphere and Oceans*, reduced minerals and volcanic gases on the early earth acted as a buffer to keep atmospheric oxygen low well after the onset of oxygenic photosynthesis. Once these were oxidized or depleted, however, the oxygen content of the atmosphere rose rapidly in what Holland called the "Great Oxidation Event." The recent estimates indicate that this Great Oxidation Event occurred between 2.4 and 2.3 billion years ago, and that atmospheric oxygen rose to 1–10% of its contemporary level—just enough to poison anaerobic organisms and support unicellular oxygen-respiring ones. Cloud assumed that the ocean had followed suit with the atmosphere. He had interpreted the banded iron formations in Archean and early Proterozoic rocks as an indication that oxygen was beginning to accumulate in shallow water and coastal environments: as long as the deep ocean was anoxic, the iron released from hydrothermal vents or eroded from rocks on land accumulated in the ocean in its reduced, soluble form. In the presence of oxygen, however, it was oxidized and formed into insoluble minerals, thus removing both the iron and free oxygen from the water. Cloud reasoned that when currents brought deep ocean water into a shallow coastal environment, the dissolved iron came into contact with oxygen, and the bands of iron oxides formed. Others postulated that the banded iron formations were produced by anaerobic iron-oxidizing photosynthetic bacteria. Either scenario required that dissolved iron be available and the deep ocean anoxic, and there was a general consensus that the disappearance of banded iron formations 1.8 billion years ago meant that the deep ocean had caught up with the atmosphere and contained enough oxygen to oxidize and precipitate out the iron, so that only small traces were left in solution. But in the 1990s, organic geochemists and geologists began to find hints that the Great Oxidation Event didn't reach the deep ocean at all, and that it may have been another billion and a half years before one of the most extensive habitats on Earth opened its doors to even the tiniest of aerobic eukaryotes.

One of the puzzles that emerged during the reconsideration of Precambrian organic matter in the 1980s was that the bitumen in many marine rocks was typically enriched in carbon-13 relative to the kerogen—exactly the opposite of the situation in rocks from later periods. Thomas Hoering had interpreted this discrepancy as a sign that the kerogen and the bitumen had not formed at the same time. He concluded that the bitumen must have migrated into the ancient rocks from more recent deposits and the molecular fossils one extracted from it were not remnants of Precambrian life at all. "That made total sense," John Hayes says. "I would have done the same thing if I'd analyzed these rocks back then!" But in the 1980s, as better preserved Precambrian rocks were discovered and uncontaminated samples became available for analysis, he and Summons realized that the relationship was *consistently* reversed: the bitumen in the Precambrian marine rocks was, in fact, enriched in carbon-13 relative to the kerogen, even when the bitumen and kerogen showed every sign of having formed in the same place at the same time. By the early 1990s they had the tools to figure out why, and Graham Logan, a new postdoc in Hayes's lab, headed for Australia to work with Summons. Hayes says he had to cajole Logan into making the cross-Pacific migration, extolling the virtues of Australian summers and the hardships of Indiana winters, but once Logan got to Australia, he didn't want to leave. "He fell for an Aussie gal," Summons says, sounding rather pleased with the outcome: a marriage and Aussie kids, someone to take over the biomarker labs in Canberra when Summons himself finally left for MIT ... not to mention a rather intriguing and still-evolving hypothesis.

The Canberra-Indiana alliance found that the short-chain *n*-alkanes in Proterozoic bitumens were consistently *enriched* in ^{13}C compared to pristane, the opposite of what is observed in Phanerozoic rocks and in the biomass of contemporary marine organisms. In the more recent rocks and sediments, the $\delta^{13}C$

values of these compound types generally reflect those of their presumed precursor compounds in photosynthetic primary producers, where isoprenoid structures like pristane are a few per mil less depleted than straight-chain compounds are, a difference that is preordained in the biosynthesis of the two types of carbon structures. Likewise, both types of lipids are more depleted than the overall biomass of the algae that makes them—and the lipids in the bitumen of Phanerozoic rocks are more depleted than those in the kerogen. But in the Proterozoic rocks it appeared that the *n*-alkanes and pristane had come from two very distinct groups of organisms that obtained their carbon from different sources or utilized different mechanisms for fixing it. But what were they? What was so different about life in the Proterozoic ocean? And why did it change when it did?

In 1995 Logan and crew published a paper that focused on a hitherto-unconsidered but fundamental difference between the Proterozoic and Phanerozoic earth: they made the inglorious observation that, popular wisdom and T-shirt pronouncements notwithstanding, "shit" has not always "happened," and they hypothesized that its advent had transformed the earth forever. Their language, of course, was slightly more eloquent, as befits a *Nature* paper, but that didn't stop their colleagues from immediately christening the idea the "shit hypothesis."

In the modern world, most organic matter that reaches the sediments is transported there by relatively large, rapidly sinking particles, most notably the sinking fecal pellets of zooplankton that feed on algae in the surface waters. But during the Precambrian, large multicellular organisms were absent, unicellular algae had not yet invented the calcium carbonate and silica encasements that endow them with ballast, and there were no zooplankton or animals with a digestive system to create fecal pellets. The organic matter produced by tiny cyanobacteria and unicellular algae in the surface waters would have stayed suspended for a very, very long time. This meant that much of the bacterial breakdown that now occurs in the sediments would have occurred in the water, most of the organic matter to reach the sediments would have derived from heterotrophic bacteria at the tail end of the food chain, and the only remnants of the surface organisms to reach the sediments would be recalcitrant isoprenoid lipids such as pristane and phytane. The relative $\delta^{13}C$ values of the molecular fossils in the Proterozoic rocks reflected this: the *n*-alkanes were enriched in ^{13}C compared to pristane, precisely as would be expected if the former came from heterotrophic organisms, whose lipids become more enriched with each step down the food chain. The $\delta^{13}C$ values of the kerogen fell between the bacterial *n*-alkanes and the photosynthetic pristane, indicating that, unlike in the Phanerozoic when the kerogen derived

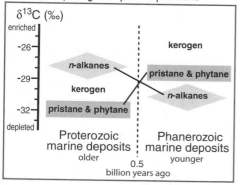

Proterozoic vs. Phanerozoic organic matter: comparing isotope compositions

mostly from algal biomass, the Proterozoic kerogen derived from a mixture of bacterial and algal biomass. More evidence came when the group analyzed a broader range of biomarkers, and when they compared results from Proterozoic rocks that had clearly formed in a shallow water environment with those from deep water marine systems: the isotopic relationship in the shallow waters was markedly different from that of deep water rocks from the same period and, in fact, mimicked the relationship found in contemporary sediments.

According to the shit hypothesis, a greater proportion of the organic matter made by photosynthetic plankton during the Precambrian either remained floating or was consumed by bacteria in the water column, with little export to the sediments. Any oxygen the phytoplankton generated during the photosynthetic reduction of CO_2 to organic matter would have been used up by heterotrophic bacteria burning the organic matter back to CO_2—with the result that, even after minerals on land had been oxidized, oxygen failed to accumulate and the deep ocean was doomed to everlasting anoxia. If the bacterial recyclers included large populations of sulfate-reducing bacteria, then the deep ocean would have been both anoxic and sulfidic—not exactly the sort of place where eukaryotic life-forms could develop. In the 1995 version of the hypothesis, it was the advent of fecal pellets and mineral-encased algal forms that changed this situation and allowed oxygen to accumulate in the ocean. Organic matter started sinking more rapidly from the surface into the deep water, and the whole organic recycling system sank with it: organic matter accumulated on the seafloor, heterotrophic bacteria concentrated in the surface sediments, methanogenic and sulfate-reducing bacteria were buried…and a larger portion of the organic matter produced by photosynthesis was preserved, leaving behind an equal measure of free oxygen. If the hypothesis were correct, then only the water near the surface of the ocean had been oxygenated, and the rest of the ocean and the seafloor—one of the most extensive of the earth's habitats—had been closed to oxygen-respiring eukaryotic life for most of life's tenure on Earth, until finally, just before the dawn of the Cambrian, shit and shells ushered in a new age! There were just two problems. One was the general consensus among geologists that the deep ocean had been at least marginally oxic throughout most of the Proterozoic. The other was that shit didn't, apparently, happen at quite the right time.

In recent years, the consensus that there was free oxygen in the Proterozoic oceans has been uprooted, and new studies of the mineral and isotopic composition of Proterozoic rocks would seem to support the Logan team's hypothesis that they were anoxic and sulfidic. But as more precise dates became available for the late Proterozoic rock formations, it was clear that the definitive isotopic shift in the biomarkers that Logan, Summons, and Hayes interpreted as evidence of a fundamental change in the way that organic matter was delivered to the sediments occurred about 570 million years ago—well *before* the first appearance of fecal pellets or any other fossil evidence of zooplankton. What, then, caused the change? The basic premise of the hypothesis, that there was a huge backlog of suspended and dissolved organic matter in the Proterozoic ocean, was robbed of explication, even as evidence accumulated to support it.

The geologic record in the late Proterozoic shows evidence of a series of ice ages that make our recent Pleistocene ones seem like summer in the Bahamas, and there has been much speculation that these had something to do with the pacing of eukaryotic evolution and a final rise in atmospheric oxygen to a level that could support oxygen-respiring multicellular animals. Three of these putative climate catastrophes are well documented in the geologic record, the first about 720 million years ago and the last about 580 million years ago, but it is difficult to date and correlate evidence in different sequences of Proterozoic rocks, so their precise timing, causes, and effects are still controversial. The most radical hypothesis, dubbed "Snowball Earth," maintains that the earth froze from pole to pole and the oceans were covered by a kilometer-thick layer of ice that blocked out all light and cut off the exchange of gases between the oceans and the atmosphere for millions of years. Eukaryotic organisms would have survived only in ill-defined isolated refuges, repopulating the world when the ice melted, and the Ediacaran expansion and Cambrian explosion were somehow the consequence of the last of these repopulations. Both fossil and biomarker evidence for such extensive die-outs is lacking, however, and most geologists and paleontologists think that the snowball was actually more of a slush ball. Meanwhile, there is mounting evidence that the expansion and evolution of eukaryotes were hampered more by the long-standing oceanic blockade on the accumulation of oxygen than by the late-Proterozoic climate catastrophes.

The best-preserved and precisely dated late-Proterozoic rock sequences come from Oman, on the Arabian Sea. The $\delta^{13}C$ values for carbonates and bulk organic matter in these rocks provide independent, but inconclusive, evidence that there was a major decrease in the oceans' organic-matter backlog 570 million years ago, contemporaneous with the isotopic shift in the biomarkers. And a certain fossil sterane that Summons's MIT group found in the same rocks may provide a more timely explanation than the advent of fecal pellets for this environmental prelude to the Cambrian explosion.

In the early 1990s, when Mike Moldowan was just making his move back to academe, the Chevron scientists published a paper in which they postulated that a distinctive sterane that they found in oils and shales of Ediacaran and early Cambrian age was recording the prevalence of marine sponges, whose fossils have been found in rocks from these periods. Isopropylcholestane has a unique Y-shaped three-carbon group on its side chain, a structure that is prominent among the sterols made by a certain class of sponges, found only in trace amounts in a handful of other organisms, and comprises a minor component of the steroid distributions in marine deposits formed after the Cambrian period. Gordon Love, a transplant from the Newcastle group, now at MIT and working with Summons, found exceptionally high abundances of isopropylcholestane in the Oman Basin oils and rocks. These oils and rocks

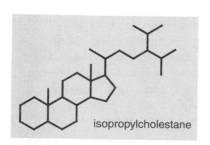

isopropylcholestane

formed from organic matter that was deposited some 657 million years ago, more than 50 million years before the first identifiable fossils of sponges or any other multicellular animals appear in the fossil record. Phylogenetic analyses indicate that sponges were the first of the still-existing animal forms to have evolved, but the biomarker evidence would seem to place them literally at the dawn of multicellular life, and their prevalence at this particular juncture in earth history may have had a profound environmental significance.

Unlike most other animal forms, sponges do not prey on algae or other eukaryotes in the surface waters, but rather live on the seafloor. Contemporary members of the isopropylcholesterol-making group feed by pumping large volumes of water through their bodies and consuming whatever dissolved or particulate organic carbon it contains, and they are often found in low-oxygen environments on the continental shelves. If, as appears to be the case from the abundance of isopropylcholestane in the Oman Basin rocks, these organisms evolved some 657 million years ago and became prevalent during the Ediacaran period, they might have been responsible for a decrease in the backlog of organic matter that was dissolved or floating in the Proterozoic oceans and the reversal of the isotopic relationship in the biomarkers that Logan and the group observed at 570 million years ago—essentially, a decrease in the relative amount of old, recycled bacterial organic matter going into the sediments as the sponges cleaned up the water.

24-Ethylcholestane is also present in anomalously high concentrations in many late Proterozoic rocks and, together with an increase in the number, variety, and complexity of acritarchs, may imply that green algae began to diversify and gain prominence at this time, perhaps responding to the hypothesized change in ocean chemistry, better able to compete with cyanobacteria as the ocean became more oxygenated and nitrate became more readily available.

Independent but still circumstantial evidence that there wasn't enough oxygen in the atmosphere to support oxygen-respiring multicellular organisms until the late Proterozoic comes from analyses of the isotopic composition of sulfide minerals in the Oman Basin sequences. These indicate that there was a stepwise increase in the amount of sulfate in the ocean—and, presumably, of oxygen in the atmosphere—between 635 and 542 million years ago. But were the deep oceans really completely anoxic throughout the entire Proterozoic, as hypothesized? A few years after the shit paper was published, Donald Canfield, who was working with microbiologists at the Bremen Max Planck Institute, offered independent evidence, this time from a broad survey of the minerals in older Proterozoic rocks, that this was the case. According to Canfield's hypothesis, most of the sulfur on the early earth was present as hydrogen sulfide from hydrothermal vents or sulfur dioxide from volcanoes, and more oxidized forms like sulfate were either nonexistent or scarce. This situation began to change after the Great Oxidation Event some 2.4 billion years ago, when the nominal buildup of oxygen in the atmosphere began oxidizing sulfide minerals on land and generating sulfate, which was delivered to the ocean in river runoff, much as it is today. As sulfate concentrations in the ocean rose, sulfate-reducing bacteria would have

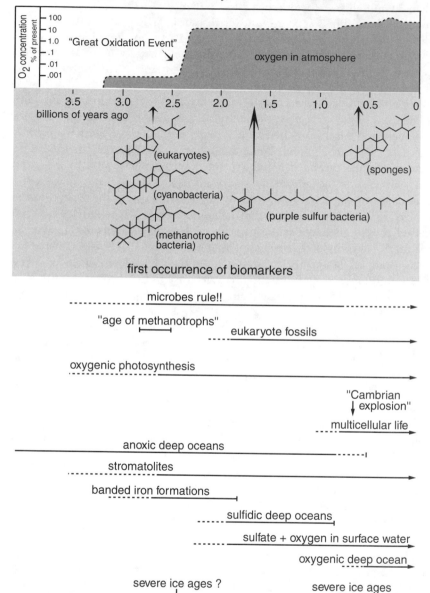

Oxygen, ocean chemistry, and life: coevolution

flourished, producing large amounts of sulfide. The sulfide would have combined with the dissolved iron in the ocean to form the mineral pyrite, thus removing both sulfide and iron from the ocean, until all of the accumulated iron in the deep ocean had been used up. This, Canfield proposed, happened about 1.8 billion years ago and was the reason the banded iron formations had disappeared: whether they were made by iron-oxidizing bacteria or formed when anoxic iron-containing deep water mixed with the oxic surface water in coastal areas, as Cloud had proposed, dissolved iron was prerequisite. Contrary to most earlier hypotheses, but in keeping with Logan's, Canfield's maintained that the deep ocean was still anoxic at this time. Indeed, he proposed that once all the iron was gone, the sulfide from the sulfate reducers accumulated in the ocean, just as Logan and his cohorts had speculated. The mineral data were too limited to be conclusive, however, and the hypothesis was controversial. Then, in 2005, Jochen Brocks and his collaborators found more clues: the lipid remains of a middle Proterozoic marine ecosystem that could only have prospered in a sulfidic, anoxic ocean—an ocean whose chemistry resembled that of the contemporary Black Sea.

Brocks, now settled into a faculty position at Australian National University, analyzed 1.64-billion-year-old rock samples from cores drilled in northern Australia's McArthur Basin, where organic-matter–rich rocks had formed at the bottom of what appeared to have been a deep marine basin during the early to middle Proterozoic. Extracts of the bitumen contained relatively large amounts of isorenieratane, chlorobactane, and a newly identified sister compound called okenane, along with various apparent breakdown products. Isorenieratane and chlorobactane, of course, are the characteristic molecular fossils of pigments found in contemporary green sulfur bacteria. Okenane bears the hydrocarbon skeleton of a red carotenoid pigment that has been found in another group of anaerobic photosynthetic bacteria, the purple sulfur bacteria.

Like green sulfur bacteria, the purple sulfur bacteria use sulfide instead of water as their electron donor, but unlike green sulfur bacteria, they require lots of light and cannot survive below a depth of 20 meters in even the clearest water. Brocks's rock cores transected the middle of the basin: if the ancient manufacturers of these pigments bore any resemblance to the living ones, then this large marine basin was entirely anoxic *and* sulfidic to within 20 meters of the surface. Significant amounts of lycopane, β-carotane, and γ-carotane, whose highly unsaturated precursors would have been rapidly destroyed in the presence of oxygen, bear further witness to the extent of the anoxia. Large amounts of 3-methyl extended hopanes indicated that aerobic methanotrophic bacteria may have been active in the oxic surface water and implied the presence of methane and methanogens, though their definitive biomarkers could not be detected.

Some geologists suspect that the McArthur Basin was periodically open to the ocean during this period, in which case it is likely that extensive areas of the world's oceans shared a similar ecology. If sulfide was in such abundance, then the formation of sulfide minerals would have removed most of the trace metals from the water—not just the iron, but also the molybdenum and copper that are essential components of enzymes that facilitate the fixation of nitrogen gas by cyanobacteria and the assimilation of nitrate by algae. Could that be what kept oxygenic photosynthesis in check for so long, an incapacity to utilize the nutrients in their environment? Or was sulfide in such high abundance that it poisoned the algae in the surface water every time a storm stirred up the water and the layers mixed? Was this the reason eukaryotic life remained in evolutionary limbo for nearly a billion years? And if the water-filtering apparatus of early sponges was ultimately responsible for blowing fresh air into the evolutionary doldrums, then what had allowed the sponges to evolve in such an environment to begin with? The timing and mechanisms of change at the end of the Proterozoic have yet to be fully deciphered, and the role of the water-filtering sponges remains hypothetical. But the emerging picture of the early and middle Proterozoic ocean is one of an anoxic, sulfidic soup of organic matter where eukaryotes were resigned to a marginal existence, photosynthetic sulfide-dependent anaerobes ruled primary production, and sulfate-reducing sulfide-producing bacteria were a defining force. One would expect methanogens to be active in such an environment, as well...but definitive molecular fossils of archaea in ancient rocks are elusive.

Summons identified crocetane in the McArthur rocks, but suspects that it has nothing to do with archaea and is, rather, a breakdown product of isorenieratene or okenone, generated during the heating and compression of these relatively mature rocks. A series of C_{21} to C_{25} acyclic isoprenoids has been found in a number of Proterozoic rocks and tentatively associated with methanotrophic archaea, based on the fact that similar compounds in methane seep environments carry the isotopic signature of methanotrophs. Unfortunately, compound-specific isotope studies are often precluded by the low concentrations of biomarkers present in these Proterozoic rocks, and none of the more distinctive biphytanyl compounds that one might expect from methanogens and methanotrophic archaea were reported. Cambrian-age oils contain regular C_{20} to C_{30} isoprenoids with an isotopic signature typical of methanogens...but, again, there are no reports of archaeal biphytanes.

Steranes are present in some of the oldest sedimentary rocks analyzed to date, and phylogenetic analyses of archaeal genes indicate that the archaea long predate the eukaryotes, so why don't we find their molecular fossils? Could it be that the archaea are not as ancient as the phylogenetic relationships would seem to imply? Or perhaps the biochemically unusual head-to-head and tail-to-tail isoprenoid linkages were a relatively late archaeal invention, and the ancestral archaea depended on regular isoprenoid chains and ether lipids like archaeol. Or maybe the biphytanes are simply eluding detection, hidden away in the kerogen and waiting for some new and more refined method of analysis. Jörn Peckmann, leader of the geobiology group at Bremen University, and Volker Thiel, now at the University of Göttingen, have found biphytanyl compounds in rocks formed in Mesozoic-era methane seep environments. They postulate that the ring-containing biphytanes may decompose when subjected to extensive thermal stress of the sort that many older rocks have been exposed to. But there's no reason why biphytane itself shouldn't be at least as sturdy as the steranes. Peckman and Thiel found it in moderately mature Jurassic rocks. In the fall of 2007, as we're preparing to ship this manuscript to its publisher in New York, we get word that a group at Woods Hole has reported finding biphytane and purported breakdown products of ring-containing tetraethers in extracts of 2.7-billion-year-old rocks from Canada. These Canadian rocks, however, have suffered more extensive metamorphosis than the contemporaneous Hamersley Basin sequences where Summons and Brocks and Eigenbrode found steranes and hopanes, and it is harder to prove that the compounds are "probably syngenetic" and seems unlikely that even biphytane could have survived intext.

As the hunt for direct, fossil molecule evidence of archaea in ancient rocks continues, the isotope compositions of carbonates, organic matter, and sulfur minerals show every indication that both methanogenic archaea and sulfate-reducing bacteria made their respective first claims to fame even before the onset of the Proterozoic eon. Indeed, phylogenetic analyses of contemporary microorganisms indicate that the genes for sulfate reduction evolved before those for photosynthesis, which poses something of a conundrum, given that sulfate was, by all estimates, in short supply until oxygenic photosynthesis evolved and the atmosphere became oxygenated. In one instance, isotopic evidence of biological sulfate reduction has been found in the minerals of a 3.8-billion-year-old evaporite deposit, and it's possible that traces of sulfate may have accumulated in some shallow basins. But in most cases, early sulfate reducers would have been isotopically invisible, not expressing their preference for the light isotope until sulfate concentrations rose to significant levels, as apparently happened on a global scale between 2.4 and 2.3 billion years ago.

The genes for methanogenesis appear to have evolved even before those for sulfate reduction. Here the mineral record poses no conundrums. All of the ingredients for methanogenesis seem to have been readily available during the Archean eon: abundant hydrogen and CO_2 for autotrophic methanogenesis, and possibly even small organic acids for fermentative methanogenesis. The complete absence of oxygen on the early earth would have allowed methanogens to thrive in the open ocean, rather than being limited to isolated waters and buried sediments as they are now. Indeed, one explanation for why the earth wasn't completely frozen over during its first two billion years is that it was wrapped in a thick blanket of methane and CO_2, which kept it warm despite the weak luminosity of a youthful sun. But as John Hayes has pointed out, the early methanogens may also have been isotopically invisible: methanogenesis leaves its distinctive isotopic signature on the methane produced as a waste product, not on the lipids of the organisms themselves. Even if they comprised the dominant life-form on

^{13}C-depleted kerogens from the Archean

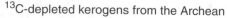

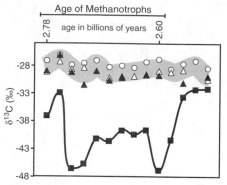

■ kerogen ▲ phytane △ pristane ○ *n*-alkanes

the early earth, methanogens would not have left an isotopic legacy in the rocks—until something evolved that could *eat* methane and incorporate its light carbon into organic matter. Hayes has proposed that this may be precisely what happened between 2.75 and 2.55 billion years ago, because the kerogen in rocks from this time is consistently 20–30 per mil more depleted in ^{13}C than that formed before and after. Such a level of depletion is well beyond the 5 per mil difference associated with changes in CO_2 concentration or with any other known pathway of carbon assimilation: it is known only from methane seep environments, where the ^{13}C-depleted methane generated by methanogens is channeled into the food chain—and thus recorded in the organic matter—by methanotrophs.

Hayes hypothesized that methanogenesis had long been the primary means of organic matter production, but the first methane munchers were aerobic methanotrophic bacteria that evolved in close association with the first oxygenic photosynthetic bacteria. He dubbed the 200-million-year stretch of the late Archean the "Age of Methanotrophs" and suggested that the methanotrophs were having their heyday, consuming huge amounts of the readily available methane. In a variation on the theme, Kai Hinrichs has proposed that, given the generally anoxic environments at the time, the methane munchers in question may have been anaerobic archaea living in association with sulfate-reducing bacteria, rather than aerobic bacteria living in association with oxygen-producing photosynthetic bacteria. The close phylogenetic relationship between methanogens and methanotrophic archaea implies that the genetic wherewithal for anaerobic methanotrophy may have been in place for almost as long as methanogenesis itself, and Hinrichs suggests that the actual onset of methanotrophy around 2.75 billion years ago was triggered by the first substantial increases in the amount of sulfate available for the sulfate reducers. Whatever the nature of the early methane munchers—and it's conceivable that both types of organisms evolved around this time—their effects on the earth's chemical cycles and climate would have been dramatic. One hypothesis holds that the associated decrease in atmospheric methane concentration facilitated the Great Oxidation Event 2.4 billion years ago, while another maintains that it triggered the earth's very first ice age, a "Snowball Earth" even more severe than those proposed for the Proterozoic ice ages.

The rocks of the Hamersley Basin sequence span the period, and though the "probably syngenetic" biomarkers identified in their bitumen don't exactly clarify the ecology of the time, they do offer some titillating bits of evidence. As in all of the Precambrian rocks studied to date, molecular fossils of archaea, including the crocetane, PMI, and biphytanyl compounds usually associated with methane-rich environments, are generally absent or eluding detection. Phytane and pristane are, however, present in enough abundance for compound-specific isotope measurements, and these would seem to weigh in against Hinrichs's version of the Age of Methanotrophs. At 30–31 per mil, $\delta^{13}C$ values for both compounds are consistent with an origin in the chlorophyll of photosynthetic bacteria, more than 10 per mil enriched compared to the contemporaneous kerogen, and completely out of the range observed for methane-consuming organisms. The presence of 3-methyl hopanes in these rocks, however, offers ambiguous evidence that methanotrophic bacteria were present. The range of steranes that Brocks and Eigenbrode and their colleagues found supports Hayes's notion that, despite the generally anoxic atmosphere and ocean, free oxygen was available, either to clusters of organisms living in direct symbiosis with oxygenic photosynthetic bacteria and algae, or in some narrow layer of surface water where they could generate oxygen faster than it was destroyed. But when Eigenbrode measured the amount of 3-methyl extended hopanes in the bitumen, she found that they were *inversely* correlated to the $\delta^{13}C$ values of the kerogen over the 200-million-year course of the putative Age of Methanotrophs—not quite what one would expect if these hopanes had come from the organisms responsible for the ^{13}C depletion. The isotope measurements that might confirm or refute a methanotroph provenance require isolating the traces of 3-methylhopanoids from huge amounts of powdered sample, and even then the chances of success with such measurements are slim—but Eigenbrode, now at the Carnegie Institute where Thomas Hoering did his groundbreaking isotopic measurements in the 1960s, says she is giving it a try.

These puzzling but provocative bits of evidence from Archean and early Proterozoic rocks call out for more study, but an ambitious NASA-supported program to drill into the ancient Australian rock sequences—for the first time, unconstrained by the mandates of mineral exploration—has been stymied by recent funding cuts. In the American political climate of the early 2000s, even NASA's adept public relations department was unable to rally support for research into the origin and evolution of life, and its astrobiology program—including plans to search for molecular fossils of life's first billion years on the earth—was targeted for more than its share of funding cuts. Nevertheless, the findings of the last few years indicate that the fossil lipids of microorganisms that ruled and transformed the earth during life's early tenure have the potential to tell us much more than "what" was living when. Many, if not most, of these fossil molecules derive from ancient cell membranes, the main boundary—guard, porter, and intermediary—between unicellular life and the water, air, and rocks of the early earth. When Geoff started analyzing Melvin Calvin's Precambrian rocks in the 1960s, he had no idea that most of the fossil molecules that would eventually be identified would derive from such a universal and critical component of life—a precondition, according to some hypotheses, for its origin, and one of the defining factors for its three domains. These are fossils that take us beyond taxonomy, beyond a family tree of relationships or even a time line of life's development, and provide insight into the nature of life's coevolution with its environment.

Given more and better samples from ancient rocks, and more extensive inquiry into the role of molecular constituents in the physiology of cell membranes, the lexicon developed over three decades of molecule collecting and natural history stands poised to tell its own story—the story of biochemical evolution. This was the story that ultimately interested Guy Ourisson, news of whose death reaches me as I write these last chapters. When I visited him in Strasbourg in 2004, ostensibly to learn about the roots of the so-called kinetics methods in petroleum exploration, he was much more interested in talking about the structure of bacterial membranes and the origin of life than about petroleum. Convinced that the key to understanding the evolution of the earliest life-forms lies in membrane function and structure, he and his longtime collaborator Yoichi Nakatani were doing experiments with very simple acyclic terpenoids, investigating the properties of the small membranelike vesicles they form in the presence of phosphate. "Why not try another look at the rock record?" I asked, ingenuously. And he shrugged, and said simply: "There's nothing left." But even with life's first footprints indisputably off the record, the molecular remnants of membrane constituents in Archean rocks may yet bring us closer to answering the questions that galvanized the scientific community in the 1950s, that lurked behind Calvin's interest in the Precambrian rocks in the 1960s, and that fed the excitement over the Martian meteorite in the 1990s: how did life originate?

11

Thinking Molecularly, Anything Goes
From Mummies to Oil Spills,
Doubts to New Directions

*To venture an idea is like moving a chess piece forward on the game
board; it may be defeated, but it initiates a game that will be won.*
—Johann Wolfgang von Goethe, 1749–1832, German writer, philosopher, and scientist
   From *Maximen und Reflexionen, Kunst und Altertum* (1821)

*We shall never cease from exploration
And the end of all our exploring
Will be to arrive where we started
And know the place for the first time.*
—T. S. Eliot, 1888–1965, British (U.S.-born) critic, dramatist, and poet
   From "Little Gidding" (1942)

Though the biomarker saga began with attempts to understand the ancient provenance of petroleum and with the concept of "fossil molecules" and search for
early forms of life, the explorations of the past 50 years have led organic geochemists far afield of these first endeavors. As Geoff and Max Blumer recognized
back in the 1960s, and as microbiologists began to realize in the early 1980s,
the usefulness of the biomarker concept is not restricted to geologic time. Most
organic geochemists have, at one time or another, applied their techniques and
expertise to the resolution of environmental problems, or found a way to address
some archaeological mystery. One of the Bristol group's most vibrant research
programs now has its chemists brushing shoulders not with geologists and
oceanographers, but with archaeologists and anthropologists concerned with the
evolution of human civilizations and societies.

   Much of the impetus for the application of biomarker concepts to archaeologists' questions in the 1970s and 1980s came from petroleum geochemists,
not least from Arie Nissenbaum, a geochemist at Israel's Weizmann Institute of
Science who developed a keen interest in the role that geological events and cir-
cumstance might have played in the history of civilizations in the Fertile Crescent
region. Nissenbaum was fascinated by the bizarre geology and chemistry of the

Dead Sea Basin area, where oil seeps and impressive raftlike chunks of asphalt floating on the surface of the lake had long tempted oil prospectors to no avail. Renewed interest in the area's oil potential in the early 1980s attracted a wave of geochemical studies, and Israeli geochemists scrambled for laboratory resources and funding from abroad. Jürgen, still with Dietrich Welte's group, did a detailed biomarker study at the behest of an Israeli colleague, and when Nissenbaum saw the results he suggested to Jürgen that they apply Jülich's considerable GC-MS capability to solving an entirely different sort of mystery.

Excavations of archaeological sites in the vicinity of the Dead Sea had turned up solid chunks of black, sticky material that was used as early as 3000 B.C., either in materials used for construction or as a glue to attach tool heads to wooden handles. Nissenbaum attempted the first geochemical analyses of this material in 1984, using the biomarker techniques that petroleum geochemists employed for correlating source rocks and oils to ascertain that it derived from the Dead Sea asphalt. Some form of asphalt use is evident in early civilizations throughout the Fertile Crescent, and indeed, the material that ranks lowest on the scale of coveted petroleum products in the modern world appears to have been a highly valued commodity for earlier civilizations. That there was an active trade in this commodity was apparent from some of the earliest written records: carved into a stone tablet by a merchant named "Lukulla" in 2039 B.C. is a list of prices for different types of asphalt, which range from "raw bitumen" to some form of processed, ready-to-use building material. But archaeologists had no way of knowing if the black, sticky stuff they often found on the surface of ancient artifacts was in fact asphalt or whether it was some sort of plant resin or wax. Both the extent of asphalt use and the ancient trade routes were subjects of much debate, as Jürgen learned when Nissenbaum contacted him in the mid-1980s and suggested that they do a biomarker analysis of samples from, of all things, Egyptian mummies.

Mummies have been the subject of conjecture and myth throughout the ages, the embalmer's art shrouded in mystery even while it was still being practiced, with secret potions developed by individual embalmers and passed down to a chosen few, but never recorded. It was an art intended only for the eyes of the gods, and, indeed, the embalmers succeeded in protecting many of their secrets for more than three millennia, notwithstanding frequent human scrutiny of their handicraft. Twentieth-century archaeologists and chemists determined that the bodies were first thoroughly dehydrated with various mixtures of salts, and then treated with some sort of organic balm that sealed them off from moisture and protected them from microbial invasion—but the nature of this balm was unclear, even in the 1980s. The most extensive historical account, written in the fifth century B.C. by the Greek Herodotus, mentions that myrrh, cassia palm wine, and cedar oil were used. The Sicilian historian Diodorus later reported that asphalt from Palestine's Dead Sea was a major component, and it was often assumed that this was why the mummies often appeared black. But archaeologists were unclear about the extent of trade between the region and Egypt, and many were convinced that wood ash and pitch, rather than asphalt, had been used to seal the bodies and were the source of the black color. Nissenbaum had obtained samples from the black balm on the coffins, baseboards, and bodies of four mummies in the British Museum and wanted Jürgen to compare their sterane and hopane fingerprints with those from the Dead Sea asphalts.

Their first analyses revealed that fresh plant materials *and* asphalt had been used in preparing the bodies. The samples contained *n*-alkanes with a typical leaf wax distribution, on the one hand, and steranes and triterpanes typical of mature geochemically altered material, on the other. The mass fragmentograms of steranes and hopanes from three of the samples, dating from 200 B.C. to 150 A.D., were nearly identical to those from extracts of Dead Sea asphalts and oils—high in gammacerane and lacking diasteranes, as is typical of oils formed in highly stratified, low-oxygen and high-salt environments. The balm from the oldest mummy, prepared around 900 B.C., clearly contained asphalt, but it hadn't come from the Dead Sea. Later analyses indicated that it may have come from a seep that was on the north shore of the Red Sea. More extensive studies by Jacques Connan, a petroleum geochemist with the French oil company Elf Aquitaine, showed that Dead Sea asphalt was used for the preparation of Egyptian mummies as early as 1100 B.C., indicating that some sort of trade between Egypt and Palestine must have been operating at the time. But his work, like Nissenbaum's and later analyses by Richard Evershed and his students in Bristol, also revealed that asphalt was not a requisite component of the preservative balms, which contain relatively large amounts of abietic acid and the related tricyclic diterpenoid compounds typical of conifer resins. Fatty acids with a distribution typical of plant oils are also common ingredients, as are the particular series of wax esters found in beeswax, and it appears that there may have been trends in mummification technology over the centuries: plant oils, resins, and beeswax were apparently à la mode during the heyday of mummification between 1350 and 1000 B.C., with asphalt coming into use toward the end of this period and gaining popularity during Roman times. The use of asphalt for more mundane purposes, however, clearly extends far back into prehistory, and far beyond the borders of ancient Egypt.

In an attempt to trace the development of asphalt technologies in early civilizations, Connan analyzed scrapings of black, potentially asphaltlike materials from hundreds of artifacts in the Louvre Museum's extensive collection. These studies revealed that some groups may have used asphalt to glue handles to their stone tools as early as the Paleolithic period, some 40,000 years ago, and such use was quite common in Neolithic times. From Neolithic through Roman times, extensive use of asphalt occurred primarily in areas where building materials such as stone and wood were scarce, and the asphalt was mixed with straw and sand to create a type of brick. The magnificent Mesopotamian city of Babylon was constructed of such bricks, but in Egypt, where stone was plentiful, asphalt seems to have been an exotic luxury item, reserved for the mummifiers' balms or for waterproofing palace baths—though according to the Bible, the basket where the Israelite slave Jochebed left her baby Moses was also lined with *heimar*, usually translated as bitumen or asphalt. Whatever the case, Connan's investigations indicate that the asphalt used in Egypt was mostly imported from the Dead Sea, while more local sources went undeveloped. By comparing hydrocarbon fingerprints and $\delta^{13}C$ values of asphalt scraped from artifacts found at different sites around Mesopotamia with those of petroleum from seeps in the region, Connan determined that a lively trade in this material had been established by 5500 B.C. The Ubaid and Uruk civilizations that thrived in what is now Iraq from the end of the Neolithic into the Bronze Age were apparently heavy users of the material, and Connan has been able to track the changes in their regional trade routes between 5800 and 3500 B.C.: first to nearby seeps in Iran, distinguished by their oleanane content and distinct sterane and hopane fingerprints, then several hundred kilometers north to seeps on the Tigris River, where oleanane is lacking and $\delta^{13}C$ values are distinctly negative; and finally, based on a subtle change in the hydrocarbon fingerprints, to seeps on the Euphrates River to the northwest.

Around the same time that Jürgen and Nissenbaum were trying to resolve the origin of the black stuff on mummies, members of the Bristol lab developed a sudden interest in maritime history. The wreck of Henry VIII's prize flagship, the *Mary Rose*, had just been recovered from the floor of the English Channel where it had lain since 1545, when, newly refitted with the most modern guns, it set sail from Portsmouth Harbor and, as the king watched from shore in horror, heeled over and sank with the first small breeze. The British media reported on the ship's contents in intimate detail, and when they mentioned that among the bows and arrows, guns, and other Tudor period relics, there were large barrels full of some sort of gooey black pitch, Geoff decided to join the fun. The black stuff coated the seams of the ship and some of the relics, and there was speculation that it was partially responsible for their excellent preservation. One couldn't tell from looking at it whether it was asphalt or some sort of wood resin, but it was an easy enough task to figure this out in the laboratory—just the thing for an undergraduate student's research project. Steranes, terpanes, and the generally unresolved hump of hydrocarbons that one typically found in asphalt were missing from the tar, but it did contain large concentrations of telltale tricyclic terpenoids. These were derivatives of the compounds that Guy Ourisson had studied so extensively in the 1950s, of interest for their antibacterial and insect-repelling qualities, and for their often genus-specific distributions—in the case of the *Mary Rose* tars, dominated by the abietic and pimaric acids indicative of a pine origin. The fourteenth-century manufacturers had apparently known how to distill and collect the volatile compounds from the wood tar, and they had produced a concentrated mixture of terpanes and mono-, di-, and triaromatic diterpenoids that had excellent waterproofing and preservative qualities. Comparative analysis of the tar on an Etruscan ship indicated that such technologies were known for thousands of years before Henry's ill-fated warship sank to its resting place on the floor of the English Channel.

The *Mary Rose* study inaugurated what would turn out to be a career-encompassing interest in the chemical analysis of archaeological relics—not for the undergraduate who prepared the extracts of tar, but for Richard Evershed, the postdoc who supervised her. Evershed had started out his career in natural products but spent several years in the Bristol lab developing analytical techniques for porphyrins and, as it turned out, analyzing the black goo from the *Mary Rose*. In 1984, the year after the *Mary Rose* was recovered, the body of the so-called "Lindow Man" was found in a bog near Manchester, again making headline news in England...and Evershed, then with a position in Liverpool, ostensibly developing analytical methods in biochemistry, somehow found himself extracting and analyzing lipids from a 2,000-year-old dead man. "I didn't actually know why I was analyzing this body," Evershed tells me. "It was sort of a nightmare." He admits that he got the idea while sitting in a pub in Liverpool, where Lindow Man was the main topic of conversation and everyone was rehashing the latest rumors about his identity, which ran the gamut from a recent murder victim to a Celtic druid who had been sacrificed to placate the gods and keep invading Romans at bay. However frivolous its inception, Evershed's study led to a quite rigorous analysis of the chemistry of tissue preservation and decomposition... and from there to analyses of the waterlogged, anoxic peat bogs where such bodies were so well preserved, and then to the application of the analytical techniques he'd developed—a mix of those learned in the Bristol lab and those of the natural products chemist—to the organic chemistry of soils, and eventually, in the late 1990s, to a call from someone at the English Heritage Foundation wondering if Evershed could do anything with the broken bits of pottery that archaeologists carefully sifted from their sites, but usually couldn't make heads or tails of.

The ceramic vessels left by former civilizations are one of the traditional mainstays of archaeological information, but determining how they were used is not always an easy task, especially when all that's left is a tiny fragment. Evershed and his colleagues found that significant amounts of lipids remained absorbed in the pores of the clay and could provide some clues. Pottery fragments found among the remains of late Saxon and early medieval settlements contained leaf wax lipids with a distinctive pattern—a simple trio of the C_{29} *n*-alkane and the corresponding mid-chain ketone and alcohol—that bore witness to the longstanding prominence of cabbage in northern European diets. Small ceramic cups found among the remains of the Minoan civilization that thrived in Crete between about 2700 and 1450 B.C. were used, according to the archaeologists, as portable lamps, and contained an array of wax esters that was clearly indicative of beeswax. Archaeologists had thought that the first portable lamps marked the early exploitation of the olive—but there was no sign of the fatty acids associated with olive oil in the cups. Here, in these artifacts that are only a few thousand years old, one can still glean some information from fatty acid distributions and wax esters. In some cases, even the ester bonds are preserved, and Evershed and his group have started using a system of high-temperature gas chromatography and mass spectroscopy that allows them to detect the intact long-chain esters of glycerol fats, providing even more information about the nature of the original waxes and fats. But just as diagenesis and thermal heating can transform the biological molecules over geologic time, cooking and the "unnatural" mixtures of substances created by even the most primitive cuisines can often obscure the sources of lipids found in archaeological remains. Evershed's team has done extensive analyses of potential foodstuffs and the isomerization and condensation reactions that occur during heating—but, as in geochemical studies, some of the most valuable molecular information in archaeological remains is obtained by combining compound-specific isotope analysis with structural information and homologue distributions.

The distributions of fatty acids and intact glycerol fats, when considered together with the compound-specific $\delta^{13}C$ values for the fatty acids, have allowed the Bristol team to distinguish the fats of ruminant animals such as cattle, sheep, and goats from those of pigs and other nonruminants, which have an entirely different diet and system of processing their food intake. The animals that a civilization consumed or made use of can thus be determined from analysis of the traces of fat in a broken pot, even when bones and other indicators are absent. The initiation of dairying is a pivotal development in human prehistory, and yet it has proven difficult to ascertain when and where humans started making use of animal milk in different parts of the world. Egyptian and Mesopotamian pictorial records from 4000–2900 B.C. offer the oldest clear evidence of dairying, but it appears that sheep were domesticated as early as 9000 B.C., and cattle and goats around 7000 B.C., so milking may have been practiced much earlier. Though the fatty acid distributions of milk and cooked meat cannot be readily distinguished, Evershed and crew have found a way to recognize traces of milk fats in pottery shards based on a subtle difference in the biosynthetic processes that animals use to produce milk fats and body fats. Animals make both their milk fats and their body fats from a mixture of fatty acids that they biosynthesize and that they obtain prefabricated from their diets. In milk fats, however, a larger proportion of the C_{18} acid comes from de novo biosynthesis, and as this generates compounds that are significantly more depleted in ^{13}C than those obtained from plants, C_{18} fatty acids from milk fats and body fats have different isotopic signatures. In the hope of determining when and where dairying was first practiced in the British Isles, Evershed's group is using this distinction to identify traces of milk fats in pottery fragments from the Celtic period.

Archaeologists' attempts to understand the development and extent of early agricultural practices have long been frustrated by a lack of structural evidence in ancient soils. But Evershed and his cohorts have been able to identify medieval farmlands where manure or some sort of compost was used from the "unnaturally" high concentrations of common lipid components left in the ancient soils. An excess of sitosterol and leaf wax alkanes indicates that plant manures were applied, whereas the reduced stanols typically produced by intestinal microbes indicate that manure was applied, with 5β-stigmastanol pointing to feces from ruminant animals that eat grasses and plants, and 5β-cholestanol pointing to feces from omnivorous animals like pigs or humans. More specific and reliable information about manure comes from the so-called bile acids, di- and trihydroxy steroid acids that are biosynthesized from sterols in animals' livers. Ruminant animals, such as cows, and omnivorous animals, such as humans, produce different sets of homologues. Bile acids are not produced by microbes or by diagenesis in the environment, and in dry conditions they are quite persistent and make excellent biomarkers on the archaeological timescale. They have been identified in the tissue of a 4,000-year-old Nubian mummy, used to map the latrines and sewage culverts in Roman ruins, and they prove that 2,000-year-old feces found in a North American cave came from human beings...among other things.

One potentially powerful but little developed method for tracking human land use patterns in prehistoric and historic time involves the same multiproxy biomarker techniques that organic geochemists have used on marine sediments to elucidate changing paleoenvironments. The thick layers of river deposits in coastal regions or the sediments at the bottom of lakes can often provide high-resolution records of the past 10,000 years, and distributions of leaf wax hydrocarbons, sterols, and triterpenoids, along with compound-specific isotope analyses, may reflect forest clearing, burning, fertilizer application, and settlement or urbanization in an associated drainage basin. Phil Meyers and his group at the University of Michigan have found that sediment cores from lakes in North America's Great Lake region contain a record of the region's changing human populations and their effects on lake ecologies since the onset of European settlement. The distributions and relative amounts of long-chain leaf wax alkanols and alkanes; algal sterols and C_{15}–C_{19} alkanols and alkanes; total organic matter, total hydrocarbon and phytol concentrations; compound-specific $\delta^{13}C$ values; and carbon to nitrogen ratios tell a story of changing vegetation type, erosion, and lake ecosystems over the past 200 years. Sediments deposited during the early nineteenth century, when large tracts of forest were cleared for agriculture, register a pronounced increase in the amount of erosion from land and algal productivity in the lakes, presumably due to the large input of nutrient-rich soils and organic matter. Between 1950 and 1975, a period that saw rapid industrial and population growth in the region's cities and widespread use of chemical fertilizers on its farms, the algae went haywire and produced so much organic matter that the bottom water became anoxic. And, finally, lake sediments from the last 30 years bear witness to the success of environmental controls imposed in the late 1970s, namely a marked decrease in algal productivity and return to oxicity in the deep water, with accordant improvement in the health of lake ecosystems.

Sediment records of changing land use patterns—and of sewage, fertilizer, or petroleum product residues—are not, of course, just of interest to archaeologists and historians. Biomarker techniques have long found application in the assessment of pollution problems and environmental management and protection, and in the past decade they have become one of the scientific mainstays of "environmental forensics," used in litigation for and against polluters. Some of these applications grew naturally out of the molecule collecting of the 1970s. Others derived from concerted efforts on the part of organic geochemists who were concerned about the spread of contaminants in the environment and noted that their techniques for identifying traces of organic compounds in complex mixtures also worked well for tracing pollutants. The studies of dust that Bernd Simoneit did while he was a student in Geoff's lab indicated that PAHs from fires, automobiles, and factories could be carried long distances on the wind and deposited in the middle of the Atlantic Ocean, just like the leaf waxes. Simoneit's recent work uses more specific biomarkers to gauge the contributions from specific sources of air pollution in populated areas: diterpenoids like abietic acid might come from a fire in a coniferous forest or from the heavy use of wood stoves in an urban area; flowering plant triterpenoids like β-amyrin or lupeol might point to a fire in a deciduous forest or to agricultural burning; the particular fingerprint of hopanes and steranes might point to the emissions of a particular petroleum-burning industrial source; and so forth.

As it happened, the development of the biomarker concept in the 1960s and 1970s had coincided with the growing public concern about environmental issues. For his part, Geoff says it was Rachel Carson's 1962 book *Silent Spring*, about the dangers of DDT, that really got him to thinking, as a chemist, about the long-term fate of environmental pollutants. But it wasn't until around 1970, after he set up shop in Bristol, that he found a chance to apply his laboratory's analytical prowess to such problems. The Severn Estuary, where he and his students did their first extensive studies of sterol diagenesis in sediments, was not only Bristol's gateway to the sea but also the outlet for waste from all the factories and cities that line the rivers of the industrial English Midlands and the coast of South Wales. So along with analyzing the steroids in the estuary's marshes, the Bristol researchers tried to analyze the petroleum hydrocarbons in the sludge, sediments, and outflow from a sewage treatment plant, and, of course, they used their GC-MS to look for DDT and other man-made chemicals. Indeed, Geoff says, some of these studies had to be curtailed because local industries got wind of what they were doing and limited their access or put pressure on the university for them to stop. Nevertheless, these and similar studies on the Clyde Estuary made it clear that the conversion of sterols to stanols during early diagenesis might be of more than academic interest. The microbes in the surface sediments began generating small amounts of stanols after a few months, and these gradually gave way to sterenes as the alcohol group was eliminated. In a natural aquatic sedimentary environment, the stanols never dominated the steroid distributions, and their existence was so ephemeral that most geochemists came to think of them as diagenetic intermediates. In the sewage sludge, however, certain stanols were present in anomalously large amounts, particularly 5β-cholestanol. As it turned out, microbes in the human intestine also reduced the sterols that the body was eliminating, mostly cholesterol, to stanols—and, unlike the sediment microbes, which produced a mixture of 5α and 5β forms, the intestinal microbes produced only 5β-stanols.

GC-MS made it easy to analyze large numbers of samples for 5β-cholestanol, and the compound, often known as coprostanol, quickly came into use as an indicator of sewage pollution. In 1990, Joan Grimalt and the environmental science group in Barcelona suggested that complete steroid distributions and ratios provided a more foolproof and source-specific indicator of sewage contamination, and Evershed's group has shown that bile acids can also provide rather specific information about sources of waste pollution, be they sewage treatment plants, a local pig farm, or an overpopulation of seagulls.

Another of the founders of the biomarker concept, Max Blumer, turned its nascent petroleum applications upside down in 1969, when he used it to track the fate of an oil spill off the coast of Cape Cod. Blumer's studies showed, for the first time, that petroleum hydrocarbons lurked in the coastal sediments long after visible signs of the oil had disappeared from the water and the beaches and tidal flats were clear of dead birds, fish, worms, and clams. Moreover, using the rudimentary GC techniques available at the time, he could determine which compounds evaporated or were broken down by microbes, and which persisted. In the years that followed, John Farrington continued and expanded on this work at Woods Hole, as did Ian Kaplan at UCLA, learning to distinguish between hydrocarbons from natural sources, like the seeps off the coast of Southern California, and those from specific types of oil pollution. Their work made it possible to track the spread of oil pollution—from an oil spill, a harbor, a shipping lane, a city, or a contaminated river—into the marine environment. Ironically, their methods developed in tandem with those that the petroleum geochemists were using for exploration: ever more sophisticated mass fragmentograms of steranes and terpanes, compound-specific and group-specific isotopic analyses, and various indices of maturity. These were the techniques that Joan Albaigés developed in his environmental chemistry group at the University of Barcelona in the late 1970s together with Pierre Albrecht in Strasbourg, that Keith Kvenvolden and others used in their analyses of Gulf of Alaska sediments after the catastrophic *Exxon Valdez* oil spill in 1989, and that Albaigés' group have employed to track the fate of oil from the Russian tanker *Prestige*, which sank off the north coast of Spain in 2002.

Chris Reddy, keeper of the long legacy of environmental research at Woods Hole, went a step further when he and his research team returned to the scene of the 1969 Cape Cod oil spill that Blumer and Farrington had studied so extensively, now equipped with a cutting-edge new instrument. They used a method that involves linking two capillary columns in sequence—so-called comprehensive two-dimensional gas chromatography, developed in the mid-1990s—and can separate many of the compounds in the unresolvable "hump" that has frustrated every chemist who ever tried to characterize the hydrocarbons in petroleum. Reddy's team obtained a core from a salt marsh that was particularly hard-hit by the 1969 spill and found that there was just as much oil present in 2003 as there had been at the time of the last study in 1975, albeit now covered by a fresh layer of uncontaminated sediments and relatively healthy marsh. Though the earlier studies indicated that a significant amount of the oil had degraded or evaporated in the first five years and the *n*-alkanes had been completely degraded by microorganisms, the new analyses revealed that some groups of branched alkanes, which earlier studies hadn't detected, as well as the cycloalkanes and aromatic compounds, remained. Apparently, the microbes had found better things to eat, or the oil degraders hadn't liked the anoxic sediment conditions that prevailed after the first few years, or they had depleted some key nutrient or energy source. Whatever the case, degradation seemed to have hit a plateau or been damped out within five years after the spill.

By the start of the twenty-first century, it would seem that the natural history of molecules had finally come of age, but it was and is still prone to remarkable spurts of growth—and self-doubt—usually inspired by the deployment of some new analytical tool, like Reddy's two-dimensional GC. The introduction of compound-specific carbon-13 measurements in the late 1980s was, as it turns out, just a first step toward including the isotopic dimension in biomarker analyses. In the late 1990s, Tim Eglinton and his colleagues found a way to measure the amount of ^{14}C—the radioactive isotope of carbon—in individual compounds...and soon found themselves mounting a new attack on one of the most insidious problems of the earth sciences. Radiocarbon dating, originally developed for use on fossil bones and archaeological relics and useful for dating anything younger than 60,000 years, had long been used to determine the ages of Pleistocene sediments. But these ages were based on the carbon in the calcium carbonate of foram shells—no one had ever tried to date individual organic compounds in the sediments. ^{14}C in the environment is less than a billionth as plentiful as ^{13}C, and there was no way to detect it with a GC-irm-MS. Instead, Tim and his group reverted to a glorified version of what Kate Freeman and John Hayes did before they developed the instrument, and what Geoff and others had done before the first GC-MS was invented: they collected the compounds from the end of a GC column until they had enough for analyses. Now, however, they used a wide-bore capillary GC column that afforded an excellent separation and allowed them to inject more sample, and after burning each compound to CO_2, they reduced the CO_2 to graphite, which could be introduced into an accelerator mass spectrometer designed for ^{14}C analysis.

The first compound-specific ^{14}C analyses showed that compounds in the same layer of sediment could have decidedly different ages, depending on whether they came from the algae in the surface waters directly above the sediments, from land plants, or from old eroded continental sediments—in other words, depending on how far they had to travel before deposition. Tim's student Ann Pearson did extensive compound-specific ^{14}C and ^{13}C analyses of recent sediments that showed the remarkable power of the combined isotopic and structural information to reveal not only the sources of biomarkers, but also the metabolic processes of recently discovered, difficult-to-study microorganisms. Pearson took advantage of an inadvertent, man-made twist in the isotopic makeup of atmospheric CO_2: intensive nuclear weapons testing in the 1950s and 1960s raised the abundance of ^{14}C in atmospheric CO_2 significantly above natural levels. Perhaps the only positive outcome of such testing is that it has provided scientists with global tracers for natural processes. Oceanographers had been tracking this pulse of so-called "bomb ^{14}C" as it moved from the atmosphere into the oceans, using it to determine the rate of uptake of atmospheric CO_2 in different regions and then to track the slow movements of deeper water masses in the great ocean conveyor. For Pearson's purposes, it was enough to know that the bomb ^{14}C had generally infiltrated the photic zone but had not yet arrived in the intermediate or deep layers of water in the Pacific where she obtained her sediment samples. She compared the ^{14}C abundance in compounds extracted from very recent sediments near the sediment–water interface with those extracted from the older, "prebomb" layers. Dinosterol, alkane diols, fatty acids, C_{24} to C_{30} *n*-alcohols, and various hopanols were significantly more enriched in ^{14}C in the contemporary sediments than in the prebomb sediments, clearly indicating that their main source organisms— algae, zooplankton, bacteria—had lived near the surface and utilized some form of carbon derived directly or indirectly from the contemporaneous atmospheric CO_2. The odd-carbon-number long-chain *n*-alkanes were likewise enriched in the postbomb sediments, indicating their provenance from contemporary land plant detritus, whereas *n*-alkanes with even carbon numbers showed no such enrichment. The latter had a distribution that indicated they derived from a geo-chemically altered source and may have come from pollution or from the seafloor petroleum seeps that pepper the region off the coast of California, where the sam-ples were taken. But it was the combined ^{13}C and ^{14}C measurements in the iso-prenoid ethers—specifically in the biphytanyl components of crenarchaeol—that gave the most surprising and provocative result. The marine crenarchaea appar-ently utilized a distinct source of carbon: they had not incorporated the dissolved CO_2 or bicarbonate from the surface waters, like the phytoplankton, nor did they recycle organic matter generated by other organisms in the surface water, like the bacteria. Rather, these marine crenarchaea made use of "old" bicarbonate from the deeper water beneath the photic zone. It was the first solid, in situ evidence that the organisms are autotrophic, as hypothesized by the NIOZ group, and it supported the observations that Ed DeLong's group in Monterey had been mak-ing, that they were most plentiful at intermediate depths.

The range of potential applications for such multidimensional biomarker analyses in both recent and ancient sediments is vast, and their utility in identi-fying the provenance of biomarkers would seem to foretell a new growth spurt for the molecular lexicon. But in 2002 when Nao Ohkouchi, a postdoc in Tim Eglinton's group at Woods Hole, determined the ^{14}C ages of alkenones in a thick layer of sediments from the Bermuda Rise in the northwest Atlantic, all hell broke loose in the geochemical community, and it seemed instead that the lexicon might end up buried in the sediments that gave rise to it. The most precise dates for late Pleistocene and Holocene marine sediments are obtained from the ^{14}C ages of planktonic foraminifera, which are relatively large and sink to the sediments quickly, recording the time of deposition of a given sediment layer. The premise was that other microfossils and biomarkers in the same layer had been depos-ited at around the same time, and the validity of correlations between proxies— between U^K_{37} and foram $\delta^{18}O$ proxy temperatures, for example—obviously depended on this fact. The Bermuda Rise was one of the few open ocean sites where the rate of deposition was high enough that one could obtain an uninter-rupted, high-resolution record of temperature change over the past ice age cycle. But Ohkouchi's results indicated, unequivocally, that the alkenones and other organic constituents were thousands of years older than the forams in the same sediment horizon. The alkenones had not been formed by coccolithophores liv-ing side by side with the foraminifera in the surface waters above the site of depo-sition, as supposed. They had, rather, been made thousands of years earlier by coccolithophores that lived, apparently, hundreds of kilometers to the north…in much colder waters. Currents or turbulence in the deep waters must have carried the tiniest particles—organic detritus that remained suspended in the water or fine clay particles with organic molecules adsorbed onto their surfaces—far from their homes and deposited them in a great pile on the Bermuda Rise, like a sort of sand dune on the seafloor.

Were biomarkers leading oceanographers and climatologists astray? Was the U^K_{37} proxy trustworthy? Tim Eglinton found himself in the unenviable position of questioning some of his father Geoff's most cherished contributions to earth science. But Ohkouchi's finding did not put an end to biomarker use or multi-proxy studies, any more than exposure of the insidious nature of hydrocarbon contamination did 40 years earlier. Rather, it raised a flag to oceanographers and geochemists to take care with their interpretations: the geochemical proxies are not mere tools to be used the way one might use a computer or an automobile or even a plastic stool, without understanding how it operates and how it was made. Context is everything, and in the case of the biomarker, the context is biological, geological, chemical, and physical—there can be no segregation of disciplines. In Pleistocene sediments, at least, ^{14}C measurements now allow one to directly determine the age of the different organic components, as well as the carbonate, in a given sediment horizon. But one must also look for other, more indirect signs of diverse timescales of organic matter deposition—physical evidence of lateral sediment transport, anomalous signs of diagenetic maturity for different compounds, and other clues to a distant origin.

Julian Sachs and his students at MIT recently made use of yet another isotopic dimension of the alkenones to gain insight into where the coccolithophores that made them had lived. Like ^{16}O and ^{18}O, natural abundances of the two stable isotopes of hydrogen, ^1H and ^2H, vary in natural waters. Alex Sessions, one of John Hayes's last Indiana students, had studied hydrogen fractionation in plants and algae and adapted the GC-irm-MS for measuring the relative amount of the rare ^2H, commonly known as deuterium, in individual lipids. Variations in deuterium enrichment, δD, reflect both biochemical and environmental effects but are also strongly dependent on the δD values of the hydrogen in the water the organism uses. Hydrogen fractionation occurs during evaporation, precipitation, and freezing, so the water in different regions of the ocean has varying δD signatures, and once biosynthetic fractionation is accounted for, these are mirrored by the δD values of lipids produced by aquatic organisms. Sachs's analyses of alkenones from coccolithophores grown in water with different δD signatures and from suspended organic matter collected in different parts of the Atlantic were consistent with this prediction, and analysis of the alkenones in the Bermuda Rise sediments pointed to a main source in the waters off the coast of Nova Scotia. Not only does the method provide a way to ensure the geographic consistency of proxy measures at a given site, but it may also offer clues about ancient current regimes. Compound-specific δD analyses of leaf wax and algal lipids in marine and lake sediments are also turning out to be a treasure trove of information about changing global patterns of rainfall and evaporation, a particularly elusive aspect of past climate systems.

Like the inclusion of isotopic dimensions in biomarker analyses, the combination of gene and lipid studies that opened the door to understanding microbial ecosystems has only just begun to make its mark. The ability to amplify and analyze tiny traces of nucleic acid fragments extracted from sediments and natural waters has allowed chemists to revisit the question of whether the ultimate biomarkers, DNA and RNA, might, after all, be of some direct use in paleontology or paleoenvironmental studies. Nucleic acids don't have nice carbon skeletons that can persist in the sediments and they do break down quickly, but even a small fragment of a DNA or RNA molecule can provide as much information as the most complex of hydrocarbons, and the new techniques can amplify and detect the tiniest trace of a fragment. Of course, they also magnify contaminants from modern organisms, and there have been many false reports of DNA preserved in ancient fossils. But in recent years, techniques for analyzing fossil DNA have been rigorously constrained, and it is now apparent that under very cold, dry, dark conditions, fragments of DNA in fossil bones can survive up to a million years, offering paleontologists and evolutionary biologists a real-time view of genetic change and its relationship to environmental change. For example, widespread extinctions of large mammals at the end of the Pleistocene epoch have been attributed to the expansion of *Homo sapiens* and increased predation, but the DNA of fossil mammoth and bison bones indicates that genetic diversity among these large mammals decreased much earlier, at the onset of the last ice age and long before the human expansion. On more recent timescales, DNA extracted from fossil fecal material has provided information about the dietary diversity of ancient humans, and DNA from fossil bones, plants, and grains from archaeological sites chronicles the spread of agriculture and domestication of crops and animals since the end of the last ice age.

Marco Coolen, now at Woods Hole, and microbiologist Jörg Overmann in Munich have been able to amplify fragments of fossil DNA and RNA extracted from Holocene and Pleistocene marine sediments. Their combined nucleic acid and lipid biomarker analyses—using the complementary information from the species-specific nucleic acids, on the one hand, and group-specific lipid biomarkers, on the other—can provide relatively reliable, detailed accounts of paleoenvironments and are viable in sediments up to about 200,000 years old, depending on conditions. Coolen and Jaap Sinninghe Damsté's NIOZ group have used such methods to track the changing populations of algae and bacteria in Antarctica's Ace Lake and were able to document the shifts in the lake's ecology over the past 10,000 years: from freshwater lake to open marine basin, and eventually to the heavily stratified enclosed basin, saline bottom water and active methane cycle of the contemporary lake. In similar studies on Mediterranean sapropels, they examined the distributions of isorenieratene and chlorobactene, and of 16S rRNA sequences from green sulfur bacteria. Surprisingly, the genes they detected were most closely related to those of green sulfur bacteria that, in contemporary times, live only in fresh and brackish water, raising the possibility that bacterial cells and debris washed into the open sea from anoxic coastal regions—in which case, the Mediterranean might not have been as anoxic during these periods of sapropel deposition as the presence of isorenieratene would otherwise seem to imply. In another study, the DNA sequences of haptophyte algae and alkenone distributions in Pleistocene sediments were used to assess the relative contributions of alkenone producers other than *Emiliania huxleyi*, a method that allows one to gauge the reliability of U^K_{37} temperatures in regions and epochs where Emily may not have been the dominant alkenone-producer.

The persistence of fossil gene fragments in the geological record appears to be limited to a million years and to fossils preserved in permafrost or sediments formed in anoxic and hypersaline environments. Genetic analyses of *living* organisms may actually make a greater contribution to tracing the activities of organisms over geologic time in that they provide information about the sources and biochemistry of lipids that leave persistent fossil molecules. Historically, natural products methods for determining biomarker source organisms have been somewhat random, and their success has been overly dependent on serendipity. But the hundreds of microbial genomes that are now available permit one to recognize genes that encode for particular enzymes and identify organisms that have the basic wherewithal to produce a given type of lipid. This facilitates a more judicious choice of organisms for lipid analyses, on the one hand. And, on the other, it allows for a more educated guess about the general nature of source organisms for fossil lipids even when specific taxonomic groups can't be identified. In other words, it makes it easier to determine who is making what. The genes can also help in elucidating the physiological roles of key lipids and provide another layer of information for interpretation of fossil lipid records. Ann Pearson's group at Harvard has begun applying some of these ideas to the study of hopanoids and steroids and found that, despite the fact that hopanoids are relatively rare constituents of anoxic sediments and absent from most of the anaerobic bacteria that had been analyzed to date, a number of groups of anaerobic bacteria that are common in anoxic sediments have the wherewithal to synthesize hopanoids—just as Volker Thiel and his colleagues in Walter Michaelis's lab began to suspect when they found ^{13}C-depleted hopanoids near a Black Sea methane seep, and Sinninghe Damsté predicted when he found they were major constituents of the anammox lipids.

A couple of years ago, sometime around his 77th birthday, Geoff was going through a presentation he'd been preparing with Richard Pancost, trying to edit it down to size, and he said to me, "A while back I was a bit dispirited, thinking that there wasn't much more to be found. I thought that we'd reached the limits." He was being awarded the Wollaston Medal, the British Geological Society's highest honor—and a sure sign that geologists had finally recognized the value of his "stamp collecting"—and the geologists had asked him to give a talk on "the future of biomarkers" as part of the award ceremony. Geoff had taken the assignment seriously and spent a couple of months talking to young scientists, brushing up on the newest developments—but the speech was only supposed to take a half hour, and now he was dismayed because he had to leave a few things out of his densely packed, 100-slide presentation.

Working with biomarkers means thinking about molecular architecture as an information source in time and space, about molecular design and structure as nature's persistent trademarks of biochemical function and biological source. But how much time? What sort of environment and what kinds of sources? Organic geochemistry's entrepreneurs have taken the concept and run with it, in every imaginable direction: as we try to complete this book in 2007, our worry is not that there is nothing left to discover, or that the book should be short and readable, but rather, that we can't keep ahead of this avalanche of new science.

Appendix

Biomarkers at a Glance

Biological molecule	Structure (example)	Known or postulated source
Straight chains		
1 *n*-Alkane (odd carbon number) $C_{25} - C_{35}$ $C_{15} - C_{19}$		Higher land plant leaf waxes Marine phytoplankton
2 *n*-Alkanol (terminal hydroxyl, even carbon number) $C_{24} - C_{34}$ $C_{16} - C_{22}$		Higher land plant leaf waxes Planktonic algae
3 *n*-Alkandiol & -keto-ol C_{28} & C_{30} 1,14-diol C_{28} & C_{30} 1-ol-14-one *n*-Alkandiol C_{30} & C_{32} 1,15-diol		Planktonic algae: some diatoms (*Proboscia*) yellow-green algae (Eustigmatophytes)
4 *n*-Fatty acid unsaturated, even carbon number, mostly $C_{14} - C_{20}$ saturated, long-chain, e.g. $C_{24} - C_{32}$		Most organisms Higher land plants
5 Alkenone $C_{37} - C_{39}$		Haptophytes
Branched chains		
6 Mono-, di- and trimethylalkane ($C_{18} - C_{20}$)		Many cyanobacteria (methyl groups at various positions)
7 *iso*- & *anteiso*-Fatty acid (C_{15} & C_{17})		Sulfate reducers and other bacteria
Acyclic isoprenoids (see also Carotenoids, Complex membrane lipids)		
8 Phytol		Chlorophylls of algae, land plants and photosynthetic bacteria
9 Tocopherol		Plants, bacteria
10 Lycopadiene		Some planktonic organisms
11 Crocetene		Anaerobic methanotrophic archaea
12 Pentamethylicosene		Methanogenic and methanotrophic archaea
13 Highly branched isoprenoid (HBI)		Some diatoms (*Rhizosolenia*)
14 Botryococcene		*Botryococcus braunii* (freshwater algae)
Diterpenoids		
15 Abietic acid		Conifers

Fossil molecule	Structure (examples)	Biogeochemical application
Straight chains		
1 *n*-Alkane (unaltered, $C_{25} - C_{35}$, odd)		Land plant organic matter $\delta^{13}C < -30$ per mil $\rightarrow C_3$ plants $\delta^{13}C > -25$ per mil $\rightarrow C_4$ plants
n-Alkane (shorter chains, odd & even)		Mature organic matter, petroleum contamination
2 *n*-Alkanol (unaltered, $C_{24} - C_{34}$) *n*-Aldehyde, *n*-ketone		Land plant organic matter $\delta^{13}C < -30$ per mil $\rightarrow C_3$ plants $\delta^{13}C > -25$ per mil $\rightarrow C_4$ plants
3 *n*-Alkandiol unaltered C_{28} & C_{30} 1,14-diol		Diatom productivity
n-Alkandiol unaltered C_{30} & C_{32} 1,15-diol		Yellow-green algae productivity
4 *n*-Fatty acid (unaltered or saturated) *n*-Alkane		Primary production $\geq C_{25}$: land plant organic matter
5 Alkenone (unaltered) *n*-Alkanes ($C_{37} - C_{39}$)		$U_{37}^{K'} \rightarrow$ temperature proxy; $\delta^{13}C \rightarrow$ proxy for paleo-pCO_2
Branched chains		
6 Mono-, di- and trimethylalkanes (unaltered)		Cyanobacteria productivity
7 *iso*- & *anteiso*-Fatty acids (unaltered) *iso*-, *anteiso*-Alkanes		Bacterial activity
Acyclic isoprenoids		
8 Phytol (unaltered)		Productivity
Phytane Pristane		Pristane/phytane $\rightarrow$ oxicity indicator
9 Pristane		Maturity parameter
10 Lycopane Acyclic isoprenoids with tail-to tail-link ($\leq C_{40}$)		Unspecific phytoplankton indicator
11 Crocetane		^{13}C-depleted $\rightarrow$ anaerobic methanotrophic archaea
12 Pentamethylicosane (PMI) Crocetane		Methanogenic and methanotrophic archaea (^{13}C-depleted)
13 Highly branched isoprenoids (HBIs, saturated)		Diatoms (*Rhizosolenia*)
14 Botryococcane (rare)		*Botryococcus braunii*
Diterpenoids		
15 Fichtelite		Conifers resinous land plant material

(*continued*)

Biological molecule	Structure (example)	Known or postulated source
Pentacyclic triterpenoids		
16 Bacteriohopanepolyols and related hopanoids		Bacteria (mostly aerobic, heterotrophic)
17 2β-Methylhopanepolyols and derivatives		Cyanobacteria, prochlorophytes
18 3β-Methylhopanepolyols and derivatives		Many aerobic methanotrophic bacteria & a few microaerophilic proteobacteria
19 Diploptene Diplopterol		Methanotrophic bacteria (^{13}C-depleted)
20 β-Amyrin		Flowering plants
21 Lupeol, betulin and derivatives		Flowering plants
22 Tetrahymanol		Protozoa, purple non-sulfur bacteria and related proteobacteria, ciliates
Steroids		
23 Cholesterol		Eucaryotes
Coprostanol		Mammalian feces
24 24-Methylcholesterol and double bond homologues		Mainly diatoms, red lineage of algae
25 24-Ethylcholesterol and double bond homologues		Mainly green algae & higher land plants
26 24-*n*-Propylsteroids		Some marine algae (Pelagophytes)
27 24-*iso*-Propylsteroids		Sponges
28 4-Methylcholesterol		Methylotrophic bacteria & a few eukaryotes

Fossil molecule	Structure (example)	Biogeochemical application
Pentacyclic triterpenoids		
16 Extended hopanes (C_{27}, C_{29} – C_{35}, rare → C_{35+})		Bacteria (mostly aerobic) epimer ratios used as maturity indicator
17 2α-Methylhopanes (C_{32} – C_{36})		Cyanobacteria productivity
18 3β-Methyl extended hopanes		Methanotrophic bacteria (^{13}C-depleted)
19 17α-Hopane		Methanotrophic bacteria (^{13}C-depleted)
20 18β- and 18α-Oleanane		Flowering plants In oils indicates early Cretaceous or younger source rock
21 18β- and 18α-Oleanane, 24,28-*dinor*-lupanes		Flowering plants, specific indicators for oil-source rock correlation
22 Gammacerane		Protozoa Indicator of stratified, low-oxygen, often saline water
Steroids		
23 Cholestane Diasterane Mono- and triaromatic steroid (all several isomers)		Carbon number distributions and ratios (C_{27}, C_{28}, C_{29}) → general source indicator; stereoisomer ratios → maturity indicators
24 24-Methylcholestanes (several isomers)		As for cholestane
25 24-Ethylcholestanes (several isomers)		As for cholestane, high relative abundance → land plant organic matter, freshwater environment or green algae
26 24-*n*-Propylcholestanes		As for cholestane; marine algae
27 24-*iso*-Propylcholestanes		As for cholestane; sponges
28 4-Methylcholestanes		Methylotrophic bacteria

(continued)

Biological molecule	Structure (example)	Known or postulated source
29 Dinosterol		Dinoflagellates (and their phylogenetic precursors?)
Chlorophylls		
30 Chlorophyll *a*		Photosynthetic plants and cyanobacteria
31 Bacteriochlorophyll *e*		Green sulfur bacteria
Carotenoids		
32 Isorenieratene		Green sulfur bacteria
33 Chlorobactene		Green sulfur bacteria
34 Okenone		Purple sulfur bacteria
Complex membrane lipids		
35 Dihexadecanoyl glycerol diester		Eukaryotes
36 Di(10-methylhexadecanoyl) glycerol diester		Sulfate-reducing bacteria
37 Diphytanyl glycerol diether (archaeol)		Halophilic archaea, methanogenic archaea, methanotrophic archaea
38 Dibiphytanyl glycerol tetraether (caldarchaeol) Ring-containing tetraethers		Methanogenic archaea, crenarchaea
Amino acids		
39 Various amino acids (free, in polypeptides or proteins)		All organisms

Fossil molecule	Structure (examples)	Biogeochemical application
29 Dinosteranes (several isomers)		Dinoflagellates
Chlorophylls		
30 Various transformation products, including chlorins and ending with metalloporphyrins		Photosynthetic eukaryotes, cyanobacteria, anoxic sediments
Phytane, maleimides		
31 Various transformation products, including chlorins and ending with metalloporphyrins		Photosynthetic bacteria, anoxic photic zone
Farnesane, maleimides		
Carotenoids		
32 Isorenieratene (unaltered) Isorenieratane		Photic zone anoxia
33 Chlorobactene (unaltered) Chlorobactane		Photic zone anoxia
34 Okenone (unaltered) Substituted alkyl benzene		Purple sulphur bacteria (unaltered); otherwise unspecific
Complex membrane lipids		
35 Hexadecanoic acid		Ubiquitous
36 10-Methylhexadecanoic acid		Sulfate-reducing bacteria
37 Archaeol (unaltered) Phytane		Archaea Methanotrophs (^{13}C-depleted)
38 Caldarchaeol (unaltered) Biphytane		Archaea Methanotrophs (^{13}C-depleted) TEX$_{86}$ (sea-surface temperature proxy)
Amino acids		
39 Various amino acids		Age determination (racemization)

Glossary

Cross-references to other glossary entries are shown in italic.

Acid   See *proton, carboxylic acid.*

Acritarch   Microfossil of uncertain biological origin, composed of insoluble, acid-resistant organic material in the form of a single cell or cyst.

Aerobic   Metabolic processes (e.g., respiration) that make use of O_2 and take place under *oxic* environmental conditions. (See *anaerobic.*)

Alcohol   A molecule containing one or more *hydroxyl groups* (–OH). Examples include methanol, ethanol, phytol, cholesterol, and bacteriohopanetetrol.

Alkane   Synonym for saturated hydrocarbon, that is, a chemical compound that contains only carbon and hydrogen, and has no *double bonds* or aromatic units. Alkanes can be straight chains of CH_2 groups (*n*-alkanes), contain one or more rings (*cyclic hydrocarbons*), or they can have one or more *alkyl* side groups (branched alkanes).

Alkenone   A common name for dialkyl *ketones* with at least one *double bond*. In geochemistry, the term usually refers specifically to a group of long-chain unsaturated ketones with 37–39 carbon atoms and 2–4 *double bonds*, which are biosynthesized by *coccolithophore* algae and used to determine past *sea surface temperatures* via the U_{37}^K index.

Alkyl   An *alkane* unit that is attached to any sort of molecule; the alkyl group is composed only of carbon and hydrogen, but the rest of the molecule can have any elemental or structural composition.

α   See *stereoisomers.*

Amino acid   A chemical compound with a carboxyl (–COOH) and an amino (–NH$_2$) group. Organisms make α-amino acids, where the amino and carboxyl groups are attached to the same carbon atom. Peptides and proteins are composed of many of these linked together by amide bonds (–CO–NH–). All but one of the amino acids used in proteins are *chiral* molecules, and they are present almost exclusively in the L configuration (see *stereoisomers*).

Anaerobic   A metabolic process that does not make use of oxygen and takes place in an *anoxic* environment.

Anammox   An acronym for *an*aerobic *amm*onium *ox*idation, a microbial process wherein ammonia (NH$_3$) reacts with nitrite (NO$_2^-$) to produce molecular nitrogen (N$_2$).

ANME   *Anaerobic* methanotrophic *archaea*. These microorganisms participate in the *anaerobic* oxidation of methane and incorporate carbon from the methane. At least three phylogenetically distinct groups of ANME have been identified; these appear to be related to several genera of *methanogens*. ANME are often found living in close association with *sulfate-reducing bacteria*, sometimes in consortia where methane is oxidized and sulfate reduced. Distinctive lipids include those listed for methanogens, but can be distinguished by their ^{13}C-depleted isotope signatures. (See *methanotroph*.)

Anoxic, anoxia   The absence of O$_2$ in a body of water or in sediments. Anoxic conditions develop in lakes, ocean basins, or *sediments* when large amounts of *organic matter* are introduced, because *aerobic* microorganisms consume the organic matter and use up all the oxygen, which is not replenished quickly enough. In natural waters, oxygen can be incorporated from the air by waves near the surface, or it can be replenished by deep currents. In sediments, the pore size and permeability determine whether oxygen can be replenished from overlying *oxic* water. Organic matter is better preserved under anoxic than oxic conditions. *Cretaceous black shales* and Mediterranean *sapropels* were deposited under anoxic conditions.

Anoxygenic phototroph   Photosynthetic *bacteria* that do not use water as an electron donor or produce molecular oxygen. Typically, they use sulfide or other reduced sulfur compounds, molecular hydrogen, or a variety of small organic molecules as electron donors. There are four known *phylogenetic* groups of anoxygenic phototrophs: the *green sulfur bacteria*, filamentous *anoxygenic phototrophs* (Chloroflexus), *purple sulfur bacteria*, and heliobacteria.

Archaea   Comprises one of the three *domains* in the *phylogenetic* tree of life. Originally called Archaebacteria, it was identified as a separate division of *prokaryotes* by Carl Woese and George E. Fox in 1977. One of their distinguishing features is that their *cell membranes* are composed of *ether*-linked *lipids* based on the *phytane* structure (*isoprenoid glycerol ethers*). Initially believed to occur only in extreme environments, archaea are now known to be widely distributed, in all kinds of habitats. Those discovered to date fall into three main phylogenetic groups, the Korarchaea, *Crenarchaea*, and *Euryarchaea*.

Aromatic ring   A hexagonal, six-carbon ring structure with the equivalent of three conjugated *double bonds*; in reality, the electrons are not localized in three separate bonds but shared among the six carbon atoms. The term "aromatic" stems from the fact that many naturally occurring compounds containing aromatic rings have distinctive scents. Polycyclic aromatic *hydrocarbons* (PAHs) have two or more fused aromatic rings and can be quite large; present in *petroleum* and formed by combustion of *organic matter*, they are common air and water pollutants.

Asphalt   A semisolid or solid form of *bitumen* with no precise chemical definition, often found at natural oil seeps. In the United States, asphalt is the residue from *petroleum* distillation used to pave roads. Sometimes referred to as "bitumen" in ancient texts and by archaeologists.

Asphaltenes   A high-*molecular-weight* polar fraction of *bitumen*. In chemical composition and polarity, asphaltenes are intermediate between *resins* and *kerogen*.

Autotroph   An organism that produces complex organic compounds from simple inorganic molecules (e.g., CO_2 or HCO_3^-) and an external source of energy, such as sunlight or chemical reactions of inorganic compounds. An autotroph is a primary producer at the base of the food chain. Photosynthetic organisms are the most widespread and productive autotrophs, but chemosynthetic organisms are important in some ecosystems.

Average chain length (ACL)   Used to compare *carbon number* distributions of unbranched compounds. The ACL of odd-carbon-number *n*-alkanes from 27 to 33 carbon atoms is given by $ACL_{27-33} = (27[C_{27}] + 29[C_{29}] + 31[C_{31}] + 33[C_{33}]) / ([C_{27}] + [C_{29}] + [C_{31}] + [C_{33}])$, where $[C_x]$ is the amount of alkane with x number of carbon atoms.

Bacteria   One of the three *domains* in the *phylogenetic* tree of life (sometimes called Eubacteria). Until *archaea* were recognized as a separate phylogenetic division, the term "bacteria" was synonymous with *prokaryote* and included both eubacteria and archaebacteria, so there is still some confusion in the literature.

Benthic   Refers to the *sediment* surface and water at the bottom of a lake or ocean, used to describe organisms that live there.

β   See *stereoisomers*.

Bicarbonate   Formed when atmospheric carbon dioxide is dissolved in water. Reaction with water yields an *equilibrium mixture* of dissolved carbon dioxide (CO_2), bicarbonate (HCO_3^-), and carbonate (CO_3^{2-}). Bicarbonate is the preferred carbon source for most aquatic photosynthetic organisms.

Biomarker   An organic *compound* in natural waters, *sediments*, soils, fossils, *crude oils*, or coal that can be unambiguously linked to specific precursor molecules made by living organisms (also: biological marker, molecular fossil, fossil molecule, geochemical fossil).

Biomass   The total weight of all living biological material in a given area.

Biphytanyl   A molecular component comprised of two phytanyl units linked together head-to-head via a covalent carbon–carbon bond. Biphytanyl groups are prominent in the *cell membrane lipids* of *archaea*, where their apparent function is to add rigidity across the membrane. In some archaea, there are cyclopentyl or cyclohexyl rings within the biphytanyl units.

Bitumen   The part of the *organic matter* in geological samples that dissolves in organic solvents; also called extractable or soluble organic matter.

Black shale   A *sedimentary rock* that contains exceptionally large amounts of *organic matter*, which makes it dark colored or black. (See *oil shale, sapropel*.)

C₃ and C₄   Two different metabolic pathways for carbon fixation during *photosynthesis*. Most broad-leafed and temperate land plants use the C_3 pathway, whereas most tropical grasses use the C_4 pathway. The two pathways produce different amounts of carbon *isotope* *fractionation*, such that *biomarkers* of the two plant groups can be distinguished by their $\delta^{13}C$ values.

Cambrian The *period* in the earth's history (542–488 million years ago) during which the first calcareous algae and exoskeletal invertebrates (e.g., trilobites, mollusks, ostracods, *foraminifera*) evolved. The radiation of animal phyla during this time is called the "Cambrian explosion."

Carbon number In this book, refers to the total number of carbon atoms in a molecule and indicated by a subscript (e.g., C_{22} indicates a 22-carbon compound). In many publications carbon number is also used to designate a specific position in a molecule as determined by the *nomenclature* of organic molecules, which unequivocally numbers each carbon atom; in this case no subscript is used (e.g., C-22 refers to the 22nd carbon atom in a molecule's numbering scheme).

Carbon Preference Index (CPI) A numerical means for determining the odd- or even-carbon-number predominance in a *homologous series* of *n*-alkanes, *n*-alcohols, or *n*-carboxylic acids based on the ratio of the amounts of odd-carbon-number to even-carbon-number compounds. Amounts are obtained by measuring compound peak heights or areas in the gas *chromatograms*. Initially defined for the long-chain C_{24} to C_{34} *n*-alkanes, it can be similarly used for other *carbon number* ranges.

Carbonyl The *functional group* (C=O) that characterizes the *ketones*.

Carboxylic acid A chemical *compound* that is characterized by the –COOH *functional group*. Carboxylic acids that have a straight or monomethyl-branched chain, with or without unsaturation, are also called fatty acids (see *proton*).

Catagenesis The thermocatalytic transformation, or maturation, of *organic matter* in *sedimentary rocks* that occurs at great depths (usually >2 km) and high temperatures (>50°C). Results in "cracking" and the generation of smaller molecules, releasing liquid and gaseous *hydrocarbons* from the *kerogen* into the *bitumen*. Under some geological conditions, these form accumulations of *crude oil* and natural gas.

Cell membrane (plasma membrane) The envelope around a cell that encloses the cytoplasm and forms a semipermeable barrier between it and the environment. In plant cells and *prokaryotes*, it is usually surrounded by a cell wall, whereas in animal cells it is the only barrier between the cytoplasm and the outside of the cell. It is composed of glycerol-based phospholipids in which a variety of protein molecules are embedded and act as channels and pumps. *Sterols*, bacteriohopanoids, or tetraethers may act to strengthen and stiffen the otherwise quite fluid membrane in *eukaryotes*, *bacteria*, and *archaea*, respectively.

Chert A white, dense, hard *sedimentary rock* composed of microcrystalline quartz (SiO_2) from the chemical or biological precipitation of silica. Chemical impurities may give rise to colored analogues.

Chiral Describes a molecule that can exist in two nonsuperimposable mirror image structures, due to the presence of one or more asymmetric carbon atoms, that is, carbon atoms with four different atoms or groups of atoms bound to them. In solution, chiral compounds rotate plane-polarized light in either clockwise or anticlockwise direction (*optical activity/rotation*). The term is also used to denote the asymmetric centers, or atoms, in a molecule.

Chromatogram The visual result of chromatographic analysis, now also used for the graphic representation of the result, where signal intensity is on the y-axis, and time on the x-axis. For example, a gas chromatogram may show the response of a flame ionization detector as a function of time. The term can also refer to the result of mass spectrometric analysis, where a mass chromatogram may show ion intensity versus time for the sum of a range of ions or for specific selected ions. (See *mass spectrometry*.)

Ciliates A group of mobile, *heterotrophic*, unicellular *eukaryotes*, typified by undulating hairlike extensions that are used for swimming, crawling, attachment, feeding, and sensation. They feed on detritus and *bacteria* and are common in natural waters. Some ciliates biosynthesize tetrahymanol, a pentacyclic *triterpenoid* that is diagenetically transformed into the *saturated hydrocarbon* gammacerane.

Coccolithophores A group of single-celled marine *planktonic* algae that secrete hubcap-shaped calcium carbonate (calcite) shields called coccoliths. Members of the Haptophyta phylum, they are widespread and abundant in the contemporary ocean. The most prominent contemporary species of coccolithophore, *Emiliania huxleyi*, produces the long-chain *alkenones* used by geologists and paleoceanographers to determine past *sea surface temperatures* (see U^K_{37}). Coccoliths comprise a major component of carbonate marine *sediments* deposited in the past 150 million years.

Column chromatography A technique for separating mixtures of organic *compounds* into pure compounds or simpler mixtures that involves dissolving them in a mobile phase and passing it through a column filled with some sort of stationary phase. Compounds in the mixture have different affinities for the mobile and stationary phases. They are adsorbed onto the stationary phase and then, as the mobile phase flows through the column, separated and sequentially released and collected. A compound's residence time in the column is determined by the particular combination of materials used for the stationary and mobile phases, and by its boiling point, molecular size, molecular shape, and polarity. The unembellished term column chromatography refers to use of a vertically positioned glass column filled with silica gel or aluminum oxide, as stationary phase; an organic solvent dripped into the top of the column, as mobile phase; slow percolation through the column under the force of gravity; and separated groups of compounds collected as they drip out the bottom. In a more general sense, *gas chromatography* and *high-performance liquid chromatography* are forms of column chromatography.

Compound General term for a material with a specific chemical structure. Compound, substance, and molecule are often used interchangeably.

Continental shelf The gently sloping seaward extension of the continent, generally covered with water during interglacial *periods* like the present, and exposed during glacial periods. Water depth on the continental shelf is less than 200 m, and the *sediments* typically contain a relatively large amount of organic and mineral matter that derives from the land. Beyond the shelf, the edge of the continent drops off more steeply and water depth increases dramatically.

Copepod A large group of small crustaceans found in the sea and most freshwater habitats. There are both *planktonic* and *benthic* species. In the ocean, copepods are the most abundant *zooplankton* and form a crucial link in the food chain between the primary producers (*phytoplankton*) and fish.

Covalent bond A chemical bond between atoms in which two atoms share a pair of electrons. Covalent bonds are what holds organic *compounds* together. They are much stronger than the ionic bonds in inorganic salts, where the anion hosts the electron pair and crystals are formed by strong electrostatic interaction between anions and cations. Other weaker types of bonds include hydrogen bonds and van der Waals interactions.

Crenarchaea A taxonomic grouping of *Archaea*. Included in the Crenarchaea are hyper-thermophiles, which can live in extremely hot environments (over 100°C); *autotrophic* organisms that are widespread in both the surface and deep waters of the ocean; and *het-erotrophic* organisms that live in *sediments* and soils. Their *cell membranes* are composed of *isoprenoid glycerol ethers* such as caldarchaeol, crenarchaeol, and ring-containing *biphyta-nyl* tetraethers. (See *TEX$_{86}$*.)

Cretaceous A *period* in the Mesozoic era of the geologic timescale (144 to 65 million years ago), characterized by a warm "greenhouse" climate and by the final breakup of the earlier supercontinent to form the contemporary continents and the Atlantic Ocean (see *seafloor spreading*), and known as the last period of the dinosaurs' reign. (See *oceanic anoxic event*.)

Crude oil A naturally occurring black or brownish liquid formed from *kerogen* in deeply buried *sediments* that have been subjected to high temperatures and pressures. Often used interchangeably with the more general term *petroleum*, which, strictly speaking, includes condensate and gas, as well as crude oil.

Culture A population of *bacteria* or other *microbes* grown in an artificial medium in the laboratory, or the process of maintaining them. Cultures may be pure, that is, restricted to one specific organism, or they may be mixed, consisting of more than one species or strain. (See *isolate*.)

Cyanobacteria A phylum of unicellular and filamentous *bacteria* that includes all oxygen-producing photosynthetic bacteria. They are important primary producers in the ocean, especially prominent in open-ocean, low-nutrient areas; many groups are able to fix molecular nitrogen, allowing them to thrive when conditions are unfavorable for other pho-tosynthetic organisms, which can only obtain their nitrogen from nitrate and ammonium. Unlike other *prokaryotes*, cyanobacteria often contain fatty acids with two or more *double bonds*. Unlike other photosynthetic bacteria, their main form of chlorophyll is the same as that in *green algae* and higher plants (chlorophyll *a*). Other distinctive *lipids* are mono-, di-, and trimethyl-branched *alkanes*, bacteriohopanoids, and, more specifically, 2β-methyl bacteriohopanoids.

Cyclic hydrocarbon A chemical *compound* made of carbon and hydrogen and having one or more ring structures.

D amino acid See *stereoisomers, amino acid*.

Dating techniques Radiometric dating techniques provide the absolute age of a fossil or rock based on the observed abundance of naturally occurring radioactive *isotopes* and their known decay rates. Different isotopes are useful for different timescales. Radioactive carbon dioxide ($^{14}CO_2$) is present in low concentrations in the atmosphere; it is incorporated into *biomass* by *photosynthesis* organisms and then passed on to other organisms in the food chain. With the death of an organism, no new carbon is introduced and radioactive decay of the ^{14}C proceeds, such that the age of a fossil (or of biogenic *sediments*) can be determined from the relative abundance of ^{14}C remaining in the fossil, the initial concentration in the atmosphere, and the decay rate. Radiocarbon dating is accurate up to about 40,000 years, and can be used to date both *organic matter* and mineral precipitates. Other radioactive isotopes that are used to date mineral precipitates in sediments and rocks over different timescales, include lead (^{210}Pb or ^{207}Pb), thorium (^{230}Th), various isotopes of uranium (^{234}U, ^{235}U, and ^{238}U), potassium (^{40}K), and rubidium (^{87}Rb). These cover timescales of 100 years to 48.8 billion years. The relative amounts of L and D *amino acids* in fossils and sediments can also be used to determine absolute ages up to 100,000 years old, based on the rate of *isomerization* between the L form initially made by the organism and the D form. (See *stereoisomers*.) Biostratigraphy uses the distributions of microfossils within rock strata to match strata that formed at the same time in different regions, that is, it correlates the strata based on the evolution of the organisms. Fossils used typically comprise pollen, *foraminifera, dinoflagellates*, radiolaria, and so forth. By comparing the relative ages of strata in one region with those in another where radiometric ages can be determined, an absolute age can be assigned.

δ¹³C   A measure of the relative amounts of the two stable *isotopes* of carbon in a substance. δ values are generally defined as δ *(per mil)* = $(R_{sample}/R_{standard} - 1)1,000$, where R is the ratio of the rare isotope to the abundant one, in this case $^{13}C/^{12}C$. Differences in isotope ratios are small, measured in parts per thousand (per mil or ‰). The original reference standard was a sample of fossils from an extinct organism called a belemnite collected decades ago near the Pee Dee River in South Carolina. This Pee Dee belemnite (PDB) has long since been used up and replaced by a new standard defined by the International Atomic Energy Agency in Vienna, the so-called Vienna PDB. The δ¹³C of carbonate mineral deposits reflect the δ¹³C of the *bicarbonate* in the water where they formed and vary depending on where and when the deposits formed. Because the ^{12}C isotope in carbon dioxide is preferentially incorporated during *photosynthesis*, biological *organic matter* is always isotopically lighter than carbonates, with distinctly negative δ¹³C values. The δ¹³C values of an organism's biomass depends on the type of organism and on the δ¹³C of its specific carbon source. (See *fractionation*, C_3 and C_4.)

δ¹⁸O   A measure of the relative amounts of the two most abundant stable *isotopes* of oxygen in a substance, determined as described for *δ¹³C* and using the same VPDB reference standard or water defined as "standard mean ocean water" (SMOW) and its updated version, Vienna standard mean ocean water (VSMOW). δ¹⁸O values of carbonate *foraminifera* shells can be used as a climate *proxy*, because the ^{18}O composition of the oxygen atoms in the *bicarbonate* and carbonate dissolved in the water reflects the ^{18}O composition of the oxygen atoms in the water, and this varies with climate. The formation of ice sheets on the continents leaves the ocean water enriched in ^{18}O during glacial times, on the one hand, and, on the other, the foraminifera shells have a preference for ^{18}O that increases at cold temperatures. Both effects are recorded in the foraminifera shells, which show the oscillating ice ages and warm periods of the past few million years.

Derivative   A *compound* that has the same basic carbon structure as its "parent" compound, but with different groups or *functional groups* attached.

Diagenesis   The transformation of *organic matter* in *sediments* by low-temperature chemical reactions and microbial activity. At temperatures above 50°C, for instance, after burial at greater depth, diagenesis is followed by *catagenesis*. (Inorganic geochemists use the term diagenesis differently.)

Diasterane   A product of *diagenesis* of *sterols*, produced by structural rearrangement of the intermediate sterenes, specifically by migration of the *methyl groups* from the upper to the lower face of the ring system (from C-10 and C-13 to C-5 and C-14, respectively). The reaction is catalyzed by acidic *sediment* components like clay minerals, so diasteranes are generally absent from pure carbonates and *evaporites*. Diasteranes are not formed by biological processes. Also called "rearranged *steranes*."

Diatoms   A large, diverse, widespread group of mostly unicellular green alga. Diatoms can be found in the surface waters of seas and lakes, or growing on the surface of *sediments* or soils. The cell's ornate skeletal structure (frustule) is made of silica and exhibits a large variety of spectacular forms. Diatom fossils are used in biostratigraphy (see *dating techniques*). Distinctive *lipids* include highly branched *isoprenoids* (HBIs), long-chain alkanediols, brassicasterol (24-methylcholesta-5,22-dien-3β-ol), and fucosterol (24-methylcholesta-5,24(28)-dien-3β-ol).

Dinoflagellates   A large, diverse group of unicellular *eukaryotes* that includes the second largest group of algae in contemporary oceans, as well as many freshwater species and nonphotosynthetic forms. *Dinoflagellate* cells are poorly preserved as fossils, but many species have a life cycle that includes a "resting phase" when they form cysts that have a more resistant cell wall and are well preserved in ancient *sediments*. A distinctive *lipid* common to most members of the group is dinosterol.

Domain   One of three branches in the tree of life: *Bacteria*, *Archaea*, or *Eukarya*.

Double bond   A *covalent bond* between atoms in which two atoms share not one but two sets of electrons.

DSDP   Deep Sea Drilling Project, a U.S.-based international project (1966–1984) designed to answer fundamental questions about the earth by drilling holes into the ocean floor and investigating the recovered *sediments*, rocks, and fluids. *Plate tectonics*, climate history, paleoceanography, and marine geology were among the main study themes. The first research ship was a converted oil drill ship called the *Glomar Challenger*. The DSDP expanded into the Ocean Drilling Project (ODP) and then, in 2003, the Integrated Ocean Drilling Project (IODP).

Eocene   A *period* in the Cenozoic era of the geological timescale (55–34 million years ago). The modern mammal fauna developed during this time of mostly warm climate. The Green River *oil shale* (Utah, Colorado, Wyoming) and the Messel oil shale (Germany) are two important *organic-matter*–rich lake deposits of Eocene age.

Equilibrium   A state in a reversible chemical reaction when the rate of the forward reaction equals that of the back reaction, so that the amounts of reactants and products cease to change with time even though both reactions are ongoing. The equilibrium ratio of products to reactants varies with temperature and depends on their thermodynamic stabilities.

Equilibrium mixture   The composition of a mixture of chemical *compounds* at the *equilibrium* state.

Ester   A chemical *compound* formed by the reaction of a *carboxylic acid* and an *alcohol*, with elimination of a molecule of water. Ester bonds (R–CO–O–R') are relatively easily cleaved (hydrolyzed) by the addition of water in the presence of an acid or base catalyst.

Ether   A chemical *compound* characterized by an oxygen atom bound to two carbon atoms, or R–O–R' where R and R' are *alkyl* or other groups. Ethers are less easily cleaved during *diagenesis* than *esters*.

Ethyl group   A two-carbon *alkyl* group in chemical structures (–CH$_2$CH$_3$).

Eukarya   One of the three *domains* in the *phylogenetic* tree of life. (See *eukaryote.*)

Eukaryote   An organism with a complex cell structure and a membrane-bound cell nucleus, which contains the genetic material. Most eukaryotes employ *aerobic* respiration. Animals, plants, and a number of unclassified unicellular organisms (protists) are Eukaryotes. Their membrane *lipids* contain glycerol diesters of straight-chain C_{16} or C_{18} fatty acids and *sterols* as rigidifiers. (See *prokaryote.*)

Euryarchaea   A taxonomic grouping within the *domain Archaea* that comprises extremely *halophilic archaea, methanogens,* and *anaerobic* methane-oxidizing archaea (*ANME*).

Evaporite   A *sedimentary rock* formed by the evaporation of water and precipitation of salt in arid marine areas. Salts are deposited according to their solubilities in water, with calcite (calcium carbonate) and gypsum (calcium sulfate) precipitated first and halite (sodium chloride) relatively late.

Extraction, extract   The process of removing *organic matter* from a geological or biological sample by dissolving it in an organic solvent, and the solution that results.

Fatty acid   See *carboxylic acid.*

Fermentation   An *anaerobic* energy-yielding process employed by a wide variety of *bacteria* and some *fungi,* wherein organic molecules are split into two smaller molecules, one oxidized and one reduced relative to the starting material. In natural waters and *sediments,* these products, usually small organic acids and *alcohols,* are used as energy and carbon sources by other bacteria. A well-known example of fermentation is the conversion of the sugar hexose by yeast (a fungus) to produce ethanol (*reduction*) and carbon dioxide (*oxidation*).

Flowering plants   The most widespread group of contemporary land plants, characterized, among other things, by flowers and by seeds that are encapsulated in an ovary, which often ripens into a fruit. The other main group of extant seed plants, the gymnosperms, have naked seeds exposed on the scale of a cone. Gymnosperms, which include conifers like pine or spruce trees, developed in the middle Paleozoic era and were the dominant land plants until the advent of flowering plants in the *Cretaceous.* Distinctive *lipids* found in both groups of land plants include long-chain *n*-alkanes, *alcohols,* and fatty acids, and 24-ethylcholesterol. Oleanoid-type pentacyclic *triterpenoids* are specific to flowering plants.

Foraminifera (forams)   A large group of unicellular *eukaryotes,* comprising *planktonic* and *benthic* species. Generally *heterotrophic,* though some species develop a symbiotic relationship with unicellular algae that they have ingested. Most extant forms are marine. Forams are distinguished by their carbonate shells, or tests, which make up a significant component of carbonate marine *sediments.*

Fractionation   1. *Isotope* fractionation: the change of isotopic composition during physical, chemical, and biochemical processes, for example, the dissolution of atmospheric carbon dioxide in water, or the photosynthetic transformation of carbon dioxide into organic molecules. 2. The separation of complex mixtures of organic *compounds* such as those in a rock *extract* into less complex mixtures, usually by *column chromatography.*

Functional group   A characteristic group of atoms in a molecule that determines its chemical reactions and associates it with a certain type of chemical *compound.* Some examples are: the *hydroxyl group* (–OH) in *alcohols,* the *carbonyl* group (C=O) in *ketones,* the carboxyl group (–COOH) in *carboxylic acids,* or the amino group ($-NH_2$) in amines.

Fungi   A group of *heterotrophic,* generally multicellular *eukaryotic* organisms character-ized by a chitinous cell wall and comprising yeasts, molds, and mushrooms. Distinctive *lipids* include the 28-carbon *sterol* ergosterol (24-methylcholesta-5,7,22-trien-3β-ol).

Gas chromatograph (GC)   An instrument used for the separation of complex mixtures by *gas chromatography.* Key components are an injector, a gas chromatographic column, and a detector. Common types of injector for biomarker analyses include split/splitless and on-column; typical columns are 30- to 50-m-long fused silica capillaries (0.2–0.3 mm inter-nal diameter) lined with a stationary phase such as silicone oil; and useful detectors include flame ionization detectors and *mass spectrometers.*

Gas chromatography   A chromatographic method in which the mobile phase is a gas, such as helium or hydrogen, and the stationary phase is a microscopically thin layer of liquid or polymer, which is either coated onto a solid support packing material in the column or on the inside of a narrow capillary tube. As the gas sweeps the molecules in the sample mixture through the column, they are retarded by absorption in the stationary phase. Larger and more *polar* compounds have a longer residence time in the stationary phase and elute from the column later than smaller and less polar constituents. The temperature of the column can be maintained constant or, when compounds with a wide range of boiling points are analyzed, increased gradually, typically a few degrees a minute. Gas chromatography can be used to identify the components in a simple mixture by comparison with *standard* mix-tures, and a flame ionization detector can be used to quantify amounts of individual com-ponents. A *mass spectrometer* is usually required as a detector for identification of unknown compounds (see *column chromatography, GC-MS*).

GC-irm-MS   Gas chromatography–isotope ratio monitoring–mass spectrometry. A *gas chromatograph* combined with a combustion interface that burns the separated compounds to CO_2 and with a special *mass spectrometer* that can then measure the relative abundance of isotopes. Used to determine the isotopic composition of individual molecular species.

GC-MS   Abbreviation for the combination of *gas chromatograph* and *mass spectrometer.*

Genetic probe   As used in geochemical applications, a tool for identifying genes in natural waters and *sediments.* A fragment of DNA or RNA from the *phylogenetic* group of organisms under study is labeled with a fluorescent chemical group and mixed with the sample, where it binds with the complimentary gene sequence and allows it to be isolated and identified.

Geochemistry   Study of the transformation of chemical substances in the geosphere. Subdisciplines are organic geochemistry and inorganic geochemistry; cosmochemistry extends the scope of interest into space.

Geologic timescale   An ordered, internationally recognized sequence of time intervals that divide the history of Earth, composed of hierarchical units, from longest to shortest: eons, eras, *periods,* and epochs.

Geothermal gradient   The rate of increase of temperature with depth in the earth. Geothermal heat flow is a physically more precise term that takes into account the different thermal conductivities of the rock and *sediment* layers. The geothermal gradient can range from 30°C/km in areas of low geothermal heat flow to 80–100°C/km in geothermally hot areas.

Green algae   A taxonomic grouping (phylum Chlorophyta) of algae with green chloro-plasts that are similar to those found in land plants. Green algae have a wide variety of forms and growth habits: unicellular plankton that grow in lakes and oceans, colonial filaments in pond scum, and leaflike seaweeds that grow along rocky and sandy intertidal areas. Some can live on tree trunks and soil, and others are symbiotic, living together with corals or *fungi* (lichens). *Isomers* of 24-ethylcholesterol, often with an additional double bond in the C-7 position, are prominent among their *lipids*.

Green sulfur bacteria   *Anaerobic* photosynthetic *bacteria* that utilize sulfide or elemental sulfur to reduce CO_2 and, in many cases, can also assimilate simple organic compounds. Their pigments include bacteriochlorophylls c, d, and e and isorenieratene and chlorobactene. They are capable of growing at relatively low light intensities and can live at greater depth in the water than most other photosynthetic organisms. (See *anoxygenic phototrophs*.)

Halophiles   A group of *archaea* that inhabit highly saline environments such as natural salt lakes and artificial evaporation ponds from desalination plants, salt mines, and even the surfaces of preserved foods, such as salted meat or fish. Most halophiles require oxygen and obtain their energy and carbon from organic compounds.

Haptophyte   Single-celled algae that are members of the Haptophyta phylum. (See *coccolithophore*.)

Heterotroph   An organism that is dependent on organic compounds biosynthesized by other organisms for its carbon and, often, its energy. A heterotroph is a consumer in the food chain. (See *autotroph*.)

High-performance liquid chromatography (HPLC)   Similar in principle to *column chromatography*, except the column is a narrow stainless steel tube (10–30 cm × 1–5 mm) that is packed with very fine grains or beads of silica, and the liquid mobile phase is pumped through the column at high pressure. Samples can be injected automatically, and for analytical HPLC are typically in the microliter range; solvents can be mixed and varied gradually during the analysis; and various detectors can be employed (ultraviolet-visible, fluorescence, mass spectrometer) for identification of the separated compounds. HPLC is particularly useful for analysis of high-molecular-weight and/or very *polar* organic compounds, but can also be used for the gross separation of *bitumen* into group-type fractions (nonaromatic *hydrocarbons*, aromatic hydrocarbons, heterocompounds) or separation of aromatic hydrocarbons into mono- di-, and tri-aromatic hydrocarbons.

Homologous series, homologue   A series of chemically related compounds that differ by the number of CH_2 units in their carbon chains. True homologues have linear additions of CH_2 groups; pseudohomologues may include *alkyl* group branches.

Hopane   A *saturated hydrocarbon* in the *hopanoid* family.

Hopanoid   A general term for a group of pentacylic *triterpenoids* with a particular fused ring structure composed of four six-membered and one five-membered ring. A wide spectrum of side groups may be attached. Bacteriohopanetetrol and *derivatives* are prominent bacterial cell constituents, and serve, among other things, to provide rigidity to the *cell membrane*. Their biosynthesis involves addition of a C_5 sugar to the side chain on the five-membered ring of a C_{30} hopanoid.

HPLC   See *high-performance liquid chromatography (HPLC)*.

Hydrocarbon   A chemical compound consisting only of the elements carbon and hydrogen. The simplest hydrocarbon is methane (CH_4).

Hydrolysis Cleavage of a chemical bond by the addition of a water molecule, usually in the presence of a base or acid catalyst. Important examples are the hydrolysis of an *ester* to yield a *carboxylic acid* and an *alcohol*, or hydrolysis of an amide to yield a carboxylic acid and an amine.

Hydropyrolysis *Pyrolysis* under pressure and in the presence of hydrogen gas and a catalyst (sometimes called hydrogenolysis). It effectively cleaves bonds at lower temperatures than traditional *pyrolysis* and is used to determine the structures and *stereochemistry* of compounds bound in *kerogen*. The kerogen may be flushed with hydrogen at even lower temperatures prior to actual bond cleavage in order to rid it of contaminants and *bitumen*.

Hydrothermal vent Fissures in the seafloor from which geothermally heated water gushes out into the ocean, found mostly in the *tectonically* active areas along mid-ocean ridges (see *seafloor spreading*) or in the vicinity of volcanoes. The dissolved metal ions and reduced chemicals in the hot water (up to 400°C) issuing from the vents provide abundant energy for chemosynthetic *bacteria* and *archaea*, which form the base of vibrant deep sea ecosystems. Sulfide-oxidizing bacteria are particularly abundant, often living in *symbiosis* with clams, mussels, and tubeworms.

Hydrous pyrolysis *Pyrolysis* in the presence of water, usually done in an autoclave over extended periods. It is often applied to organic-matter–rich rocks to simulate *petroleum* formation in the laboratory, as it functions at lower temperatures than traditional pyrolysis.

Hydroxyl group The *functional group* (–OH) that characterizes *alcohols*.

Ice cores Obtained by drilling into the continental ice sheets of Greenland and Antarctica, ice cores extend thousands of meters into the ice sheets and comprise ice formed over 100,000 years or more. Tiny amounts of atmospheric gases were trapped in the pores of the ice when it crystallized and can be used to investigate how the earth's atmosphere and climate have changed during the past several ice age cycles.

Infrared spectroscopy An analytical technique used to investigate the structure of organic molecules. Infrared light is passed through a sample of a pure organic substance (gas, liquid, or solid) inducing vibrations in the molecules, and examination of the transmitted light reveals how much energy was absorbed at each wavelength. The resulting absorption spectrum reveals details about the compound's *functional groups* and overall structure.

Isolate (bacterial) The separation and enrichment of a single strain or selected mixture of bacteria in the laboratory; also refers to the strain isolated. (See *culture*.)

Isolation, isolate (chemical) The separation and purification of a single chemical *compound* from a mixture; also refers to the purified compound.

Isomerization The transformation of a molecule into its *isomer*. In this book, it refers to conversions between *stereoisomers*.

Isomers Chemical *compounds* with the same elemental composition but different arrangement of the atoms. (See *stereoisomers*.)

Isoprenoid A general term for chemical compounds that are composed of repeating five-carbon-atom isoprene (2-methylbuta-1,3-diene) units. Also called terpenes or terpenoids. Monoterpenes have two isoprene units (C_{10}), diterpenes have four (C_{20}), triterpenes have six (C_{30}), and so forth. Sesquiterpenes have three isoprene units (C_{15}), sesterterpenes have five (C_{25}), and polyterpenes consist of long chains of many isoprene units. The individual isoprene units are most commonly joined at opposite ends, or head-to-tail (regular isoprenoids), but may also be joined head-to-head or tail-to-tail. They can form open chains (acyclic isoprenoids) or ring compounds (cyclic terpenoids).

Isoprenoid glycerol ethers   A group of *lipid* structures that occur in the membranes of *archaea*. Includes diphytanyl glycerol diethers (e.g., archaeol) and dibiphytanyl diglycerol tetraethers (often referred to in the geochemical literature as GDGTs for glycerol diphytanyl glycerol tetraethers).

Isopropyl   An *alkyl* group wherein the three carbon atoms are arranged in a Y shape ($-CH(CH_3)_2$).

Isotopes   The isotopes of an element have the same number of *protons* but a different number of neutrons in their atomic nuclei. Thus they have different atomic weights, but the same chemical affinities. For example, carbon-12 (^{12}C) has 6 protons and 6 neutrons, whereas carbon-13 (^{13}C) has six protons and seven neutrons, and carbon-14 (^{14}C) has six protons and eight neutrons. (See $\delta^{13}C$, $\delta^{18}O$, *fractionation*.)

Jurassic   A *period* in the Mesozoic era of the geologic timescale (206–142 million years ago) when dinosaurs dominated and the first birds appeared.

Kerogen   The insoluble organic component in *sediments*. Kerogen is a macromolecular geochemical transformation product derived, over time, from decaying *biomass* in the sediments.

Ketone   A class of chemical *compounds* characterized by the *carbonyl* group.

Lipid   A chemical *compound* that is insoluble in water but soluble in organic solvents. Lipids are the main components of cell *membranes* and among the most plentiful compounds in *biomass*, together with carbohydrates and proteins. During *diagenesis* the decaying biomass becomes selectively enriched in lipids, which are likewise enriched in the kerogen, and become the most important starting material for the formation of *petroleum* during *catagenesis*.

Macromolecule   A very large molecule that may be composed of diverse structural subunits, unlike a polymer, which consists entirely of repeating units.

Mass extinction   The sudden extinction of a large number of types of organisms, as recorded by the fossil record in *sedimentary rocks*; in most cases, believed to be the result of rapid climate change associated with a catastrophic event such as a meteor impact or extreme volcanic activity.

Mass spectrometer (MS)   The instrument used to perform *mass spectrometry*.

Mass spectrometry   An analytical technique used to obtain information on molecular structures. There are many variations on the theme, but the most common application in organic geochemistry involves introduction of a pure substance or mixture, as a gas, into the high-vacuum system of a *mass spectrometer (MS)*, either directly or as the effluent of a *gas chromatograph*. Here, the molecules are ionized by a high-energy beam of electrons that breaks the molecule into charged ions, including a *molecular ion* and charged fragments. These are separated according to their mass-to-charge ratios (*m/z* where *z* is 1 except for *aromatic rings*, which can carry more than one charge), and their abundance is registered by a detector. The results are displayed as a mass spectrum (ion intensity versus *m/z*) or mass *chromatogram* (total, summed selected, or selected single ion intensities versus time). Two mass spectrometers can be linked in series (MS-MS) to obtain more detailed information. Here, the ions formed in the first mass spectrometer are bombarded with atoms from an inert gas, such as argon, to induce further fragmentation, and this is recorded in the second mass spectrometer.

Maturity   The degree to which the *organic matter* in a *sediment* or rock has been transformed under the influence of increasing burial depth and temperature. A certain level of maturity is required before *petroleum* generation by thermal cracking of *kerogen* can commence. Maturity is measured by various physical and chemical parameters, for example, vitrinite reflectance, *carbon preference index* of *n*-alkanes, and *biomarker* compound ratios.

Metabolite   The generally small molecules produced as intermediates or products of biochemical reactions in a cell (metabolism). A primary metabolite is directly involved in normal growth, development, and reproduction, and common to many types of organisms, for example, sugars, *amino acids*, and nucleic acids. A secondary metabolite has some less universal function that nevertheless may be important to a particular organism's long-term survival, for instance, pigments and antibiotics.

Methane hydrate   A solid crystalline material composed of methane molecules trapped inside a cage-like lattice of water molecules that are held together by hydrogen bonds. Methane hydrates can form from bacterial or thermogenic methane, are stable at particular combinations of low temperature and high pressure, and are often found beneath the sediment surface along the continental margins, in some marine basins, and beneath the permafrost in polar regions.

Methane seep, methane vent   A place where methane gas is seeping out of the *sediments*, often found along continental margins.

Methanogens   A group of *anaerobic archaea* that obtain their energy from the *reduction* of carbon, rather than oxygen, and produce methane, rather than CO_2, as a waste product. The most prominent known forms of methanogenesis employ CO_2 or small organic molecules as their electron acceptors, and H_2 or organic molecules, respectively, as electron donors. Both forms obtain their carbon from *organic matter*. Methanogens generally perform the last step in the decay of organic matter and are ubiquitous in sediments, soils, and natural waters where oxygen is absent.

Methanotrophic bacteria   *Aerobic bacteria* with a wide range of different morphologies (e.g., *Methylomonas*, *Methylobacter*, *Methylococcus*). They are widespread in soils, *sediments*, and natural waters. Common bacterial *lipids* such as iso-C_{17} and anteiso-C_{18} fatty acids and diploptene can serve as *biomarkers* when their $\delta^{13}C$ values also point to a methanotrophic source. More specific biomarkers for methanotrophic *bacteria* include 3β-methyl bacteriohopanoids and 4α-methyl *sterols*. They are the only *prokaryotes* known to biosynthesize sterols in quantity.

Methanotrophs   *Bacteria* or *archaea* (*ANME*) that obtain their energy and carbon from the *oxidation* of methane. The *lipids* of methanotrophs tend to be markedly depleted in carbon-13 due to their consumption of carbon-13–depleted methane.

Methyl group   A one-carbon *alkyl* branch in a chemical structure ($-CH_3$).

Methylene group   A structural unit in a molecule consisting of a CH_2 group either as part of a *saturated* chain ($-CH_2-$) or as an *unsaturated* group at the end of a chain ($=CH_2$).

Microbe, microorganism General terms for unicellular organisms, used to refer collectively to *prokaryotes* and/or unicellular *eukaryotes*.

Molecular ion The ion formed in *mass spectrometry* that is the starting molecule minus an electron (or, occasionally, plus an electron). Pseudomolecular ions are sometimes formed by the addition of a small charged particle (e.g., NH_4^+ or H^-) in the *mass spectrometer* (secondary ionization), something that is particularly common in *HPLC-MS*.

Molecular sieve A substance that contains tiny pores of uniform size (e.g., 5 Å) used to separate the straight-chain compounds from the branched and cyclic ones in a solution. Straight-chain compounds such as *n*-alkanes and *n*-alkanols are trapped in the pore spaces, and the bulkier branched and cyclic compounds remain in solution. Various methods are used to purge the straight-chain compounds from the sieve. Minerals such as zeolite, porous glass, charcoals, and other substances are used. Crystals of urea can be used in a similar fashion (urea adduction).

Molecular weight The sum of the atomic weights of the elemental constituents of a molecule. In general chemistry the weighted average of the *isotopes* of each element is used, but when dealing with *mass spectrometry* the atomic weights of specific isotopes are used.

MS See *mass spectrometer (MS)*.

n-Alkane See *alkane*.

Nomenclature A set of conventions used to unequivocally describe the structures of chemical *compounds*, as determined by the International Union of Pure and Applied Chemistry (IUPAC).

Oceanic anoxic event A period when the world's oceans were depleted in oxygen and *organic matter* preservation in the *sediments* was enhanced, resulting in widespread deposition of *black shales*. Several such events occurred during the *Cretaceous*.

ODP See *DSDP*.

Oil shale A *sedimentary rock* that is rich in fossil *organic matter* but has not been buried deeply enough to generate *petroleum*. Oil can be produced from an oil shale by artificial heating (retorting).

Optical activity/rotation The tendency of certain substances to bend, or rotate, a ray of monochromatic plane-polarized light to the right or left. Optical activity is a property of *chiral* organic molecules such as *amino acids, steroids,* and *triterpenoids*.

Organic geochemistry See *geochemistry*.

Organic matter Remnants of decayed *biomass*.

Oxic Describes an environment where molecular oxygen is present. (See *anoxic*.)

Oxidation A chemical reaction involving the addition of an oxygen atom to a molecule, the removal of two hydrogen atoms from a molecule (as in the formation of a double bond) or, more generally, the loss of one or more electrons from a molecule. In a chemical reaction system, oxidation is always coupled with *reduction*; that is, of two reaction partners, one is oxidized whereas the other one is reduced (redox reaction).

PAH Polycyclic aromatic hydrocarbon. (See *aromatic ring*.)

pCO₂   Partial pressure of carbon dioxide in the atmosphere. In paleoclimate research, past pCO_2 levels are measured by analysis of the gases trapped in old layers of ice in the Greenland and Antarctica ice sheets. (See *ice cores.*)

Period   See *geologic timescale.*

Per mil, ‰   Parts per thousand. Measurement scale for low numbers or concentrations like in *isotope* analysis.

Petroleum   A general term for the products of the thermocatalytic transformation of *kerogen* during *catagenesis.* The most important products are natural gas and *crude oil,* which accumulate in large quantities in natural reservoirs in the subsurface. (See *asphalt.*)

Photic zone   The upper layer of water in a lake or ocean where there is enough light for *photosynthesis.* The depth of the photic zone depends on the turbidity of the water and may range from a few meters to about 200 m in extremely clear water.

Photosynthesis   The process by which carbon dioxide or *bicarbonate* are converted to *biomass* in plants, algae, and some *bacteria,* using light as an energy source.

Phylogenetic   A way of classifying groups of organisms based on their evolutionary relationships, usually based on comparisons of their genetic information (phylogenetic relationship). A phylogenetic tree shows organisms that developed early in the history of life positioned near the roots and those that evolved later at higher positions of the tree. The main branches signify the three *domains* of life on Earth.

Phytane   The diterpenoid (C_{20} acyclic *isoprenoid*) formed from phytol during diagenesis, probably the most abundant hydrocarbon on Earth. Phytol is readily cleaved from chlorophyll *a,* which is common to all plants, algae, and many bacteria, and produced in great quantities upon death of the organisms. Another common source may be the *isoprenoid* membrane *lipids* of *archaea.*

Phytoplankton   Small or microscopic photosynthetic organisms (plants, algae, and bacteria) that live near the surface of lakes and oceans (e.g., *coccolithophores, diatoms, dinoflagellates*). They form the base of the marine food chain, its primary producers. (See *primary productivity.*)

Plankton, planktonic   Refers to organisms that float or drift in the water of a lake or ocean, generally small organisms that serve as food for fish and other larger animals. Planktonic describes anything relating to or characteristic of such organisms. (See *benthic.*)

Plate tectonics   A theory describing movements in the earth's crust that accounts for the changing positions of the continents and oceans over geologic time. The earth's lithosphere (crust plus rigid part of the mantle) is broken up into jigsaw-like pieces called *tectonic* plates, which move slowly about the planet. These plates move away from each other at mid-ocean ridges. Where oceanic plates converge, one plate slides beneath the other; where two continental plates collide, mountains form. Earthquakes, volcanic activity, mountain building, and oceanic trenches occur along plate boundaries.

PMI   Pentamethylicosane; a *saturated* acyclic C_{25} *isoprenoid hydrocarbon* found in *sediments,* believed to derive from the corresponding *unsaturated* hydrocarbons (pentamethylicosenes, with up to five double bonds) made by *methanogenic* and *methanotrophic archaea.*

pO₂   Partial pressure of oxygen in the atmosphere.

Polar   Describes molecules with a high affinity for water or solvents like methanol (which are likewise composed of polar molecules). Polarity stems from the presence of atoms with free electrons (e.g., oxygen, nitrogen, sulfur) in compounds like *alcohols* and *carboxylic acids* or, to a lesser extent, the delocalized electrons in aromatic *hydrocarbons.*

Polycyclic aromatic hydrocarbons (PAHs)   See *aromatic ring.*

Pore water   The water that fills the spaces between grains of sediments. Special techniques must be used to "squeeze" it out of the sediments for analysis.

Porphyrin   An organic *compound,* derived from the *diagenetic* transformation of chlorophyll, with four nitrogen-bearing pyrrol rings and a fully *aromatic ring* system that makes it very stable. Porphyrins can survive thermocatalytic cracking of *kerogen* and are found in deeply buried *sedimentary rocks* and *crude oil.*

ppm   Parts per million. Measurement scale for very low numbers or concentrations.

Primary productivity   The total amount of photosynthetic *biomass* production by *autotrophic* organisms. Often refers to the production of biomass by *phytoplankton* in the *photic zone* of the ocean.

Prokaryote   An organism with an uncompartmentalized cell structure, lacking a cell nucleus or any other membrane-enclosed structure. A general term used to refer to all non-eukaryotic life, that is, a collective term for *bacteria* and *archaea.* (See *eukaryote.*)

Protokerogen   A chemically ill-defined precursor of *kerogen* formed during early *diagenesis.* The term "humic matter" is often used in the same sense, to refer to structurally ill-characterized high-molecular-weight organic material in soils or in the ocean.

Proton   A subatomic particle with an electric charge of one positive fundamental unit. The nucleus of an atom is made up of protons and neutrons together. A proton is formed when an atom of hydrogen, whose nucleus contains one proton and no neutrons, loses an electron. A molecule that releases protons is an acid. Protons released from acids can catalyze many chemical reactions. (See *carboxylic acid.*)

Proxy   A parameter that makes use of variables that can be empirically determined to estimate some other variable that cannot be measured directly. Proxies are often used to assess and quantify conditions in the geological past. Examples include the U^{K}_{37} or TEX_{86} proxies for *sea surface temperatures,* and the $\delta^{18}O$ proxy for global ice mass and ocean temperatures.

Purple sulfur bacteria   *Anaerobic* photosynthetic *bacteria* that utilize sulfide or elemental sulfur to reduce CO_2 and, in many cases, can also assimilate simple organic compounds. They are commonly found in sulfur springs or in the *anoxic* zones of lakes, where their large blooms give the water a conspicuous purple color. Their pigments include okenone and bacteriochlorophylls *a* and *b.* (See *anoxygenic phototrophs.*)

Pyrolysis   A method for analyzing *kerogen* or other recalcitrant organic materials that involves breaking them up into their chemical constituents by heating in the absence of any chemical reactants, in particular molecular oxygen. Flash pyrolysis usually involves rapid heating in a stream of helium. (See *hydrous pyrolysis, hydropyrolysis.*)

R   See *stereoisomers.*

Racemic   Denotes a 1:1 mixture of the two mirror-image forms of a *chiral* molecule.

Reduction A chemical reaction involving the addition of hydrogen (e.g., to a double bond to form a single bond) or removal of oxygen, or, more generally, the acceptance of one or more electron(s). (See *oxidation*.)

Resin 1. A hydrocarbon excretion of many plants, particularly conifers. Usually a viscous liquid containing a variety of terpenes, often used in varnishes and adhesives. 2. The *polar* heterocompounds in *bitumen*.

Rubisco Ribulose-1,5-*bis*phosphate carboxylase/oxygenase. An enzyme that catalyzes the first step of the Calvin cycle of photosynthesis, incorporating CO_2. It is the first step in carbon fixation in C_3 plants.

S See *stereoisomers*.

Sapropel An *organic-matter–rich sedimentary rock*, often used synonymously with *black shale*, but generally used for younger, less consolidated sediments. Sapropels are found in the sedimentary records of the Mediterranean Sea and the Black Sea, as well as in many lake deposits.

Saturated Refers to chemical compounds with no double bonds or *aromatic rings*. (See *unsaturated*.)

Scanning electron microscope (SEM) An instrument that produces high-resolution images of very small items (e.g., fossils of unicellular organisms). Its focus has a great depth of field so that the images appear three-dimensional.

Seafloor spreading The movement of two *tectonic* plates away from each other as new basaltic ocean floor (oceanic crust) forms at mid-oceanic ridges such as the mid-Atlantic Ridge. (See *plate tectonics*.)

Sea surface temperature (SST) An important parameter in oceanography and climate research. Proxies used to estimate past sea surface temperatures include the U^K_{37} index, TEX_{86}, and, to some extent, $\delta^{18}O$.

Sediment The solid material that collects at the bottom of lakes, rivers, and oceans. It is generally composed of the carbonate or siliceous shells of aquatic organisms, clay, mineral rock fragments, and *organic matter*.

Sedimentary rock The lithified form of *sediment* produced by the pressure and heat experienced during the deep burial that results from continuing sedimentation and *subsidence*.

Shale A fine-grained *sedimentary rock* that easily fractures along bedding planes and consists of clay minerals or, more generally, clay-sized particles that may also include carbonate.

Sponge An animal of the phylum Porifera. Sponges are sessile, mostly marine filter feeders that pump water through their bodies to filter particles out of the water for feeding. Their *lipids* include C_{30} *sterols* with a propyl or *isopropyl* group at C-24 in the side chain.

Standard In analytical chemistry, a chemical substance that has an authentic composition and structure obtained by chemical synthesis or by *isolation* from a natural source (e.g., a plant or a geological sample). Standards are used to unequivocally identify an unknown chemical compound.

Sterane The *alkane* form of a *steroid.*

Stereochemistry The study of the spatial orientation of atoms in molecules and the reactions that change that orientation; often used synonymously with configuration, to refer to the particular orientation of atoms in a molecule.

Stereoisomers Two or more forms of a molecule that differ only in the relative spatial orientation of the atoms. The specific orientations at a given atom are designated *R* or *S* according to a convention that defines the relative hierarchy of its attached atoms and groups of atoms. If the atom in question is part of a ring structure, then the two configurations are designated α or β, depending on whether the attached hydrogen atom or group of atoms points down from or up from the plane of the ring structure, which is again defined by a chemical convention related to the hierarchy of the attached groups. The configurations of *amino acids* are traditionally designated L or D based on an older chemical convention.

Steroid A class of polycyclic chemical compounds with a particular fused ring structure composed of three six-membered rings and a five-membered ring. They may have a wide range of side groups and *functional groups* attached. Steroids are biosynthesized by plants (phytosterols) and animals and may enter the geological record after the death of these organisms. Sterols from organisms are converted to sterenes during early *diagenesis,* and *organic matter* in deeply buried *sediments* and *crude oils* contains *steranes* and aromatic steroid *hydrocarbons.*

Sterol A *steroid alcohol.*

Stratified See *water column.*

Stratigraphy The characterization of *sedimentary rock* layers, or strata. Many different parameters are used, including rock type, fossils, age, magnetic properties, *isotope* signatures, and biomarkers. (See *dating techniques.*)

Subsidence The downward movement of the earth's surface due to an increasing load of sediments or *tectonic* movements such as earthquakes and faulting.

Sulfate-reducing bacteria A morphologically and phylogenetically diverse group of *anaerobic bacteria* that obtain their energy from the *reduction* of sulfate and *oxidation* of organic acids, *alcohols,* or molecular hydrogen. They are widespread in aquatic and *terrestrial* environments.

Symbiosis A mutually beneficial association of different organisms.

Syngenetic Formed contemporaneously with sediment deposition. Used to differentiate organic matter that derives from the environment where the sediments formed and is approximately the same age, from that which may have become incorporated in a rock much later.

Tectonic Denotes a process related to movements of the earth's crust, for example, earthquakes and volcanic activity. (See *plate tectonics.*)

Terpene, terpenoid See *isoprenoid.*

Terrestrial Originating on the planet Earth (used in some texts to distinguish continental from marine organisms and material, but not here).

Tetraether See *isoprenoid glycerol ethers.*

TEX$_{86}$ A measure of the relative abundances of ring-containing *tetraethers* that correlates with the temperature of the water where the *crenarchaea* that produced them lived. It is used as a *proxy* parameter for the determination of past *sea surface temperatures*.

Thin-layer chromatography (TLC) An analytical technique used to separate mixtures of organic compounds, based on the same principles as *column chromatography*, except that the stationary phase (silica gel or aluminum oxide) is a coating on an aluminum or glass plate. A drop or two of the mixture to be analyzed is applied near the bottom of the plate, which is then placed standing in a shallow pool of solvent (mobile phase) inside a glass chamber. The solvent then moves slowly up the plate by capillary action, separating the mixture according to polarity into single compounds or groups of compounds. These can be viewed by spraying the plate with some substance that reacts with the organic compounds to dye them, or by pretreating the stationary phase such that they can be viewed with a UV lamp. The separated compounds can be scraped off the plate and extracted from the silica for further analysis.

Total organic carbon (TOC) Typically used as a measure of the total amount of organic matter in a rock or sediment, determined by combustion at high temperature and measurement of the amount of CO_2 produced.

Triterpenoid See *isoprenoid*. Pentacyclic triterpenoids of the oleanane, lupane, and ursane families are *biomarkers* of *flowering plants*; those of the *hopane* family are biomarkers of *bacteria*.

U$_{37}^{K}$ A measure of the degree of unsaturation of C_{37} *alkenones* that is directly correlated with the temperature of the water where the *coccolithophores* that produced them lived. The slightly modified U$_{37}^{K}$′ index most widely used disregards the tetraunsaturated alkenone and is simply the ratio of diunsaturated to di- and triunsaturated alkenones. Both indices serve as a proxy for past *sea surface temperatures*.

Unsaturated Refers to a chemical compound containing one or more double bond(s) or *aromatic rings*. (See *saturated*.)

Upwelling The upward movement of cold, deep, usually nutrient-rich water. It is produced in areas where the winds create offshore or divergent surface currents, such as along the western boundaries of continents and along the equator. Coastal upwelling occurs along the coasts of Southwest Africa, Western Australia, California, and West Africa. Seasonal upwelling driven by the monsoon occurs off the coast of Oman and in the Arabian Sea. The influx of nutrients leads to high *primary productivity* in these regions.

Water column A term used by oceanographers and geochemists to refer to the conceptual column of water, from the surface of the water to the surface of the sediments, often divided into layers or zones of different densities depending on temperature and salinity. When such layers are pronounced and don't mix, the water column is said to be stratified.

Wax ester An informal name used for any compound that is composed of a long-chain alcohol linked by an ester bond to a long-chain fatty acid, often of similar length. Wax esters typically range from 32 to 56 carbon atoms in length, and are found in animal, plant, and microbial tissues, where they may serve as energy stores, waterproofing, or lubrication.

Younger Dryas   A relatively brief period (about 1,300 years) of exceptionally cold climate that suddenly interrupted the warming phase from the last glacial to the present interglacial about 11,000 years ago.

Zooplankton   *Planktonic* animals that graze on *phytoplankton*. The most abundant zooplankton are microscopic in size, but the term includes everything from *copepods* to shrimp and jellyfish.

Figure List

Biosynthesis of unbranched carbon chains:
odd versus even numbers of carbon atoms   15

Ancient rocks, 1960s   19

Microfossils in the Gunflint chert: 1960s microphotographs   20
After Barghoorn, E. S., and S. A. Tyler. 1965. Microorganisms from the
Gunflint chert. *Science* 147:563–577.

Analysis of hydrocarbons in ancient shales   22

Identifying pristane in the 50-million-year-old Green River shale   23
After Eglinton, G., and M. Calvin. 1967. Chemical fossils.
Scientific American 216:32–43.

Head-to-tail isoprenoid link   24

Geological transformation of chlorophyll a   25

The one-billion-year-old Nonesuch shale: gas chromatograms of alkanes,
packed column, 1964   26
After Eglinton, G., P. M. Scott, T. Belsky, A. L. Burlingame, W. Richter, and M. Calvin.
1966. Occurrence of isoprenoid alkanes in a Precambrian sediment. In: Hobson,
G. D., and M. C. Louis, eds., *Advances in Organic Geochemistry 1964*, Pergamon
Press (London): 41–74.

Cell membrane   29
Based on a figure in Ourisson, G., and Y. Nakatani. 1994. The terpenoid theory
of the origin of cellular life: The evolution of terpenoids to cholesterol. *Chemistry*
and Biology 1:11–23.

Fatty acid structures   31

Optical stereoisomers   34

Phytol, and the stereoisomers of phytanic acid   35

Isomerization of α-amino acids   37

Oil migration and accumulation   51
After Tissot, B. P., and D. H. Welte. 1984. *Petroleum Formation and Occurrence*,
2nd ed., Springer (Heidelberg, Germany): 294.

Oil drop with hydrocarbon structures   52

Biosynthesis of sterols and pentacyclic triterpenoids:
a simplified scheme   54

Some sterols and triterpenoids found in organisms   55

Kerogen and biomarkers: the journey from organisms to petroleum 95
A schematic representation, after Tegelaar, E. W., J. W. de Leeuw, S. Derenne, and
C. Largeau. 1989. A reappraisal of kerogen formation. *Geochimica et Cosmochimica
Acta* 53:3103–3106.

Some organic sulfur compounds and possible precursors 97

Deep sea drilling 102

Diagenesis of steroids and hopanoids 105

Molecular analysis of alcohols in immature marine sediments 107

Dinosterol, and some dinoflagellates 108
The scanning electron micrographs of dinoflagellates are by Maria A. Faust at the
Smithsonian National Museum of Natural History, used by permission. The top left
image is of *Ornithocercus magnificus*, bottom left is *Ceratochoris horrida*, and bottom
right is *Protoperidinium crassipes*. © Smithsonian Institution.

Alkane diols 109

Highly branched isoprenoid (HBI) 110

Emiliania huxleyi cells, *scanning electron micrograph* 112
Courtesy of Jane Lewis, University of Westminster.

Alkenones in Emiliania huxleyi 114
After Prahl, F. G., L. A. Muehlhausen, and D. L. Zahnle. 1988. Further evaluation of
long-chain alkenones as indicators of paleoceanographic conditions. *Geochimica et
Cosmochimica Acta* 52:2303–2310.

Orbital variations 116

Alkenone distributions: Japan Trench and Middle American Trench 119
Based on Brassell, S. C. 1993. Applications of biomarkers for delineating marine paleocli-
matic fluctuations during the Pleistocene. In: Engel, M. H., and S. A. Macko, eds., *Organic
Geochemistry—Principles and Applications*, Plenum Press (New York): 699–738.

*Average degree of unsaturation versus algal growth temperature
for long-chain unsaturated lipids in* Emiliania huxleyi, *1982* 121
After Marlowe, I. T. 1984. *Lipids as Paleoclimatic Indicators*. Ph.D. thesis,
University of Bristol.

*Alkenone unsaturation (U_{37}^K) and oxygen isotope ($\delta^{18}O$) stratigraphy:
Kane Gap core, east equatorial Atlantic Ocean, 1984* 122
After Brassell, S. C., G. Eglinton, I. T. Marlowe, U. Pflaumann, and M. Sarnthein.
1986. Molecular stratigraphy: A new tool for climatic assessment. *Nature* 320:129–133.

Calibration of $U_{37}^K{}'$ from suspended and sinking particulate matter, 1987 124
After Prahl, F. G., and S. G. Wakeham. 1987. Calibration of unsaturation patterns in
long-chain ketone compositions for paleotemperature assessment. *Nature* 330:367–369.

Geophysics Geosystems 7:Q08010 (doi:10.1029/2005GC001223); and Rommerskirchen, F., A. Plader, G. Eglinton, Y. Chikaraishi, and J. Rullkötter. 2006. Chemotaxonomic significance of distribution and stable carbon isotopic composition of long-chain alkanes and alkan-1-ols in C_4 grass waxes. *Organic Geochemistry* 37:1303–1332.

Mapping with biomarkers and plant pollen: comparing interglacial and glacial African ecosystems 172

After Rommerskirchen, F., G. Eglinton, L. Dupont, U. Güntner, C. Wenzel, and J. Rullkötter. 2003. A north to south transect of Holocene southeast Atlantic continental margin sediments: Relationship between aerosol transport and compound-specific $\delta^{13}C$ land plant biomarker and pollen records. *Geochemistry Geophysics Geosystems* 4(12):1101 (doi:10.1029/2003GC000541); and Rommerskirchen, F., G. Eglinton, L. Dupont, and J. Rullkötter. 2006. Glacial/interglacial changes in southern Africa: Compound-specific $\delta^{13}C$ land plant biomarker and pollen records from southeast Atlantic continental margin sediments. *Geochemistry Geophysics Geosystems* 7:Q08010 (doi:10.1029/2005GC001223).

Did Cretaceous oceanic anoxic events allow C_4 land plants to become more prevalent? 174

Evidence from a DSDP core off the coast of northwest Africa. After Kuypers, M. M. M., R. D. Pancost, and J. S. Sinninghe Damsté. 1999. A large and abrupt fall in atmospheric CO_2 concentration during Cretaceous times. *Nature* 399:342–345.

A hydrothermal vent landscape (1980, East Pacific Rise) 178

From Haymon, R. M. 1982. *Hydrothermal Deposition on the East Pacific Rise at 21°N.* Ph.D. dissertation, University of California, San Diego. Courtesy of Rachel M. Haymon.

Woese's universal phylogenetic tree, 1987 179

After Woese, C. R. 1987. Bacterial evolution. *Microbiological Reviews* 51:221–271.

Examples of unusual lipids in archaea 180

Microbial mats from around the world 184

The photos of cyanobacteria and of the Solar Lake core are courtesy of Bo Barker Jørgensen at the Max Planck Institute for Marine Biology. The photo of the Yellowstone mat is courtesy of David Ward at Montana State University Bozeman.

A microbial mat community from an alkaline hot spring: schematic representation of a vertical profile 187

Based on Ward, D. M., J. Jentaie, Y. Bing Zeng, G. Dobson, S. C. Brassell, and G. Eglinton. 1989. Lipic biochemical markers and the composition of microbial mats. In *Microbial Mats: Physiological Ecology of Benthic Microbial Communities*, American Society for Microbiology (Washington, DC); Zeng, Y. B., D. M. Ward, S. C. Brassell, and G. Eglinton. 1992. Biogeochemistry of hot spring environments 3. *Chemical Geology* 95:347–360; and David Ward, personal communication.

Microbial breakdown of organic matter in marine sediments 189

Based on Jørgensen, B. B. 2006. Bacteria and marine biogeochemistry. In: Schulz, H. D., and M. Zabel, eds., *Marine Geochemistry*, 2nd ed., Springer-Verlag (Heidelberg, Germany): 169–206.

anaerobic methane oxidation in Mediterranean sediments by a consortium of methanogenic archaea and bacteria. *Applied and Environmental Microbiology* 66:1126–1132; and
Egorov, A. V., and M. K. Ivanov. 1998. Hydrocarbon gases in sediments and mud brec-
cia from the central and eastern Part of the Mediterranean Ridge. *Geo-Marine Letters*
18:127–138.

*Biomarkers of methanotrophs in ice age sediments from the Santa Barbara
Basin: evidence of methane hydrate decomposition?* 208
After Hinrichs, K.-U., L. R. Hmelo, and S. P. Sylva. 2003. Molecular fossil record of
elevated methane levels in Late Pleistocene coastal waters. *Science* 299:1214–1217.

Crenarchaeol 215

Tetraether distributions: North Sea versus Arabian Sea 216
After Schouten, S., E. C. Hopmans, E. Schefuß, and J. S. Sinninghe Damsté. 2002.
Distributional variations in marine crenarchaeotal membrane lipids: A new tool for recon-
structing ancient sea water temperatures. *Earth and Planetary Science Letters* 204:265–274.

Calibration of the Tetraether Index (TEX$_{86}$) in surface sediments, 2002 217
After Schouten, S., E. C. Hopmans, E. Schefuß, and J. S. Sinninghe Damsté. 2002.
Distributional variations in marine crenarchaeotal membrane lipids: A new tool for recon-
structing ancient sea water temperatures. *Earth and Planetary Science Letters* 204:265–274.

Examples of ladderane lipids from anammox bacteria 222

*Elevated levels of 2-methyl hopanoids in Cretaceous black shales: did
cyanobacteria take over during oceanic anoxic events?* 225
After Kuypers, M. M. M., Y. van Breugel, S. Schouten, E. Erba, and J. S. Sinninghe Damsté.
2004. N_2-fixing cyanobacteria supplied nutrient N for Cretaceous oceanic anoxic events.
Geology 32:853–856.

Geologic records of oleanane in ancient sediments and fossil plants 231
After Moldowan, J. M., J. Dahl, B. J. Huizinga, F. J. Fago, L. J. Hickey, T. M. Peakman,
and D. W. Taylor. 1994. The molecular fossil record of oleanane and its relation to
angiosperms. *Science* 265:768–771.

*Evolution and diversification of marine algae: microfossils and molecular
fossils* 237
After Schwark, L., and P. Empt. 2006. Sterane biomarkers as indicators of Palaeozoic
algal evolution and extinction events. *Palaeogeography Palaeoclimatology Palaeoecology*
240:225–236; and Knoll, A. H., R. E. Summons, J. R. Waldbauer, and J. E. Zumberge.
2007. The geological succession of primary producers in the oceans. In: Falkowski, P.,
and A. H. Knoll, eds., *The Evolution of Primary Producers in the Sea*, Elsevier
(Amsterdam): 133–163.

*Proterozoic versus Phanerozoic organic matter: comparing isotope
compositions* 248
After Brocks, J. J., R. Buick, R. E. Summons, and G. A. Logan. 2003. A reconstruction
of Archean biological diversity based on molecular fossils from the 2.78 to 2.45 billion-
year-old Mount Bruce Supergroup, Hamersley Basin, Western Australia. *Geochimica et
Cosmochimica Acta* 67:4321–4335.

Isopropylcholestane   250

Oxygen, ocean chemistry, and life: coevolution   252

Data synthesized from various sources, including Catling, D. C., and M. W. Claire. 2005. How Earth's atmosphere evolved to an oxic state: A status report. *Earth and Planetary Science Letters* 237:1–20; Brocks, J. J., R. Buick, R. E. Summons, and G. A. Logan. 2003. A reconstruction of Archean biological diversity based on molecular fossils from the 2.78 to 2.45 billion-year-old Mount Bruce Supergroup, Hamersley Basin, Western Australia. *Geochimica et Cosmochimica Acta* 67(22):4321–4335; Brocks, J. J., G. D. Love, R. E. Summons, A. H. Knoll, G. A. Logan, and S. A. Bowden. 2005. Biomarker evidence for green and purple sulphur bacteria in a stratified Palaeoproterozoic sea. *Nature* 437:866–870; Peterson, K. J., R. E. Summons, and P. C. J. Donoghue. 2007. Molecular palaeobiology. *Palaeontology* 50:775–809; and Roger Summons, personal communication.

Okenone and okenane structures   253

^{13}C-depleted kerogens from the Archean   256

After Brocks, J. J., R. Buick, R. E. Summons, and G. A. Logan. 2003. A reconstruction of Archean biological diversity based on molecular fossils from the 2.78 to 2.45 billion-year-old Mount Bruce Supergroup, Hamersley Basin, Western Australia. *Geochimica et Cosmochimica Acta* 67:4321–4335.

A biomarker-centric tree of life   356

In this tree, we have included the common names and groupings of organisms that are discussed in *Echoes of Life*, along with a few familiar groups as reference points, and their characteristic biomarkers. Some of the prokaryote groups, such as sulfate-reducing bacteria and methanotrophs, are turning out to have representatives in phylogenetically disparate branches—we have not tried to represent this on our tree but rather have showed them in the group associated with the illustrated biomarkers. The lengths of the branches do not accurately represent the distances between branch points, but we have attempted to give some conceptual sense of those distances, based on both phylogenetic knowledge and the geologic record.

A synthesis of information presented throughout this book and contained in the bibliography, the tree was initially inspired by Jochen Brocks and Roger Summons. The relative placement of organisms on the tree is based on the best consensus we could find of current phylogenetic knowledge. We also, of course, were constrained by space and format. See, in particular: Brocks, J. J., and R. E. Summons. 2003. Sedimentary hydrocarbons, biomarkers for early life. In: Holland, H. D., and K. K. Turekian, eds., *Treatise on Geochemistry*, Vol. 8, *Biogeochemistry*, Elsevier (Amsterdam): 63–115. See also Brocks, J. J., and A. Pearson. 2005. Building the biomarker tree of life. *Molecular Geomicrobiology Reviews in Mineralogy and Geochemistry* 59:233–258; Pace, N. R. 2001. The universal nature of biochemistry. *Proceedings of the National Academy of Sciences* 98: 805–808; Pennisi, E. 2003. Drafting a Tree. *Science* 300: 1694; and the Tree of Life Web Project, http://www.tolweb.org.

Selected Bibliography

More than 1,500 original research papers and reviews in many different fields were consulted during the writing of *Echoes of Life*. The chapter bibliographies are, of necessity, lists of selected references. Choosing what to include and what to leave out was a difficult task, which seemed, at times, to involve an almost random decision process. We focused on the biomarker papers per se and did not include the many fine works from peripheral research fields. We generally did not include natural products surveys or studies of lipids biosynthesis, unless they were combined with sediment studies. Such work is essential for interpretation of biomarkers found in natural environments, but it ranges far and wide in the literature and would make this bibliography untenable. Choosing among the biomarker papers was even more difficult. We included a few articles of historical significance, works that we feel had a profound impact at the time they appeared. We included papers that we discussed explicitly in the text. In most cases, our discussion is a synthesis of many research papers by various groups of authors. Here, we have tried to frame the work by choosing key early papers where a particular line of inquiry is introduced, follow-up papers that give a good overview of the topic and a feel for the spectrum of applications, and a few contemporary papers that will lead readers to current research. In order to maintain a readable and accessible narrative, we have not cited sources in the text, but references for each chapter are listed in chronological order and the loosely historical narrative framework should allow readers to pick out most of the papers discussed. The listing of complete titles allows readers to scan them for subjects of interest.

Chapter 1: Molecular Informants

Eglinton, G., R. J. Hamilton, R. A. Raphael, and A. G. Gonzalez. 1962. Hydrocarbon constituents of the wax coatings of plant leaves: A taxonomic survey. *Nature* 193:739–742.

Eglinton, G., and R. J. Hamilton. 1963. The distribution of alkanes. *Chemical Plant Taxonomy,* Academic Press (London): 187–217.

Eglinton, G., and R. J. Hamilton. 1967. Leaf epicuticular waxes. *Science* 156:1322–1335.

Chapter 2: Looking to the Rocks

Cummins, J. J., and W. E. Robinson. 1964. Normal and isoprenoid hydrocarbons isolated from oil-shale bitumen. *Journal of Chemical and Engineering Data* 9(2):304–307.

Eglinton, G., P. M. Scott, T. Belsky, A. L. Burlingame, and M. Calvin. 1964. Hydrocarbons of biological origin from a one-billion-year-old sediment. *Science* 145:263–264.

Meinschein, W. G., E. S. Barghoorn, and J. W. Schopf. 1964. Biological remnants in a Precambrian sediment. *Science* 145:262–263.

Blumer, M., and W. D. Snyder. 1965. Isoprenoid hydrocarbons in recent sediments. *Science* 150:1588–1589.

Eglinton G., A. G. Douglas, J. R. Maxwell, J. N. Ramsay, and S. Ställberg-Stenhagen. 1966. Occurrence of isoprenoid fatty acids in the Green River Shale. *Science* 153:1133–1135.

Kvenvolden, K. 1966. Molecular distributions of normal fatty acids and paraffins in some lower Cretaceous sediments. *Nature* 209:573–577.

Douglas, A. G., G. Eglinton, and W. Henderson. 1967. Thermal alteration of the organic matter in sediments. In: *Advances in Organic Geochemistry 1966,* Pergamon Press (Oxford, UK): 181–207.

Eglinton, G., and M. Calvin. 1967. Chemical fossils. *Scientific American* 216:32–43.

Hoering, T. C. 1967. The organic geochemistry of Precambrian rocks. In: P. H. Abelson, ed., *Researches in Geochemistry,* 2nd ed., Wiley (New York): 87–111.

Calvin, M. 1969. *Chemical Evolution.* Clarendon Press (Oxford, UK): 88.

Kvenvolden, K. A., E. Peterson, and G. E. Pollock. 1969. Optical configuration of amino-acids in pre-Cambrian Fig Tree chert. *Nature* 221:141–143.

Sever, J., and P. L. Parker. 1969. Fatty alcohols (normal and isoprenoid) in sediments. *Science* 164(3883):1052–1054.

Winters, K., P. L. Parker, and C. van Baalen. 1969. Hydrocarbons of blue-green algae: Geochemical significance. *Science* 163:467–468.

Kvenvolden, K., J. Lawless, K. Pering, E. Peterson, J. Flores, C. Ponnamperuma, I. R. Kaplan, and C. Moore. 1970. Evidence for extraterrestrial amino acids and hydrocarbons in the Murchison Meteorite. *Nature* 228:923–926.

Schopf, J. W., J. M. Hayes, and M. R. Walter. 1983. Evolution of Earth's earliest ecosystems: Recent progress and unsolved problems. In: Schopf, J. W., ed., *Earth's Earliest Biosphere,* Princeton University Press (Princeton, NJ): 359–384.

Hoering, T. C., and V. Navale. 1987. A search for molecular fossils in the kerogen of Precambrian sedimentary rocks. *Precambrian Research* 34:247–267.

Chapter 3: From the Moon to Mars

Hayes, J. M. 1967. Organic constituents of meteorites—a review. *Geochimica et Cosmochimica Acta* 31:1395–1440.

Abelson, P. H., ed. 1970. The moon issue. *Science* 167(3918).

Bada, J. L. 1991. Amino acid cosmogeochemistry. *Philosophical Transactions of the Royal Society of London Series B* 333:349–358.

Bada, J. L., D. P. Glavin, G. D. McDonald, and L. Becker. 1998. A search for endogenous amino acids in martian meteorite ALH84001. *Science* 279(5349):362–365.

Simoneit, B. R. T., R. E. Summons, and L. L. Jahnke. 1998. Biomarkers as tracers for life on early earth and mars. *Origins of Life and Evolution of the Biosphere* 28:475–483.

Botta, O., and J. L. Bada. 2002. Extraterrestrial organic compounds in meteorites. *Surveys in Geophysics* 23:411–467.

Simoneit, B. R. T. 2004. Biomarkers (molecular fossils) as geochemical indicators of life. *Advances in Space Research* 33:1255–1261.

Chapter 4: Black Gold

Burlingame, A. L., P. Haug, T. Belsky, and M. Calvin. 1965. Occurrence of biogenic steranes and pentacyclic triterpanes in an Eocene shale (52 million years) and in an early Precambrian shale (2.7 billion years): A preliminary report. *Proceedings of the National Academy of Sciences of the United States of America* 54:1406–1412.

Hills, I. R., and E. V. Whitehead. 1966. Triterpanes in optically active petroleum distillates. *Nature* 209:977–979.

Hills, I. R., E. V. Whitehead, D. E. Anders, J. J. Cummins, and W. E. Robinson. 1966. An optically active triterpane, gammacerane in Green River, Colorado oil shale bitumen. *Journal of the Chemical Society Chemical Communications* 20:752–754.

Albrecht, P., and G. Ourisson. 1968. Diagénèse des hydrocarbures saturés dans une série sédimentaire épaisse (Doula, Cameroun). *Geochimica et Cosmochimica Acta* 33:142–147.

Henderson, W., V. Wollrab, and G. Eglinton. 1968. Identification of steroids and triterpenes from a geological source by capillary gas-liquid chromatography and mass spectrometry. *Journal of the Chemical Society Chemical Communications* 13:710–712.

Hills, I. R., G. W. Smith, and E. V. Whitehead. 1968. Optically active spirotriterpane in petroleum distillates. *Nature* 219:243–246.

Maclean, I., G. Eglinton, K. Douraghi-Zadeh, R. G. Ackman, and S. N. Hooper. 1968. Correlation of stereoisomerism in present day and geologically ancient isoprenoid fatty acids. *Nature* 218:1019–1023.

Albrecht, P., and G. Ourisson. 1969. Triterpene alcohol isolation from oil shale. *Science* 163:1192–1193.

Anderson, P. C., P. M. Gardner, E. V. Whitehead, D. E. Anders, and W. E. Robinson. 1969. The isolation of steranes from Green River oil shale. *Geochimica et Cosmochimica Acta* 33:1304–1307.

Henderson, W., V. Wollrab, and G. Eglinton. 1969. The identification of steranes and triterpanes from a geological source by capillary gas liquid chromatography and mass spectrometry. In: Schenck, P. A., and I. Havenaar, eds., *Advances in Organic Geochemistry 1968,* Pergamon Press (Oxford, UK): 181–207.

Cox, R. E., J. R. Maxwell, G. Eglinton, C. T. Pillinger, R. G. Ackman, and S. N. Hooper. 1970. The geological fate of chlorophyll: The absolute stereochemistries of a series of acyclic isoprenoid acids in a 50 million year old lacustrine sediment. *Journal of the Chemical Society Chemical Communications:*1639–1641.

Hills, I. R., G. W. Smith, and E. V. Whitehead. 1970. Hydrocarbons from fossil fuels and their relationship with living organisms. *Journal of the Institute of Petroleum* 56:127–137.

Hills, I. R., and E. V. Whitehead. 1970. Pentacyclic triterpanes from petroleum and their significance. In: Hobson, G. D., and G. C. Speers, eds., *Advances in Organic Geochemistry 1966,* Pergamon Press (Oxford, UK): 89–110.

Albrecht, P., and G. Ourisson. 1971. Biogenic substances in sediments and fossils. *Angewandte Chemie International Edition* 10:209–225.

Gallegos, E. J. 1971. Identification of new steranes, terpanes, and branched paraffins in Green River shale by combined capillary gas chromatography and mass spectrometry. *Analytical Chemistry* 43:1151–1160.

Rhead, M. M., G. Eglinton, and G. H. Draffan. 1971. Hydrocarbons produced by the thermal alteration of cholesterol under conditions simulating the maturation of sediments. *Chemical Geology* 8:277–297.

Arpino, P., P. Albrecht, and G. Ourisson. 1972. Studies on the organic constituents of lacustrine Eocene sediments. Possible mechanisms for the formation of some geolipids related to biologically occurring terpenoids. In: Van Gaertner, H. R., and H. Wehner, eds., *Advances in Organic Geochemistry 1971,* Pergamon Press (Oxford, UK): 173–187.

Ensminger, A., P. Albrecht, G. Ourisson, B. J. Kimble, J. R. Maxwell, and G. Eglinton. 1972. Homohopane in Messel oil shale: First identification of a C_{31} pentacyclic triterpane in nature. Bacterial origin of some triterpanes in ancient sediments? *Tetrahedron Letters* 36:3861–3864.

Maxwell, J. R., R. E. Cox, G. Eglinton, and C. T. Pillinger. 1973. Stereochemical studies of acyclic isoprenoid compounds—II. The role of chlorophyll in the derivation of isoprenoid-type acids in a lacustrine sediment. *Geochimica et Cosmochimica Acta* 37:297–313.

Powell, T. G., and D. M. McKirdy. 1973 Relationship between ratio of pristane to phytane, crude oil composition and geologic environment in Australia. *Nature* 243:37–39.

Ensminger, A., A. van Dorsselaer, Ch. Spyckerelle, P. Albrecht, and G. Ourisson. 1974. Pentacyclic triterpenes of the hopane type as ubiquitous geochemical markers: Origin and significance. In: Tissot, B., and F. Bienner, eds., *Advances in Organic Geochemistry 1973,* Editions Technip (Paris): 245–260.

Kimble, B. J., J. R. Maxwell, and G. Eglinton. 1974. Identification of steranes and triterpanes in geolipid extracts by high-resolution gas chromatography and mass spectrometry. *Chemical Geology* 14:173–198.

Kimble, B. J., J. R. Maxwell, R. P. Philp, G. Eglinton, P. Albrecht, A. Ensminger, P. Arpino, and G. Ourisson. 1974. Tri- and tetraterpenoid hydrocarbons in the Messel oil shale. *Geochimica et Cosmochimica Acta* 38:1165–1181.

van Dorsselaer, A., A. Ensminger, C. Spyckerelle, M. Dastillung, O. Sieskind, P. Arpino, P. Albrecht, G. Ourisson, P. W. Brooks, S. J. Gaskell, et al. 1974. Degraded and extended hopane derivatives (C_{27} to C_{35}) as ubiquitous geochemical markers. *Tetrahedron Letters* 14:1349–1352.

Whitehead, E. V. 1974. The structure of petroleum pentacyclanes. In: Tissot, B., and F. Bienner, eds., *Advances in Organic Geochemistry 1973*, Editions Technip (Paris): 226–243.

Mulheirn, L. J., and G. Ryback. 1975. Stereochemistry of some steranes from geological sources. *Nature* 256:301–302.

Rubinstein, I., and P. Albrecht. 1975. The occurrence of nuclear methylated steranes in a shale. *Journal of the Chemical Society Chemical Communications* 25:957–958.

Rubinstein, I., O. Sieskind, and P. Albrecht. 1975. Rearranged sterenes in a shale: Occurrence and simulated formation. *Journal of the Chemical Society Perkin Transactions 1* 19:1833–1836.

Albrecht, P., M. Vandenbroucke, and M. Mandengué. 1976. Geochemical studies on the organic matter from the Douala Basin (Cameroon)—I. Evolution of the extractable organic matter and the formation of petroleum. *Geochimica et Cosmochimica Acta* 40:791–799.

Gallegos, E. J. 1976 Analysis of organic mixtures using metastable transition spectra. *Analytical Chemistry* 48:1348–1351.

Petrov, A. A., S. D. Pustil'nikova, N. N. Abriutina, and G. R. Kagramanova. 1976. Petroleum steranes and triterpanes. *Neftekhimiya* 16:411–427.

Dastillung, M., and P. Albrecht. 1977. Δ^2-Sterenes as diagenetic intermediates in sediments. *Nature* 269:678–679.

Ensminger, A., P. Albrecht, G. Ourisson, and B. Tissot. 1977. Evolution of polycyclic alkanes under the effect of burial (early Toarcian shales, Paris Basin). In: Campos, R., and J. Goñi, eds., *Advances in Organic Geochemistry 1975*, ENADISMA (Madrid): 46–52.

van Dorsselaer, A., P. Albrecht, and G. Ourisson. 1977. Changes in composition of polycyclic alkanes by thermal maturation (Yallourn lignite, Australia). In: Campos, R., and J. Goñi, eds., *Advances in Organic Geochemistry 1975*, ENADISMA (Madrid): 53–59.

Ensminger, A., G. Joly, and P. Albrecht. 1978. Rearranged steranes in sediments and crude oils. *Tetrahedron Letters*:1575–1578.

Gagosian, R. B., and J. W. Farrington. 1978. Sterenes in surface sediments from the southwest African shelf and slope. *Geochimica et Cosmochimica Acta* 42:1091–1101.

Seifert, W. K. 1978. Steranes and terpanes in kerogen pyrolysis for correlation of oils and source rocks. *Geochimica et Cosmochimica Acta* 42:473–484.

Seifert, W. K., and J. M. Moldowan. 1978. Applications of steranes, terpanes and monoaromatics to the maturation, migration and source of crude oils. *Geochimica et Cosmochimica Acta* 42:77–95.

Ourisson, G., P. Albrecht, and M. Rohmer. 1979. The hopanoids: Palaeochemistry and biochemistry of a group of natural products. *Pure and Applied Chemistry* 51:709–729.

Seifert, W., and J. M. Moldowan. 1979. The effect of biodegradation on steranes and terpanes in crude oils. *Geochimica et Cosmochimica Acta* 43:111–126.

Sieskind, O., G. Joly, and P. Albrecht. 1979. Simulation of the geochemical transformations of sterols: Superacid effect of clay minerals. *Geochimica et Cosmochimica Acta* 41:1675–1679.

Mackenzie, A. S., R. L. Patience, J. R. Maxwell, M. Vandenbroucke, and B. Durand. 1980. Molecular parameters of maturation in the Toarcian shales, Paris Basin, France—I. Changes in the configuration of acyclic isoprenoid alkanes, steranes and triterpanes. *Geochimica et Cosmochimica Acta* 44:1709–1721.

Seifert, W. K., and Moldowan, J. M. 1980. The effect of thermal stress on source-rock quality as measured by hopane stereochemistry. In: Douglas, A. G., and J. R. Maxwell, eds., *Advances in Organic Geochemistry 1979,* Pergamon Press (Oxford, UK): 229–237.

Seifert, W. K., J. M. Moldowan, and R. W. Jones. 1980. The application of biological markers to petroleum exploration. *Proceedings of the Tenth World Petroleum Congress* 2:425–440.

Mackenzie, A. S., C. F. Hoffmann, and J. R. Maxwell. 1981. Molecular parameters of maturation in the Toarcian shales, Paris Basin, France—III. Changes in aromatic steroid hydrocarbons. *Geochimica et Cosmochimica Acta* 45:1345–1355.

Mackenzie, A. S., A. C. Lewis, and J. R. Maxwell. 1981. Molecular parameters of maturation in the Toarcian shales, Paris Basin, France—IV. Laboratory thermal alteration studies. *Geochimica et Cosmochimica Acta* 45:2369–2376.

Mackenzie, A. S., J. M. E. Quirke, and J. R. Maxwell. 1981. Molecular parameters of maturation in the Toarcian shales, Paris Basin, France—II. Evolution of metalloporphyrins. In: Douglas, A. G., and J. R. Maxwell, eds., *Advances in Organic Geochemistry 1979,* Pergamon Press (Oxford, UK): 239–248.

Seifert, W. K., and Moldowan, J. M. 1981. Paleoreconstruction by biological markers. *Geochimica et Cosmochimica Acta* 45:783–794.

Mackenzie, A. S., N. A. Lamb, and J. R. Maxwell. 1982. Steroid hydrocarbons and the thermal history of sediments. *Nature* 295:223–226.

Mackenzie, A. S., R. L. Patience, D. A. Yon, and J. R. Maxwell. 1982. The effect of maturation on the configuration of acyclic isoprenoid acids in sediments. *Geochimica et Cosmochimica Acta* 46:782–792.

van Graas, G., J. M. A. Baas, B. van de Graaf, and J. W. de Leeuw. 1982. Theoretical organic geochemistry. I. The thermodynamic stability of several cholestane isomers calculated by molecular mechanics. *Geochimica et Cosmochimica Acta* 46:2399–2402.

Brassell, S. C., G. Eglinton, and J. R. Maxwell. 1983. The geochemistry of terpenoids and steroids. *Biochemical Society Transactions* 11:575–586.

Mackenzie, A. S., and D. McKenzie. 1983. Isomerization and aromatization of hydrocarbons in sedimentary basins formed by extension. *Geological Magazine* 120:417–470.

Abbott, G. D., C. A. Lewis, and J. R. Maxwell. 1984. Laboratory simulation studies of steroid aromatization and alkane isomerization. *Organic Geochemistry* 6:31–38.

Mackenzie, A. S. 1984. Application of biological markers in petroleum geochemistry. In: Brooks, J., and D. H. Welte, eds., *Advances in Petroleum Geochemistry,* Vol. 1, Academic Press (London): 115–214.

Moldowan, J. M. 1984. C_{30} steranes, novel markers for marine petroleums and sedimentary rocks. *Geochimica et Cosmochimica Acta* 48:2767–2768.

Abbott, G. D., C. A. Lewis, and J. R. Maxwell. 1985. Laboratory models for aromatization and isomerization of hydrocarbons in sedimentary basins. *Nature* 318:651–653.

Mackenzie, A. S., C. Beaumont, R. Boutilier, and J. Rullkötter. 1985. The aromatization and isomerization of hydrocarbons and the thermal and subsidence history of the Nova Scotia margin. *Philosophical Transactions of the Royal Society of London Series A* 315:203–232.

ten Haven, H. L., J. W. de Leeuw, T. M. Peakman, and J. R. Maxwell. 1986. Anomalies in steroid and hopanoid maturity indices. *Geochimica et Cosmochimica Acta* 50:853–855.

Caccialanza, P. G., and A. Riva. 1987. Separation and identification of a new biological marker, 18β(H)-oleanane, in crude oil and ancient sediments, by high resolution gas chromatography-mass spectrometry. *La Revista dei Combustibili* 42:3–8.

Summons, R. E., J. K. Volkman, and C. J. Boreham. 1987. Dinosterane and other steroidal hydrocarbons of dinoflagellate origin in sediments and petroleum. *Geochimica et Cosmochimica Acta* 51:3075–3082.

Eglinton, T. I., and A. G. Douglas. 1988. Quantitative study of biomarker hydrocarbons released from kerogens during hydrous pyrolysis. *Energy and Fuels* 2:81–88.

Grantham, P. J., and Wakefield, L. L. 1988. Variations in the sterane carbon number distributions of marine source rock derived crude oils through geological time. *Organic Geochemistry* 12:61–73.

Riva, A., P. G. Caccialanza, and F. Quagliaroli. 1988. Recognition of 18β(H)-oleanane in several crudes and Tertiary-Upper Cretaceous sediments. Definition of a new maturity parameter. *Organic Geochemistry* 13:671–675.

Summons, R. E., and R. J. Capon. 1988. Fossil steranes with unprecedented methylation in ring-A. *Geochimica et Cosmochimica Acta* 52:2733–2736.

Ourisson, G. 1989. The evolution of terpenes to sterols. *Pure and Applied Chemistry* 61:345–348.

Abbott, G. D., G. Y. Wang, T. I. Eglinton, A. K. Home, and G. S. Petch. 1990. The kinetics of sterane biological marker release and degradation processes during the hydrous pyrolysis of vitrinite kerogen. *Geochimica et Cosmochimica Acta* 54:2451–2461.

Marzi, R., J. Rullkötter, and W. S. Perriman. 1990. Application of the change of sterane isomer ratios to the reconstruction of geothermal histories: Implications of the results of hydrous pyrolysis experiments. *Organic Geochemistry* 16:91–102.

Moldowan, J. M., F. J. Fago, C. Y. Lee, S. R. Jacobson, D. S. Watt, N.-E. Slougui, A. Jeganathan, and D. C. Young. 1990. Sedimentary 24-*n*-propylcholestanes, molecular fossils diagnostic of marine algae. *Science* 247:309–312.

Rullkötter, J., and W. Michaelis. 1990. The structure of kerogen and related materials. A review of recent progress and future trends. *Organic Geochemistry* 16:829–852.

Summons, R. E., and L. L. Jahnke. 1990. Identification of methylhopanes in sediments and petroleum. *Geochimica et Cosmochimica Acta* 54:247–251.

Ourisson, G., and P. Albrecht. 1992. Hopanoids. 1. Geohopanoids: The most abundant natural products on earth? *Accounts of Chemical Research* 25:398–402.

de Leeuw, J. W., and C. Largeau. 1993. A review of macromolecular organic compounds that comprise living organisms and their role in kerogen, coal, and petroleum formation. In: Engel, M. H., and S. A. Macko, eds., *Organic Geochemistry,* Plenum Press (New York): 23–72.

Abbott, G. D., B. Bennett, and G. S. Petch. 1995. The thermal degradation of 5α(H)-cholestane during closed-system pyrolysis. *Geochimica et Cosmochimica Acta* 59:2259–2264.

van Duin, A. C. T., J. S. Sinninghe Damsté, M. P. Koopmans, B. van de Graaf, and J. W. de Leeuw. 1997. A kinetic calculation method of homohopanoid maturation: Applications in the reconstruction of burial histories of sedimentary basins. *Geochimica et Cosmochimica Acta* 61:2409–2429.

Farrimond, P., A. Taylor, and N. Telnæs. 1998. Biomarker maturity parameters: The role of generation and thermal degradation. *Organic Geochemistry* 29:1181–1197.

Sugden, M. A., and G. D. Abbott. 2002. The stereochemistry of bound and extractable pentacyclic triterpenoids during closed system pyrolysis. *Organic Geochemistry* 33:1515–1521.

de Leeuw, J. W., G. J. M. Versteegh, and P. F. van Bergen. 2006. Biomacromolecules of algae and plants and their fossil analogues. *Plant Ecology* 182:209–233.

Chapter 5: Deep Sea Mud

Attaway, D., and P. L. Parker. 1970. Sterols in recent marine sediments. *Science* 169:674–675.

Simoneit, B. R. T., R. Chester, and G. Eglinton. 1977. Biogenic lipids in particulates from the lower atmosphere over the eastern Atlantic. *Nature* 267:682–685.

Didyk, B. M., B. R. T. Simoneit, S. C. Brassell, and G. Eglinton. 1978. Organic geochemical indicators of palaeoenvironmental conditions of sedimentation. *Nature* 272:216–222.

Boon, J. J., and J. W. de Leeuw. 1979. The analysis of wax esters, very long mid-chain ketones and sterol ethers isolated from Walvis Bay diatomaceous ooze. *Marine Chemistry* 7:117–132.

Boon, J. J., W. I. C. Rijpstra, F. de Lange, J. W. de Leeuw, M. Yoshioka, and Y. Shimizu. 1979. Black Sea sterol—a molecular fossil for dinoflagellate blooms. *Nature* 277:125–127.

Brassell, S. C., P. A. Comet, G. Eglinton, P. J. Isaacson, J. McEvoy, J. R. Maxwell, I. D. Thomson, P. J. C. Tibbetts, and J. K. Volkman. 1980. The origin and fate of lipids in the Japan Trench. In: Douglas, A. G., and J. R. Maxwell, eds., *Advances in Organic Geochemistry 1979*, Pergamon Press (Oxford, UK): 375–392.

de Leeuw, J. W., F. W. van der Meer, W. I. C. Rijpstra, and P. A. Schenck. 1980. On the occurrence and structural identification of long chain unsaturated ketones and hydrocarbons in sediments. In: Douglas, A. G., and J. R. Maxwell, eds., *Advances in Organic Geochemistry 1979*, Pergamon Press (Oxford, UK): 211–217.

Eglinton, G., S. C. Brassell, V. Howell, and J. R. Maxwell. 1980. The role of organic geochemistry in the Deep Sea Drilling Project (DSDP/IPOD). In: Douglas, A. G., and J. R. Maxwell, eds., *Advances in Organic Geochemistry 1979*, Pergamon Press (Oxford, UK): 391–400.

Mackenzie, A. S., R. L. Patience, J. R. Maxwell, M. Vandenbroucke, and B. Durand. 1980. Molecular parameters of maturation in the Toarcian shales, Paris Basin, France I. Changes in the configuration of acyclic isoprenoid alkanes, steranes and triterpanes. *Geochimica et Cosmochimica Acta* 44:1709–1721.

Quirke, J. M. E., G. J. Shaw, P. D. Soper, and J. R. Maxwell. 1980. Petroporphyrins II. The presence of porphyrins with extended alkyl substituents. *Tetrahedron* 36:3261–3267.

Volkman, J. K., G. Eglinton, E. D. S. Corner, and J. R. Sargent. 1980. Novel unsaturated straight-chain C_{37}–C_{39} methyl and ethyl ketones in marine sediments and a coccolithophore *Emiliania huxleyi*. In: Douglas, A. G., and J. R. Maxwell, eds., *Advances in Organic Geochemistry 1979*, Pergamon Press (Oxford, UK): 219–227.

de Leeuw, J. W., W. I. C. Rijpstra, and P. A. Schenck. 1981. The occurrence and identification of C_{30}, C_{31} and C_{32} alkan-1,15-diols and alkan-15-one-1-ols in Unit I and Unit II Black Sea sediments. *Geochimica et Cosmochimica Acta* 45:2281–2285.

Mackenzie, A. S., S. C. Brassell, G. Eglinton, and J. R. Maxwell. 1982. Chemical fossils: The geological fate of steroids. *Science* 217:491–504.

Klok, J., M. Baas, H. C. Cox, J. W. de Leeuw, and P. A. Schenck. 1984. Loliolides and dihydroactinidiolide in a recent marine sediment probably indicate a major transformation pathway of carotenoids. *Tetrahedron Letters* 25:5577–5580.

Brassell, S. C., G. Eglinton, I. T. Marlowe, U. Pflaumann, and M. Sarnthein. 1986. Molecular stratigraphy: A new tool for climatic assessment. *Nature* 320:129–133.

Farrimond, P., G. Eglinton, and S. C. Brassell. 1986. Alkenones in Cretaceous black shales, Blake-Bahama Basin, western North Atlantic. *Organic Geochemistry* 10:897–903.

Robson, J. N., and S. J. Rowland. 1986. Identification of novel widely distributed sedimentary acyclic sesterterpenoids. *Nature* 324:561–563.

Prahl, F. G., and S. G. Wakeham. 1987. Calibration of unsaturation patterns in long-chain ketone compositions for palaeotemperature assessment. *Nature* 330:367–369.

ten Haven, H. L., and J. Rullkötter. 1988. The diagenetic fate of taraxer-14-ene and oleanene isomers. *Geochimica et Cosmochimica Acta* 52:2543–2548.

ten Haven, H. L., J. Rullkötter, and D. H. Welte. 1989. Steroid biological marker hydrocarbons as indicators of organic matter diagenesis in deep sea sediments: Geochemical reactions and influence of different heat flow regimes. *Geologische Rundschau* 78:841–850.

Marlowe, I. T., S. C. Brassell, G. Eglinton, and J. C. Green. 1990. Long-chain alkenones and alkyl alkenoates and the fossil coccolith record of marine sediments. *Chemical Geology* 88:349–375.

Sinninghe Damsté, J. S., and J. W. de Leeuw. 1990. Analysis, structure and geochemical significance of organically-bound sulphur in the geosphere: State of the art and future research. *Organic Geochemistry* 16:1077–1101.

Sinninghe Damsté, J. S., M. E. L. Kohnen, and J. W. de Leeuw. 1990. Thiophenic biomarkers for palaeoenvironmental assessment and molecular stratigraphy. *Nature* 345:609–611.

Wakeham, S. G. 1990. Algal and bacterial hydrocarbons in particulate matter and interfacial sediment of the Cariaco Trench. *Geochimica et Cosmochimica Acta* 54:1325–1336.

Eckardt, C. B., B. J. Keely, J. R. Waring, M. I. Chicarelli, and J. R. Maxwell. 1991. Preservation of chlorophyll-derived pigments in sedimentary organic matter. *Philosophical Transactions of the Royal Society of London Series B* 333:339–348.

Keely, B. J., and J. R. Maxwell. 1991. Structural characterization of the major chlorins in a recent sediment. *Organic Geochemistry* 17:663–669.

Eglinton, G., S. A. Bradshaw, A. Rosell, M. Sarnthein, U. Pflaumann, and R. Tiedemann. 1992. Molecular record of secular sea surface temperature changes on 100-year timescales for glacial terminations I, II and IV. *Nature* 356:423–426.

Kohnen, M. E. L., S. Schouten, J. S. Sinninghe Damsté, J. W. de Leeuw, D. A. Merritt, and J. M. Hayes. 1992. Recognition of paleobiochemicals by a combined molecular sulfur and isotope geochemical approach. *Science* 256:358–362.

ten Haven, H. L., T. M. Peakman, and J. Rullkötter. 1992. Early diagenetic transformation of higher-plant triterpenoids in deep-sea sediments from Baffin Bay. *Geochimica et Cosmochimica Acta* 56:2001–2024.

Rosell-Melé, A., G. Eglinton, U. Pflaumann, and M. Sarnthein. 1995. Atlantic core-top calibration of U^K_{37} index as a sea-surface palaeotemperature indicator. *Geochimica et Cosmochimica Acta* 59:3099–3107.

Sinninghe Damsté, J. S., F. Kenig, M. P. Koopmans, J. Köster, S. Schouten, J. M. Hayes, and J. W. de Leeuw. 1995. Evidence for gammacerane as an indicator of water column stratification. *Geochimica et Cosmochimica Acta* 59:1895–1900.

Volkman, J. K., S. M. Barrett, S. I. Blackburn, and E. L. Sikes. 1995. Alkenones in *Gephyrocapsa oceanica:* Implications for studies of paleoclimate. *Geochimica et Cosmochimica Acta* 59:513–520.

Zhao, M., N. A. S. Beveridge, N. J. Shackleton, M. Sarnthein, and G. Eglinton. 1995. Molecular stratigraphy of cores off northwest Africa: Sea surface temperature history over the last 80 ka. *Paleoceanography* 10:661–675.

Grice, K., R. Gibbison, J. E. Atkinson, L. Schwark, C. B. Eckardt, and J. R. Maxwell. 1996. Maleimides (1H-pyrrole-2,5-diones) as molecular indicators of anoxygenic photosynthesis in ancient water columns. *Geochimica et Cosmochimica Acta* 60:3913–3924.

Harris, P. G., M. Zhao, A. Rosell-Melé, R. Tiedemann, M. Sarnthein, and J. R. Maxwell. 1996. Chlorin accumulation rate as a proxy for Quaternary marine primary productivity. *Nature* 383:63–65.

Hedges, J. I., R. G. Keil, and R. Benner. 1997. What happens to terrestrial organic matter in the ocean? *Organic Geochemistry* 27:195–212.

Versteegh, G. J. M., H.-J. Bosch, and J. W. de Leeuw. 1997. Potential palaeoenvironmental information on C_{24} to C_{36} mid-chain diols, keto-ols and mid-chain hydroxyl fatty acids; a critical review. *Organic Geochemistry* 27:1–13.

Villanueva, J., J. O. Grimalt, E. Cortijo, L. Vidal, and L. Labeyrie. 1997. A biomarker approach to the organic matter deposited in the North Atlantic during the last climatic cycle. *Geochimica et Cosmochimica Acta* 61:4633–4646.

Villanueva, J., J. O. Grimalt, L. D. Labeyrie, E. Cortijo, and L. Vidal. 1998. Precessional forcing of productivity in the North Atlantic Ocean. *Paleoceanography* 13:561–571.

Volkman, J. K., S. M. Barrett, S. I. Blackburn, M. P. Mansour, E. L. Sikes, and F. Gelin. 1998. Microalgal biomarkers: A review of recent research developments. *Organic Geochemistry* 29:1163–1179.

Bouloubassi, I., J. Rullkötter, and P. A. Meyers. 1999. Origin and transformation of organic matter in Pliocene-Pleistocene Mediterranean sapropels: Organic geochemical evidence reviewed. *Marine Geology* 153:177–197.

Cacho, I., J. O. Grimalt, C. Pelejero, M. Canals, F. J. Sierro, J. A. Flores, and N. Shackleton. 1999. Dansgaard-Oeschger and Heinrich event imprints in Alboran Sea paleotemperatures. *Paleoceanography* 14:698–705.

Hinrichs, K.-U., R. R. Schneider, P. J. Müller, and J. Rullkötter. 1999. A biomarker perspective on paleoproductivity variations in two Late Quaternary sediment sections from the Southeast Atlantic Ocean. *Organic Geochemistry* 30:341–366.

Schulte, S., F. Rostek, E. Bard, J. Rullkötter, and O. Marchal. 1999. Variations of oxygen-minimum and primary productivity recorded in sediments of the Arabian Sea. *Earth and Planetary Science Letters* 173:205–221.

Belt, S. T., W. G. Allard, G. Massé, J.-M. Robert, and S. J. Rowland. 2000. Highly branched isoprenoids (HBIs): Identification of the most common and abundant sedimentary isomers. *Geochimica et Cosmochimica Acta* 64:3839–3851.

Cacho, I., J. O. Grimalt, F. J. Sierro, N. Shackleton, and M. Canals. 2000. Evidence for enhanced Mediterranean thermohaline circulation during rapid climatic coolings. *Earth and Planetary Science Letters* 183:417–429.

Mix, A. C., E. Bard, G. Eglinton, L. D. Keigwin, A. C. Ravelo, and Y. Rosenthal. 2000. Alkenones and multiproxy strategies in paleoceanographic studies. *Geochemistry, Geophysics, Geosystems* 1:2000GC000056.

Sachs, J. P., R. R. Schneider, T. I. Eglinton, K. H. Freeman, G. Ganssen, J. F. McManus, and D. W. Oppo. 2000. Alkenones as paleoceanographic proxies. *Geochemistry, Geophysics, Geosystems* 1 (doi:10.1029/2000GC000059).

Volkman, J. K. 2000. Ecological and environmental factors affecting alkenone distributions in seawater and sediments. *Geochemistry, Geophysics, Geosystems* 1 (doi:10.1029/2000GC000061).

Werne, J. P., D. J. Hollander, A. Behrens, P. Schaeffer, P. Albrecht, and J. S. Sinninghe Damsté. 2000. Timing of early diagenetic sulfurization of organic matter: A precursor-product relationship in Holocene sediments of the anoxic Cariaco Basin, Venezuela. *Geochimica et Cosmochimica Acta* 64:1741–1751.

Werne, J. P., D. J. Hollander, T. W. Lyons, and L. C. Peterson. 2000. Climate-induced variations in productivity and planktonic ecosystem structure from the Younger Dryas to Holocene in the Cariaco Basin, Venezuela. *Paleoceanography* 15:19–29.

Villanueva, J., E. Calvo, C. Pelejero, J. O. Grimalt, A. Boelaert, and L. Labeyrie. 2001. A latitudinal productivity band in the central North Atlantic over the last 270 kyr: An alkenone perspective. *Paleoceanography* 16:617–626.

Rinna, J., B. Warning, P. A. Meyers, H.-J. Brumsack, and J. Rullkötter. 2002. Combined organic and inorganic geochemical reconstruction of paleodepositional conditions of a Pliocene sapropel from the eastern Mediterranean Sea. *Geochimica et Cosmochimica Acta* 66:1969–1986.

Menzel, D., P. F. van Bergen, S. Schouten, and J. S. Sinninghe Damsté. 2003. Reconstruction of changes in export productivity during Pliocene sapropel deposition: A biomarker approach. *Palaeogeography, Palaeoclimatology, Palaeoecology* 190:273–287.

Rimbu, N., G. Lohmann, J.-H. Kim, H. W. Arz, and R. Schneider. 2003. Arctic/North Atlantic Oscillation signature in Holocene sea surface temperature trends as obtained from alkenone data. *Geophysical Research Letters* 30(6) (doi:10.1029/2002GL016570).

Sinninghe Damsté J. S., M. M. M. Kuypers, S. Schouten, S. Schulte, and J. Rullkötter. 2003. The lycopane/C_{31} *n*-alkane ratio as a proxy to assess palaeoxicity during sediment deposition. *Earth and Planetary Science Letters* 209:215–226.

Sinninghe Damsté, J. S., S. Rampen, W. I. C. Rijpstra, B. Abbas, G. Muyzer, and S. Schouten. 2003. A diatomaceous origin for long-chain diols and mid-chain hydroxy methyl alkanoates widely occurring in Quaternary marine sediments: Indicators for high-nutrient conditions. *Geochimica et Cosmochimica Acta* 67:1339–1348.

Calvo, E., C. Pelejero, G. A. Logan, and P. de Deckker. 2004. Dust-induced changes in phytoplankton composition in the Tasman Sea during the last four glacial cycles. *Paleoceanography* 19:PA2020 (doi:10.1029/2003PA000992).

Kuypers, M. M. M., L. J. Lourens, W. I. C. Rijpstra, R. D. Pancost, I. A. Nijenhuis, and J. S. Sinninghe Damsté. 2004. Orbital forcing of organic carbon burial in the proto-North Atlantic during oceanic anoxic event 2. *Earth and Planetary Science Letters* 228:465–482.

Martrat, B., J. O. Grimalt, C. López-Martínez, I. Cacho, F. J. Sierro, J. A. Flores, R. Zahn, M. Canals, J. H. Curtis, and D. A. Hodell. 2004. Abrupt temperature changes in the Western Mediterranean over the past 250,000 years. *Science* 306:1762–1765.

Menzel, D., S. Schouten, P. F. van Bergen, and J. S. Sinninghe Damsté. 2004. Higher plant vegetation changes during Pliocene sapropel formation. *Organic Geochemistry* 35:1343–1353.

Moreno, A., I. Cacho, M. Canals, J. O. Grimalt, and A. Sanchez-Vidal. 2004. Millennial-scale variability in the productivity signal from the Alboran Sea record, western Mediterranean Sea. *Palaeogeography, Palaeoclimatology, Palaeoecology* 211:205–219.

Pancost, R. D., and C. S. Boot. 2004. The palaeoclimatic utility of terrestrial biomarkers in marine sediments. *Marine Chemistry* 92:239–261.

Pancost, R. D., N. Crawford, S. Magness, A. Turner, H. C. Jenkyns, and J. R. Maxwell. 2004. Further evidence for the development of photic-zone euxinic conditions during Mesozoic oceanic anoxic events. *Journal of the Geological Society, London* 161:353–364.

Wagner, T., J. S. Sinninghe Damsté, P. Hofmann, and B. Beckmann. 2004. Euxinia and primary production in Late Cretaceous eastern equatorial Atlantic surface waters fostered orbitally driven formation of marine black shales. *Paleoceanography* 19:PA3009 (doi:10.1029/2003PA000898).

Dumitrescu, M., and S. C. Brassell. 2005. Biogeochemical assessment of sources of organic matter and paleoproductivity during the early Aptian Oceanic Anoxic Event at Shatsky Rise, ODP Leg 198. *Organic Geochemistry* 36:1002–1022.

Wakeham, S. G., M. L. Peterson, J. I. Hedges, and C. Lee. 2005. Lipid biomarker fluxes in the Arabian Sea, with a comparison to the equatorial Pacific Ocean. *Deep-Sea Research II* 49:2265–2301.

López-Martínez, C., J. O. Grimalt, B. Hoogakker, J. Gruetzner, M. J. Vautraverfs, and I. N. McCave. 2006. Abrupt wind regime changes in the North Atlantic Ocean during the past 30,000—60,000 years. *Paleoceanography* 21:PA4215 (doi:10.1029/2006PA001275).

Zhao, M., J. L. Mercer, G. Eglinton, M. J. Higgins, and C.-Y. Huang. 2006. Comparative molecular biomarker assessment of phytoplankton paleoproductivity for the last 160 kyr off Cap Blanc, NW Africa. *Organic Geochemistry* 37:72–97.

Belt, S. T., G. Massé, S. J. Rowland, M. Poulin, C. Michel, and B. LeBlanc. 2007. A novel chemical fossil of palaeo sea ice: IP_{25}. *Organic Geochemistry* 38:16–27.

Chapter 6: More Molecules, More Mud, and the Isotopic Dimension

Summons, R. E., and T. G. Powell. 1986. *Chlorobiaceae* in Palaeozoic seas revealed by biological markers, isotopes and geology. *Nature* 319:763–765.

Hayes, J. M., R. Takigiku, R. Ocampo, H. J. Callot, and P. Albrecht. 1987. Isotopic compositions and probable origins of organic molecules in the Eocene Messel shale. *Nature* 329:48–51.

Popp, B. N., R. Takigiku, J. M. Hayes, J. W. Louda, and E. W. Baker. 1989. The post-Paleozoic chronology and mechanism of ^{13}C depletion in primary marine organic matter. *American Journal of Science* 289:436–454.

Freeman, K. H., J. M. Hayes, J.-M. Trendel, and P. Albrecht. 1990. Evidence from carbon isotope measurements for diverse origins of sedimentary hydrocarbons. *Nature* 343:254–256.

Hayes, J. M., K. H. Freeman, B. N. Popp, and C. H. Hoham. 1990. Compound-specific isotopic analyses: A novel tool for reconstruction of ancient biogeochemical processes. *Organic Geochemistry* 16:1115–1128.

Jasper, J. P., and J. M. Hayes. 1990. A carbon isotope record of CO_2 levels during the late Quaternary. *Nature* 347:462–464.

Rieley, G., R. J. Collier, D. M. Jones, G. Eglinton, P. A. Eakin, and A. E. Fallick. 1991. Sources of sedimentary lipids deduced from stable carbon-isotope analyses of individual compounds. *Nature* 352:425–426.

Freeman, K. H., and J. M. Hayes. 1992. Fractionation of carbon isotopes by phytoplankton and estimates of ancient CO_2 levels. *Global Biogeochemical Cycles* 6:185–198.

Hayes, J. M. 1993. Factors controlling ^{13}C contents of sedimentary organic compounds: Principles and evidence. *Marine Geology* 113:111–125.

Jasper, J. P., J. M. Hayes, A. C. Mix, and F. G. Prahl. 1994. Photosynthetic fractionation of ^{13}C and concentrations of dissolved CO_2 in the central equatorial Pacific during the last 255,000 years. *Paleoceanography* 9:781–798.

Street-Perrott, F. A., Y. Huang, R. A. Perrott, G. Eglinton, P. Barker, L. B. Khelifa, D. D. Harkness, and D. O. Olago. 1997. Impact of lower atmospheric carbon dioxide on tropical mountain ecosystems. *Science* 278:1422–1426.

Popp, B. N., F. Kenig, S. G. Wakeham, E. A. Laws, and R. R. Bidigare. 1998. Does growth rate affect ketone unsaturation and intracellular carbon isotopic variability in *Emiliania huxleyi*? *Paleoceanography* 13:35–41.

Kuypers, M. M. M., R. D. Pancost, and J. S. Sinninghe Damsté. 1999. A large and abrupt fall in atmospheric CO_2 concentration during Cretaceous times. *Nature* 399:342–345.

Pagani, M., K. H. Freeman, and M. A. Arthur. 1999. Late Miocene atmospheric CO_2 concentrations and the expansion of C_4 grasses. *Science* 285:876–879.

Pancost, R. D., K. H. Freeman, and S. G. Wakeham. 1999. Controls on the carbon-isotope compositions of compounds in Peru surface waters. *Organic Geochemistry* 30:319–340.

Huang, Y., L. Dupont, M. Sarnthein, J. M. Hayes, and G. Eglinton. 2000. Mapping of C_4 plant input from North West Africa into North East Atlantic sediments. *Geochimica et Cosmochimica Acta* 64:3505–3513.

Huang, Y., F. A. Street-Perrott, S. E. Metcalfe, M. Brenner, M. Moreland, and K. H. Freeman. 2001. Climate change as the dominant control on glacial-interglacial variations in C_3 and C_4 plant abundance. *Science* 293:1647–1651.

Benthien, A., N. Andersen, S. Schulte, P. J. Müller, R. R. Schneider, and G. Wefer. 2002. Carbon isotopic composition of the $C_{37:2}$ alkenone in core top sediments of the South Atlantic Ocean: Effects of CO_2 and nutrient concentrations. *Global Biogeochemical Cycles* 16 (doi:10.1029/2001GB001433).

Pagani, M. 2002. The alkenone-CO_2 proxy and ancient atmospheric carbon dioxide. *Philosophical Transactions of the Royal Society of London Series A* 360:609–632.

Rommerskirchen, F., G. Eglinton, L. Dupont, U. Güntner, C. Wenzel, and J. Rullkötter. 2003. A north to south transect of Holocene Southeast Atlantic continental margin sediments: Relationship between aerosol transport and compound-specific $\delta^{13}C$ land plant biomarker and pollen records. *Geochemistry, Geophysics, Geosystems* 4 (doi:10.1029/2003GC000541).

Schefuß, E., S. Schouten, J. H. F. Jansen, and J. S. Sinninghe Damsté. 2003. African vegetation controlled by tropical sea surface temperatures in the mid-Pleistocene. *Nature* 422:418–421.

Hayes, J. M. 2004. Isotopic order, biogeochemical processes, and earth history. Goldschmidt Lecture, Davos, Switzerland, August 2002. *Geochimica et Cosmochimica Acta* 68:1691–1700.

Kuypers, M. M. M., L. J. Lourens, W. I. C. Rijpstra, R. D. Pancost, I. A. Nijenhuis, and J. S. Sinninghe Damsté. 2004. Orbital forcing of organic carbon burial in the proto-North Atlantic during oceanic anoxic event 2. *Earth and Planetary Science Letters* 228:465–482.

Benthien, A., N. Andersen, S. Schulte, P. J. Müller, R. R. Schneider, and G. Wefer. 2005. The carbon isotopic record of the C37:2 alkenone in the South Atlantic: Last Glacial Maximum (LGM) vs. Holocene. *Palaeogeography, Palaeoclimatology, Palaeoecology* 221:123–140.

Engel, A., I. Zondervan, K. Aerts, L. Beaufort, A. Benthien, L. Chou, B. Delille, J.-P. Gattuso, J. Harlay, C. Heemann, et al. 2005. Testing the direct effect of CO_2 concentration on a bloom of the coccolithophorid *Emiliania huxleyi* in mesocosm experiments. *Limnology and Oceanography* 50:493–507.

Kolonic, S., T. Wagner, A. Forster, J. S. Sinninghe Damsté, B. Walsworth-Bell, E. Erba, S. Turgeon, H.-J. Brumsack, E. H. Chellai, H. Tsikos, et al. 2005. Black shale deposition on the Northwest African shelf during the Cenomanian/Turonian oceanic anoxic event: Climate coupling and global organic carbon burial. *Paleoceanography* 20:PA1006 (doi:10.1029/2003PA000950).

Schefuß, E., S. Schouten, and R. R. Schneider. 2005. Climatic controls on central African hydrology during the past 20,000 years. *Nature* 437:1003–1006.

Rommerskirchen, F., G. Eglinton, L. Dupont, and J. Rullkötter. 2006. Glacial/interglacial changes in southern Africa: Compound-specific $\delta^{13}C$ land plant biomarker and pollen records from southeast Atlantic continental margin sediments. *Geochemistry, Geophysics, Geosystems* 7 (doi:10.1029/2005GC001223).

Chapter 7: Microbiologists (Finally) Climb on Board

Moldowan, J. M., and W. K. Seifert. 1979. Head-to-head linked isoprenoid hydrocarbons in petroleum. *Science* 204:169–171.

Brassell, S. C., A. M. K. Wardroper, I. D. Thomson, J. R. Maxwell, and G. Eglinton. 1981. Specific acyclic isoprenoids as biological markers of methanogenic bacteria in marine sediments. *Nature* 290:693–696.

Chappe, B., P. Albrecht, and W. Michaelis. 1982. Polar lipids of archaebacteria in sediments and petroleums. *Science* 217:65–66.

Ourisson, G., P. Albrecht, and M. Rohmer. 1982. Predictive microbial biochemistry—from molecular fossils to procaryotic membranes. *Trends in Biochemical Sciences* 7:236–239.

Rowland, S. J., N. A. Lamb, C. F. Wilkinson, and J. R. Maxwell. 1982. Confirmation of 2,6,10,15,19-pentamethyleicosane in methanogenic bacteria and sediments. *Tetrahedron Letters* 23:101–104.

Edmunds, K. L. H., and G. Eglinton. 1984. Microbial lipids and carotenoids and their early diagenesis in the Solar Lake laminated microbial mat sequence. In: Cohen, Y., R. W. Castenholtz, and H. O. Halvorson, eds., *Microbial Mats: Stromatolites*, Alan R. Liss (New York): 343–389.

Ward, D. M., S. C. Brassell, and G. Eglinton. 1985. Archaebacterial lipids in hot-spring microbial mats. *Nature* 318:656–659.

Summons, R. E., and T. G. Powell. 1987. Identification of aryl isoprenoids in source rocks and crude oils: Biological markers for the green sulphur bacteria. *Geochimica et Cosmochimica Acta* 51:557–566.

Ward, D. M., J. Shiea, Y. B. Zeng, G. Dobson, S. Brassell, and G. Eglinton. 1989. Lipid biochemical markers and the composition of microbial mats. In: Cohen, Y., and E. Rosenberg. *Microbial Mats. Physiological Ecology of Benthic Microbial Communities*, American Society for Microbiology (Washington, DC): 439–454.

Summons, R. E., and L. L. Jahnke. 1990. Identification of the methylhopanes in sediments and petroleum. *Geochimica et Cosmochimica Acta* 54:247–251.

Shiea, J., S. C. Brassell, and D. M. Ward. 1991. Comparative analysis of extractable lipids in hot spring microbial mats and their component photosynthetic bacteria. *Organic Geochemistry* 17:309–319.

Zeng, Y. B., D. M. Ward, S. C. Brassell, and G. Eglinton. 1992. Biogeochemistry of hot spring environments 2. Lipid compositions of Yellowstone (Wyoming, U.S.A.) cyanobacterial and *Chloroflexus* mats. *Chemical Geology* 95:327–345.

Summons, R. E., L. L. Jahnke, and Z. Roksandic. 1994. Carbon isotopic fractionation in lipids from methanotrophic bacteria: Relevance for interpretation of the geochemical record of biomarkers. *Geochimica et Cosmochimica Acta* 58:2853–2863.

Kenig, F., J. S. Sinninghe Damsté, A. C. Kock-van Dalen, W. I. C. Rijpstra, A. Y. Huc, and J. W. de Leeuw. 1995. Occurrence and origin of mono-, di-, and trimethylalkanes in modern and Holocene cyanobacterial mats from Abu Dhabi, United Arab Emirates. *Geochimica et Cosmochimica Acta* 59:2999–3015.

Summons, R. E., P. D. Franzmann, and P. D. Nichols. 1998. Carbon isotopic fractionation associated with methylotrophic methanogenesis. *Organic Geochemistry* 28:465–475.

Elvert, M., E. Suess, and M. J. Whiticar. 1999. Anaerobic methane oxidation associated with marine gas hydrates: Superlight C-isotopes from saturated and unsaturated C_{20} and C_{25} irregular isoprenoids. *Naturwissenschaften* 86:295–300.

Hinrichs, K.-U., J. M. Hayes, S. P. Sylva, P. G. Brewer, and E. F. DeLong. 1999. Methane-consuming archaebacteria in marine sediments. *Nature* 398:802–805.

Summons, R. E., L. L. Jahnke, J. M. Hope, and G. A. Logan. 1999. 2-Methylhopanoids as biomarkers for cyanobacterial oxygenic photosynthesis. *Nature* 400:554–557.

Thiel, V., J. Peckmann, R. Seifert, P. Wehrung, J. Reitner, and W. Michaelis. 1999. Highly isotopically depleted isoprenoids: Molecular markers for ancient methane venting. *Geochimica et Cosmochimica Acta* 63:3959–3966.

Boetius, A., K. Ravenschlag, C. J. Schubert, D. Rickert, F. Widdel, A. Gieseke, R. Amann, B. B. Jørgensen, U. Witte, and O. Pfannkuche. 2000. A marine microbial consortium apparently mediating anaerobic oxidation of methane. *Nature* 407:623–625.

Hinrichs, K.-U., R. E. Summons, V. Orphan, S. P. Sylva, and J. M. Hayes. 2000. Molecular and isotopic analysis of anaerobic methane-oxidizing communities in marine sediments. *Organic Geochemistry* 31:1685–1701.

Pancost, R. D., J. S. Sinninghe Damsté, S. de Lint, M. J. E. C. van der Maarel, J. C. Gottschal, and the Medinaut Shipboard Scientific Party. 2000. Biomarker evidence for widespread anaerobic methane oxidation in Mediterranean sediments by a consortium of methanogenic archaea and bacteria. *Applied and Environmental Microbiology* 66:1126–1132.

Bian, L., K.-U. Hinrichs, T. Xie, S. C. Brassell, N. Iversen, H. Fossing, B. B. Jørgensen, and J. M. Hayes. 2001. Algal and archaeal polyisoprenoids in a recent marine sediment: Molecular isotopic evidence for anaerobic oxidation of methane. *Geochemistry, Geophysics, Geosystems* 2 (doi:2000GC000112).

Elvert, M., J. Greinert, E. Suess, and M. J. Whiticar. 2001. Carbon isotopes of biomarkers derived from methane-oxidizing microbes at Hydrate Ridge, Cascadia convergent margin. In: Paull, C. K., and W. P. Dillon, eds., *Natural Gas Hydrates: Occurrence, Distribution, and Dynamics,* Monograph Series 124, American Geophysical Union (Washington, DC): 115–129.

Hinrichs, K.-U., and A. Boetius. 2001. The anaerobic oxidation of methane: New insights in microbial ecology and biogeochemistry. In: Wefer, G., D. Billett, D. Hebbeln, B. B. Jørgensen, M. Schlüter, and T. van Weering, eds., *Ocean Margin Systems,* Springer (Heidelberg, Germany): 457–477.

Jahnke, L. L., W. Eder, R. Huber, J. M. Hope, K.-U. Hinrichs, J. M. Hayes, D. J. des Marais, S. L. Cady, and R. E. Summons. 2001. Signature lipids and stable carbon isotope analyses of Octopus Spring hyperthermophilic communities compared with those of *Aquificales* representatives. *Applied and Environmental Microbiology* 67:5179–5189.

Orphan, V. J., C. H. House, K.-U. Hinrichs, K. D. McKeegan, and E. F. DeLong. 2001. Methane-consuming archaea revealed by directly coupled isotopic and phylogenetic analysis. *Science* 293:484–487.

Pancost, R. D., E. C. Hopmans, J. S. Sinninghe Damsté, and the Medinaut Shipboard Scientific Party. 2001. Archaeal lipids in Mediterranean cold seeps: Molecular proxies for anaerobic methane oxidation. *Geochimica et Cosmochimica Acta* 65:1611–1627.

Hinrichs, K.-U., L. R. Hmelo, and S. P. Sylva. 2003. Molecular fossil record of elevated methane levels in late Pleistocene coastal waters. *Science* 299:1214–1217.

Pancost, R. D., and J. S. Sinninghe Damsté. 2003. Carbon isotopic compositions of prokaryotic lipids as tracers of carbon cycling in diverse settings. *Chemical Geology* 195:29–58.

Schouten, S., S. G. Wakeham, E. C. Hopmans, and J. S. Sinninghe Damsté. 2003. Biogeochemical evidence that thermophilic archaea mediate the anaerobic oxidation of methane. *Applied and Environmental Microbiology* 69:1680–1686.

Blumenberg, M., R. Seifert, J. Reitner, T. Pape, and W. Michaelis. 2004. Membrane lipid patterns typify distinct anaerobic methanotrophic consortia. *Proceedings of the National Academy of Sciences of the United States of America* 101:11111–11116.

Orphan, V. J., W. Ussler III, T. H. Naehr, C. H. House, K.-U. Hinrichs, and C. K. Paull. 2004. Geological, geochemical, and microbiological heterogeneity of the seafloor around methane vents in the Eel River Basin, offshore California. *Chemical Geology* 205:265–289.

Peckmann, J., and V. Thiel. 2004. Carbon cycling at ancient methane-seeps. *Chemical Geology* 205:443–467.

Wakeham, S. G., E. C. Hopmans, S. Schouten, and J. S. Sinninghe Damsté. 2004. Archaeal lipids and anaerobic oxidation of methane in euxinic water columns: A comparative study of the Black Sea and Cariaco Basin. *Chemical Geology* 205:427–442.

Elvert, M., E. C. Hopmans, T. Treude, A. Boetius, and E. Suess. 2005. Spatial variations of methanotrophic consortia at cold methane seeps: Implications from a high-resolution molecular and isotopic approach. *Geobiology* 3:195–209.

Orcutt, B., A. Boetius, M. Elvert, V. Samarkin, and S. B. Joye. 2005. Molecular biogeochemistry of sulfate reduction, methanogenesis and the anaerobic oxidation of methane at Gulf of Mexico cold seeps. *Geochimica et Cosmochimica Acta* 69:4267–4281.

Pape, T., M. Blumenberg, R. Seifert, V. N. Egorov, S. B. Gulin, and W. Michaelis. 2005. Lipid geochemistry of methane-seep-related Black Sea carbonates. *Palaeogeography, Palaeoclimatology, Palaeoecology* 227:31–47.

Stadnitskaia, A., G. Muyzer, B. Abbas, M. J. L. Coolen, E. C. Hopmans, M. Baas, T. C. E. van Weering, M. K. Ivanov, E. Poludetkina, and J. S. Sinninghe Damsté. 2005. Biomarker and 16S rDNA evidence for anaerobic oxidation of methane and related carbonate precipitation in deep-sea mud volcanoes of the Sorokin Trough, Black Sea. *Marine Geology* 217:67–96.

Biddle, J. F., J. S. Lipp, M. A. Lever, K. G. Lloyd, K. B. Sørensen, R. Anderson, H. F. Fredricks, M. Elvert, T. J. Kelly, D. P. Schrag, et al. 2006. Heterotrophic archaea dominate sedimentary subsurface ecosystems off Peru. *Proceedings of the National Academy of Sciences of the United States of America* 103:3846–3851.

Birgel, D., V. Thiel, K.-U. Hinrichs, M. Elvert, K. A. Campbell, J. Reitner, J. D. Farmer, and J. Peckmann. 2006. Lipid biomarker patterns of methane-seep microbialites from the Mesozoic convergent margin of California. *Organic Geochemistry* 37:1289–1302.

Blumenberg, M., R. Seifert, and W. Michaelis. 2007. Aerobic methanotrophy in the oxic-anoxic transition zone of the Black Sea water column. *Organic Geochemistry* 38:84–91.

Chapter 8: Weird Molecules, Inconceivable Microbes, and Unlikely Environmental Proxies

Hoefs, M. J. L., S. Schouten, J. W. de Leeuw, L. L. King, S. G. Wakeham, and J. S. Sinninghe Damsté. 1997. Ether lipids of planktonic archaea in the marine water column. *Applied and Environmental Microbiology* 63:3090–3095.

Schouten, S., M. J. L. Hoefs, M. P. Koopmans, H.-J. Bosch, and J. S. Sinninghe Damsté. 1998. Structural characterization, occurrence and fate of archaeal ether-bound acyclic and cyclic biphytanes and corresponding diols in sediments. *Organic Geochemistry* 29:1305–1319.

Hopmans, E. C., S. Schouten, R. D. Pancost, M. T. J. van der Meer, and J. S. Sinninghe Damsté. 2000. Analysis of intact tetraether lipids in archaeal cell material and sediments by high performance liquid chromatography/atmospheric pressure chemical ionization mass spectrometry. *Rapid Communications in Mass Spectrometry* 14:585–589.

Schouten, S., E. C. Hopmans, R. D. Pancost, and J. S. Sinninghe Damsté. 2000. Widespread occurrence of structurally diverse tetraether membrane lipids: Evidence for the ubiquitous presence of low-temperature relatives of hyperthermophiles. *Proceedings of the National Academy of Sciences of the United States of America* 97:14421–14426.

Schouten, S., E. C. Hopmans, E. Schefuss, and J. S. Sinninghe Damsté. 2002. Distributional variations in marine crenarchaeotal membrane lipids: A new tool for reconstructing ancient sea water temperatures? *Earth and Planetary Science Letters* 204:265–274.

Sinninghe Damsté, J. S., W. I. C. Rijpstra, E. C. Hopmans, F. G. Prahl, S. G. Wakeham, and S. Schouten. 2002. Distribution of membrane lipids of planktonic Crenarchaeota in the Arabian Sea. *Applied and Environmental Microbiology* 68:2997–3002.

Sinninghe Damsté, J. S., S. Schouten, E. C. Hopmans, A. C. T. van Duin, and J. A. J. Geenevasen. 2002. Crenarchaeol: The characteristic core glycerol dibiphytanyl glycerol tetraether membrane lipid of cosmopolitan pelagic crenarchaeota. *Journal of Lipid Research* 43:1641–1651.

Sinninghe Damsté, J. S., M. Strous, W. I. C. Rijpstra, E. C. Hopmans, J. A. J. Geenevasen, A. C. T. van Duin, L. A. van Niftrik, and M. S. M. Jetten. 2002. Linearly concatenated cyclobutane lipids form a dense bacterial membrane. *Nature* 491:708–712.

Kuypers, M. M. M., A. O. Sliekers, G. Lavik, M. Schmid, B. B. Jørgensen, J. G. Kuenen, J. S. Sinninghe Damsté, M. Strous, and M. S. M Jetten. 2003. Anaerobic ammonium oxidation by anammox bacteria in the Black Sea. *Nature* 422:608–611.

Schouten, S., E. C. Hopmans, A. Forster, Y. van Breugel, M. M. M. Kuypers, and J. S. Sinninghe Damsté. 2003. Extremely high sea-surface temperatures at low latitudes during the middle Cretaceous as revealed by archaeal membrane lipids. *Geology* 31:1069–1072.

Jenkyns, H. C., A. Forster, S. Schouten, and J. S. Sinninghe Damsté. 2004. High temperatures in the Late Cretaceous Arctic Ocean. *Nature* 432:888–892.

Kuypers, M. M. M., Y. van Breugel, S. Schouten, E. Erba, and J. S. Sinninghe Damsté. 2004. N_2-fixing cyanobacteria supplied nutrient N for Cretaceous oceanic anoxic events. *Geology* 32:853–856.

Powers, L. A., J. P. Werne, T. C. Johnson, E. C. Hopmans, J. S. Sinninghe Damsté, and S. Schouten. 2004. Crenarchaeotal membrane lipids in lake sediments: A new paleotemperature proxy for continental paleoclimate reconstruction? *Geology* 32:613–616.

Powers, L. A., T. C. Johnson, J. P. Werne, I. S. Castañeda, E. C. Hopmans, J. S. Sinninghe Damsté, and S. Schouten. 2005. Large temperature variability in the southern African tropics since the Last Glacial Maximum. *Geophysical Research Letters* 32 (doi:10.1029/2004GL022014).

Schmid, M. C., B. Maas, A. Dapena, K. van de Pas-Schoonen, J. van de Vossenberg, B. Kartal, L. van Niftrik, I. Schmidt, I. Cirpus, J. G. Kuenen, et al. 2005. Biomarkers for in situ detection of anaerobic ammonium-oxidizing (anammox) bacteria. *Applied and Environmental Microbiology* 71:1677–1684.

Wuchter, C., S. Schouten, S. G. Wakeham, and J. S. Sinninghe Damsté. 2005. Temporal and spatial variation in tetraether membrane lipids of marine Crenarchaeota in particulate organic matter: Implications for TEX_{86} paleothermometry. *Paleoceanography* 20 (doi:10.1029/2004PA001110).

Huguet, C., J.-H. Kim, J. S. Sinninghe Damsté, and S. Schouten. 2006. Reconstruction of sea surface temperature variations in the Arabian Sea over the last 23 kyr using organic proxies (TEX$_{86}$ and U$^K_{37}$'). *Paleoceanography* 21 (doi:10.1029/2005PA001215).

Sluijs, A., S. Schouten, M. Pagani, M. Woltering, H. Brinkhuis, J. S. Sinninghe Damsté, G. R. Dickens, M. Huber, G.-J. Reichart, R. Stein, et al. 2006. Subtropical Arctic ocean temperatures during the Palaeocene/Eocene thermal maximum. *Nature* 441:610–613.

Chapter 9: Molecular Paleontology and Biochemical Evolution

Summons, R. E., J. Thomas, J. R. Maxwell, and C. J. Boreham. 1992. Secular and environmental constraints on the occurrence of dinosterane in sediments. *Geochimica et Cosmochimica Acta* 56:2437–2444.

Brassell, S. C. 1994. Isopentenoids and geochemistry. In: Nes, W. D., ed., *Isopentenoids and Other Natural Products: Evolution and Function,* Symposium Series 562, American Chemical Society (Washington, DC): 2–30.

Moldowan, J. M., J. Dahl, B. J. Huizinga, F. J. Fago, L. J. Hickey, T. M. Peakman, and D. W. Taylor. 1994. The molecular fossil record of oleanane and its relation to angiosperms. *Science* 265:768–771.

Moldowan, J. M., J. Dahl, S. R. Jacobson, B. J. Huizinga, F. J. Fago, R. Shetty, D. S. Watt, and K. E. Peters. 1996. Chemostratigraphic reconstruction of biofacies: Molecular evidence linking cyst-forming dinoflagellates with pre-Triassic ancestors. *Geology* 24:159–162.

Moldowan, J. M., and N. M. Talyzina. 1998. Biogeochemical evidence for dinoflagellate ancestors in the early Cambrian. *Science* 281:1168–1170.

Kuypers, M. M. M., P. Blokker, J. Erbacher, H. Kinkel, R. D. Pancost, S. Schouten, and J. S. Sinninghe Damsté. 2001. Massive expansion of marine archaea during a mid-Cretaceous oceanic anoxic event. *Science* 293:92–94.

Kuypers, M. M. M., P. Blokker, E. C. Hopmans, H. Kinkel, R. D. Pancost, S. Schouten, and J. S. Sinninghe Damsté. 2002. Archaeal remains dominate marine organic matter from the early Albian oceanic anoxic event 1b. *Palaeogeography, Palaeoclimatology, Palaeoecology* 185:211–234.

Sinninghe Damsté, J. S., G. Muyzer, B. Abbas, S. W. Rampen, G. Massé, W. G. Allard, S. T. Belt, J.-M. Robert, S. J. Rowland, J. M. Moldowan, et al. 2004. The rise of the rhizosolenid diatoms. *Science* 304:584–587.

Xie, S., R. D. Pancost, H. Yin, H. Wang, and R. P. Evershed. 2005. Two episodes of microbial change coupled with Permo/Triassic faunal mass extinction. *Nature* 434:494–497.

Taylor, D. W., H. Li, J. Dahl, F. J. Fago, D. Zinniker, and J. M. Moldowan. 2006. Biogeochemical evidence for the presence of the angiosperm molecular fossil oleanane in Paleozoic and Mesozoic non-angiospermous fossils. *Paleobiology* 32:179–190.

Zhang, C. L., A. Pearson, Y.-L. Li, G. Mills, and J. Wiegel. 2006. Thermophilic temperature optimum for crenarchaeol synthesis and its implication for archaeal evolution. *Applied and Environmental Microbiology* 72:4419–4422.

Chapter 10: Early Life Revisited

Ourisson, G., M. Rohmer, and K. Poralla. 1987. Prokaryotic hopanoids and other polyterpenoid sterol surrogates. *Annual Review of Microbiology* 41:301–333.

Summons, R. E., S. C. Brassell, G. Eglinton, E. Evans, R. J. Horodyski, N. Robinson, and D. M. Ward. 1988. Distinctive hydrocarbon biomarkers from fossiliferous sediment of the Late Proterozoic Walcott Member, Chuar Group, Grand Canyon, Arizona. *Geochimica et Cosmochimica Acta* 52:2625–2637.

Summons, R. E., T. G. Powell, and C. J. Boreham. 1988. Petroleum geology and geochemistry of the Middle Proterozoic McArthur Basin, Northern Australia: III. Composition of extractable hydrocarbons. *Geochimica et Cosmochimica Acta* 52:1747–1763.

Summons, R. E., and M. R. Walter. 1990. Molecular fossils and microfossils of prokaryotes and protists from Proterozoic sediments. *American Journal of Science* 290:212–244.

des Marais, D. J., H. Strauss, R. E. Summons, and J. M. Hayes. 1992. Carbon isotope evidence for the stepwise oxidation of the Proterozoic environment. *Nature* 359:605–609.

Hayes, J. M. 1994. Global methanotrophy at the Archean-Proterozoic transition. In: Bengtson, S., ed., *Early Life on Earth,* Nobel Symposium 84, Columbia University Press (New York): 220–236.

McCaffrey, M. A., J. M. Moldowan, P. A. Lipton, R. E. Summons, K. E. Peters, A. Jeganathan, and D. S. Watt. 1994. Paleoenvironmental implications of novel C_{30} steranes in Precambrian to Cenozoic age petroleum and bitumen. *Geochimica et Cosmochimica Acta* 58:529–532.

Logan, G. A., J. M. Hayes, G. B. Hieshima, and R. E. Summons. 1995. Terminal Proterozoic reorganization of biogeochemical cycles. *Nature* 376:53–56.

Logan, G. A., R. E. Summons, and J. M. Hayes. 1997. An isotopic biogeochemical study of Neoproterozoic and Early Cambrian sediments from the Centralian Superbasin, Australia. *Geochimica et Cosmochimica Acta* 61:5391–5409.

Brocks, J. J., G. A. Logan, R. Buick, and R. E. Summons. 1999. Archean molecular fossils and the early rise of eukaryotes. *Science* 285:1033–1036.

Summons, R. E., L. L. Jahnke, J. M. Hope, and G. A. Logan. 1999. 2-Methylhopanoids as biomarkers for cyanobacterial oxygenic photosynthesis. *Nature* 400:554–557.

Shen, Y., R. Buick, and D. E. Canfield. 2001. Isotopic evidence for microbial sulphate reduction in the early Archaean era. *Nature* 410:77–81.

Hinrichs, K.-U. 2002. Microbial fixation of methane carbon at 2.7 Ga: Was an anaerobic mechanism possible? *Geochemistry, Geophysics, Geosystems* 3 (doi:10.1029/2001GC000286).

Brocks, J. J., R. Buick, G. A. Logan, and R. E. Summons. 2003. Composition and syngeneity of molecular fossils from the 2.78 to 2.45 billion-year-old Mount Bruce Supergroup, Pilbara Craton, Western Australia. *Geochimica et Cosmochimica Acta* 67:4289–4319.

Brocks, J. J., R. Buick, R. E. Summons, and G. A. Logan. 2003. A reconstruction of Archean biological diversity based on molecular fossils from the 2.78 to 2.45 billion-year-old Mount Bruce Supergroup, Hamersley Basin, Western Australia. *Geochimica et Cosmochimica Acta* 67:4321–4335.

Brocks, J. J., G. D. Love, C. E. Snape, G. A. Logan, R. E. Summons, and R. Buick. 2003. Release of bound aromatic hydrocarbons from late Archean and Mesoproterozoic kerogens via hydropyrolysis. *Geochimica et Cosmochimica Acta* 67:1521–1530.

Brocks, J. J., and R. E. Summons. 2003. Sedimentary hydrocarbons, biomarkers for early life. In: Holland, H. D., and K. K. Turekian, eds., *Treatise on Geochemistry*, Vol. 8, *Biogeochemistry*, Elsevier (Amsterdam): 63–115.

Greenwood, P. F., and R. E. Summons. 2003. GC-MS detection and significance of crocetane and pentamethylicosane in sediments and crude oils. *Organic Geochemistry* 34:1211–1222.

Rothman, D. H., J. M. Hayes, and R. E. Summons. 2003. Dynamics of the Neoproterozoic carbon cycle. *Proceedings of the National Academy of Sciences of the United States of America* 100:8124–8129.

Brocks, J. J., G. D. Love, R. E. Summons, A. H. Knoll, G. A. Logan, and S. Bowden. 2005. Biomarker evidence for green and purple sulphur bacteria in a stratified Palaeoproterozoic sea. *Nature* 437:866–870.

Brocks, J. J., and A. Pearson. 2005. Building the biomarker tree of life. *Molecular Geomicrobiology Reviews in Mineralogy and Geochemistry* 59:233–258.

Olcott, A. N., A. L. Sessions, F. A. Corsetti, A. J. Kaufmann, and T. F. de Oliviera. 2005. Biomarker evidence for photosynthesis during Neoproterozoic glaciation. *Science* 310:471–474.

Volkman, J. K. 2005. Sterols and other triterpenoids: Source specificity and evolution of biosynthetic pathways. *Organic Geochemistry* 36:139–159.

Eigenbrode, J. L., and K. H. Freeman. 2006. Late Archean rise of aerobic microbial ecosystems. *Proceedings of the National Academy of Sciences of the United States of America* 103:15759–15764.

Fike, D. A., J. P. Grotzinger, L. M. Pratt, and R. E. Summons. 2006. Oxidation of the Ediacaran Ocean. *Nature* 444:744–747.

Hayes, J. M., and J. R. Waldbauer. 2006. The carbon cycle and associated redox processes through time. *Philosophical Transactions of the Royal Society of London Series B* 361:931–950.

McKirdy, D. M., L. J. Webster, K. R. Arouri, K. Grey, and V. A. Gostin. 2006. Contrasting sterane signatures in Neoproterozoic marine rocks of Australia before and after the Acraman asteroid impact. *Organic Geochemistry* 37:189–207.

Summons, R. E., A. S. Bradley, L. L. Jahnke, and J. R. Waldbauer. 2006. Steroids, triterpenoids and molecular oxygen. *Philosophical Transactions of the Royal Society of London Series B* 361:951–968.

Knoll, A. H., R. E. Summons, J. R. Waldbauer, and J. E. Zumberge. 2007. The geological succession of primary producers in the oceans. In: Falkowski, P., and A. H. Knoll, eds., *The Evolution of Primary Producers in the Sea,* Academic Press (London): 133–163.

Peterson, K. J., R. E. Summons, and P. C. J. Donoghue. 2007. Molecular palaeobiology. *Palaeontology* 50:775–809.

Ventura, G. T., F. Kenig, C. M. Reddy, J. Schieber, G. S. Frysinger, R. K. Nelson, E. Dinel, R. B. Gaines, and P. Schaeffer. 2007. Molecular evidence of Late Archean archaea and the presence of a subsurface hydrothermal biosphere. *Proceedings of the National Academy of Sciences of the United States of America* 104:14260–14265.

Chapter 11: Thinking Molecularly, Anything Goes

Blumer, M., and J. Sass. 1972. Oil pollution: Persistence and degradation of spilled fuel oil. *Science* 176:1120–1122.

Farrington, J. W., and B. W. Tripp. 1977. Hydrocarbons in western North Atlantic surface sediments. *Geochimica et Cosmochimica Acta* 41:1627–1641.

Albaigés, J., and P. Albrecht. 1978. Fingerprinting marine pollutant hydrocarbons by computerized gas chromatography-mass spectrometry. In: Pfafflin, J. R., ed., *Encyclopedia of Environmental Science and Engineering,* Gordon and Brech (Philadelphia, PA): 171–190.

Crisp, P. T., S. Brenner, M. I. Venkatesan, E. Ruth, and I. R. Kaplan. 1979. Organic chemical characterization of sediment-trap particulates from San Nicolas, Santa Barbara, Santa Monica and San Pedro Basins, California. *Geochimica et Cosmochimica Acta* 43:1791–1801.

Nissenbaum, A., A. Serban, R. Amiran, and O. Ilan. 1984. Dead Sea asphalt from the excavations in Tel Arad and Small Tel Malhata. *Paléorient* 10:157–161.

Evershed, R. P., K. Jerman, and G. Eglinton. 1985. Pine wood origin for pitch from the Mary Rose. *Nature* 314:528–530.

Rullkötter, J., and A. Nissenbaum. 1988. Dead Sea asphalt in Egyptian mummies: Molecular evidence. *Naturwissenschaften* 75:618–621.

Grimalt, J. O., P. Fernandez, J. M. Bayona, and J. Albaigés. 1990. Assessment of fecal sterols and ketones as indicators of urban sewage inputs to coastal waters. *Environmental Science and Technology* 24:357–363.

Evershed, R. P., C. Heron, and L. J. Goad. 1991. Epicuticular wax components preserved in potsherds as chemical indicators of leafy vegetables in ancient diets. *Antiquity* 65:540–544.

Nissenbaum, A. 1992. Molecular archaeology: Organic geochemistry of Egyptian mummies. *Journal of Archaeological Science* 19:1–6.

Bence, A. E., K. A. Kvenvolden, and M. C. Kennicutt. 1996. Organic geochemistry applied to environmental assessments of Prince William Sound, Alaska, after the Exxon Valdez oil spill—a review. *Organic Geochemistry* 24:7–42.

Boeda, E., J. Connan, D. Dessort, S. Muhesen, N. Mercier, H. Valladas, and N. Tisnerat. 1996. Bitumen as a hafting material on Middle Palaeolithic artefacts. *Nature* 380:336–338.

Eglinton, T. I., B. C. Benitez-Nelson, A. Pearson, A. P. McNichol, J. E. Bauer, and E. R. M. Druffel. 1997. Variability in radiocarbon ages of individual organic compounds from marine sediments. *Science* 277:796–799.

Elhmmali, M. M., D. J. Roberts, and R. P. Evershed. 1997. Bile acids as a new class of sewage pollution indicator. *Environmental Science and Technology* 31:3663–3668.

Evershed, R. P., H. R. Mottram, S. N. Dudd, S. Charters, A. W. Stott, G. J. Lawrence, A. M. Gibson, A. Conner, P. W. Blinkhorn, and V. Reeves. 1997. New criteria for the identification of animal fats preserved in archaeological pottery. *Naturwissenschaften* 84:402–406.

Evershed, R. P., S. J. Vaughan, S. N. Dudd, and J. S. Soles. 1997. Fuel for thought? Beeswax in lamps and conical cups from Late Minoan Crete. *Antiquity* 71:979–985.

Coolen, M. J. L., and J. Overmann. 1998. Analysis of subfossil molecular remains of purple sulfur bacteria in a lake sediment. *Applied and Environmental Microbiology* 64:4513–4521.

Dudd, S. N., and R. P. Evershed. 1998. Direct demonstration of milk as an element of archaeological economies. *Science* 282:1478–1481.

Brown, T. A. 1999. How ancient DNA may help in understanding the origin and spread of agriculture. *Philosophical Transactions of the Royal Society of London Series B* 354:89–97.

Bull, I. D., I. A. Simpson, P. F. van Bergen, and R. P. Evershed. 1999. Muck'n molecules: Organic geochemical methods for detecting ancient manuring. *Antiquity* 73:86–96.

Connan, J. 1999. Use and trade of bitumen in antiquity and prehistory: Molecular archaeology reveals secrets of past civilizations. *Philosophical Transactions of the Royal Society of London Series B* 354:33–50.

Evershed, R. P., S. N. Dudd, S. Charters, H. Mottram, A. W. Stott, A. Raven, P. F. van Bergen, and H. A. Bland. 1999. Lipids as carriers of anthropogenic signals from prehistory. *Philosophical Transactions of the Royal Society of London Series B* 354:19–31.

Buckley, S. A., and R. P. Evershed. 2001. Organic chemistry of embalming agents in Pharaonic and Graeco-Roman mummies. *Nature* 413:837–841.

Pearson, A., A. P. McNichol, B. C. Benitez-Nelson, J. M. Hayes, and T. I. Eglinton. 2001. Origins of lipid biomarkers in Santa Monica Basin surface sediment: A case study using compound-specific $\delta^{14}C$ analysis. *Geochimica et Cosmochimica Acta* 65:3123–3137.

Poinar, H. N., M. Kuch, K. D. Sobolik, I. Barnes, A. B. Stankiewicz, T. Kuder, W. G. Spaulding, V. M. Bryant, A. Cooper, and S. Paabo. 2001. A molecular analysis of dietary diversity for three archaic Native Americans. *Proceedings of the National Academy of Sciences of the United States of America* 98:4317–4322.

Sauer, P. E., T. I. Eglinton, J. M. Hayes, A. Schimmelmann, and A. L. Sessions. 2001. Compound-specific D/H ratios of lipid biomarkers from sediments as a proxy for environmental and climatic conditions. *Geochimica et Cosmochimica Acta* 65:213–222.

Bull, I. D., M. J. Lockheart, M. M. Elhmmali, D. J. Roberts, and R. P. Evershed. 2002. The origin of faeces by means of biomarker detection. *Environment International* 27:647–654.

Evershed, R. P., S. N. Dudd, M. S. Copley, R. Berstan, A. W. Stott, H. Mottram, S. A. Buckley, and Z. Crossman. 2002. Chemistry of archaeological animal fats. *Accounts of Chemical Research* 35:660–668.

Ohkouchi, N., T. I. Eglinton, L. D. Keigwin, and J. M. Hayes. 2002. Spatial and temporal offsets between proxy records in a sediment drift. *Science* 298:1224–1227.

Simoneit, B. R. T. 2002. Biomass burning—a review of organic tracers for smoke from incomplete combustion. *Applied Geochemistry* 17:129–162.

Meyers, P. A. 2003. Applications of organic geochemistry to paleolimnological reconstructions: A summary of examples from the Laurentian Great Lakes. *Organic Geochemistry* 34:261–289.

Dawson, C., K. Grice, S. X. Wang, R. Alexander, and J. Radke. 2004. Stable hydrogen isotopic composition of hydrocarbons in torbanites (Late Carboniferous to Late Permian) deposited under various climatic conditions. *Organic Geochemistry* 35:189–197.

Noblet, J. A., D. L. Young, E. Y. Zeng, and S. Ensari. 2004. Use of fecal steroids to infer the sources of fecal indicator bacteria in the lower Santa Ana River watershed, California: Sewage is unlikely a significant source. *Environmental Science and Technology* 38:6002–6008.

Shapiro, B., A. J. Drummond, A. Rambaut, M. C. Wilson, P. E. Matheus, A. V. Sher, O. G. Pybus, M. T. P. Gilbert, I. Barnes, J. Binladen, et al. 2004. Rise and fall of the Beringian steppe bison. *Science* 306:1561–1565.

Copley, M. S., R. Berstan, A. J. Mukherjee, S. N. Dudd, V. Straker, S. Payne, and R. P. Evershed. 2005. Dairying in antiquity. III. Evidence from absorbed lipid residues dating to the British Neolithic. *Journal of Archaeological Science* 32:523–546.

Englebrecht, A. C., and J. P. Sachs. 2005. Determination of sediment provenance at drift sites using hydrogen isotopes and unsaturation ratios in alkenones. *Geochimica et Cosmochimica Acta* 69:4253–4265.

Peacock, E. E., R. K. Nelson, A. R. Solow, J. D. Warren, J. L. Baker, and C. M. Reddy. 2005. The West Falmouth oil spill: ~100 kg of oil found to persist decades later. *Environmental Forensics* 6:273–281.

Willerslev, E., and A. Cooper. 2005. Ancient DNA. *Proceedings of the Royal Society of London Series B* 272:3–16.

Connan, J., A. Nissenbaum, K. Imbus, J. Zumberge, and S. Macko. 2006. Asphalt in iron age excavations from the Philistine Tel Miqne-Ekron city (Israel): Origin and trade routes. *Organic Geochemistry* 37:1768–1786.

Coolen, M. J. L., A. Boere, B. Abbas, M. Baas, S. G. Wakeham, and J. S. Sinninghe Damsté. 2006. Ancient DNA derived from alkenone-biosynthesizing haptophytes and other algae in Holocene sediments from the Black Sea. *Paleoceanography* 21 (doi:10.1029/2005PA001188).

Barnes, I., B. Shapiro, A. Lister, T. Kuznetsova, A. Sher, D. Guthrie, and M. G. Thomas. 2007. Genetic structure and extinction of the woolly mammoth, *Mammuthus primigenius*. *Current Biology* 17:1072–1075.

Coolen, M. J. L., and J. Overmann. 2007. 217,000-year-old DNA sequences of green sulfur bacteria in Mediterranean sapropels and their implications for the reconstruction of the paleoenvironment. *Environmental Microbiology* 9:238–249.

Diez, S., E. Jover, J. M. Bayona, and J. Albaigés. 2007. Prestige oil spill. III. Fate of a heavy oil in the marine environment. *Environmental Science and Technology* 41:3075–3082.

Index

Page numbers in italic refer to figures.

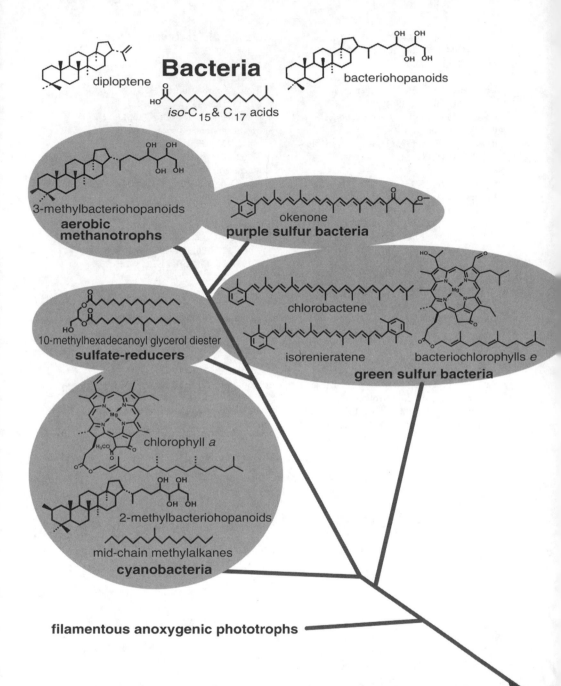

A Biomarker-centric Tree of Life

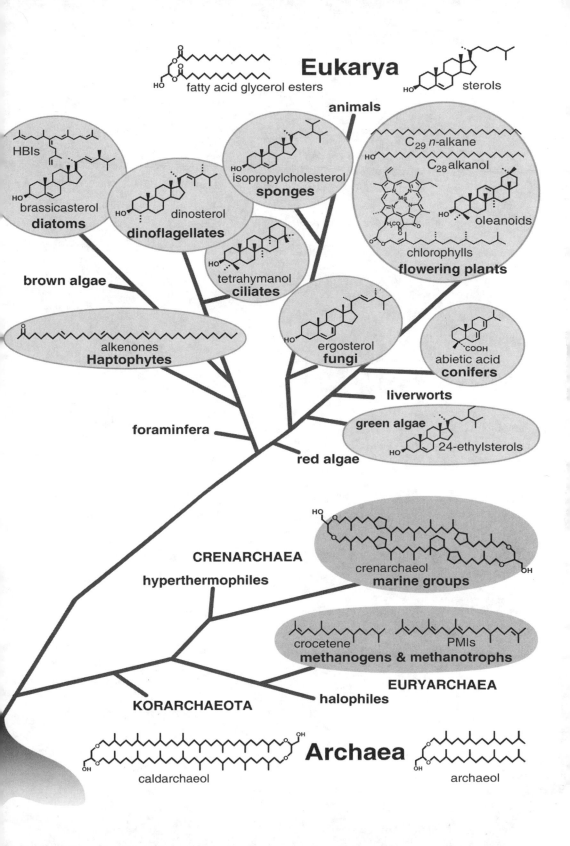